MICROBIOLOGICAL EXAMINATION of WATER and WASTEWATER

MICROBIOLOGICAL EXAMINATION of WATER and WASTEWATER

Maria Csuros
Csaba Csuros

With Contributions By
Klara Ver

LEWIS PUBLISHERS
Boca Raton London New York Washington, D.C.

Library of Congress Cataloging-in-Publication Data

Csuros, Maria.
Microbiological examination of water and wastewater / Maria Csuros and Csaba Csuros.
p. cm.
Includes bibliographical references and index.
ISBN 1-56670-179-1 (alk. paper)
1. Sanitary microbiology. I. Csuros, Csaba. II. Title.
QR48.C78 1999
579—dc21 98-48684
CIP

International Standard Book Number 1-56670-179-1
Library of Congress Card Number 98-48684
Printed in the United States of America 1 2 3 4 5 6 7 8 9 0
Printed on acid-free paper

Preface

Microbiological tests in environmental pollution control are intended to indicate the degree of contamination by the detection and enumeration of indicator organisms.

Many excellent microbiology textbooks and laboratory manuals are available to teach microbiology courses. A great number of published scientific materials and government agency releases are also available for selected areas of environmental microbiology. However, these texts are difficult to use as laboratory guides and they are not designed for training purposes.

There is a great need for information related to the general operation of environmental microbiology laboratories as well as for training laboratory technicians to detect indicator microorganisms in environmental samples. This book completely satisfies these demands and provides those requirements. It is designed as an important tool for environmental laboratory technicians and to serve as a complete guide in their training and in their work.

With the presentation of the fundamentals of microbiology, microbial metabolism, microbial growth and control, and the classification and introduction of microorganisms, the text helps to better understand the purpose and significance of the mechanisms of microbiological examination. Discussion of the general operations and techniques used in environmental microbiology laboratories, sample collection and preservation, detailed QA/QC requirements, and selected step-by-step methodologies, are provided to make this book a useful guide for laboratory technicians working in this field. The text is divided into four parts:

1. General microbiology,
2. Environmental microbiology,
3. Safety and quality control in environmental microbiology laboratories,
4. Techniques and methods in the routine environmental microbiology examination.

The authors of this book hope that as a practical handbook, it will be a valuable contribution to environmental education and special training programs. The aim is to assist environmental laboratory technicians in their work. This handbook also gives a helping hand to environmental laboratory managers, environmental consultants, water and wastewater treatment plant operators and managers, industrial hygienists and health officials.

Authors

Maria Csuros is an environmental chemist, she received her Ph.D. in Environmental Chemistry from the Janus Pannonius University, Hungary. Most of her professional life has revolved around environmental testing laboratories and teaching.

Her first encounter with environmental science occurred at the Environmental and Public Health Laboratory in Hungary, where she worked as the supervisor of the Water Department.

After moving to the United States, she continued to dedicate her knowledge and time to environmental analytical work and education. She designed and developed an environmental science program for Pensacola Junior College that focused on environmental sampling and analysis. An associate professor and coordinator of this program, she teaches chemistry and environmental courses.

She has published two books *Environmental Sampling and Analysis for the Technician* (1994) and *Environmental Sampling and Analysis Laboratory Manual* (1996) by CRC/LEWIS Publishers. She has also authored a number of environmental publications in Hungary.

Csaba Csuros is a professor of science at Pensacola Junior College and has received the excellence in teaching award.

He earned his Ph.D. at Jozsef Attila University, Hungary, in Microbiology. His professional life focused on medical and public health laboratory work and he teaches microbiology, anatomy, and physiology.

He was a delegate to the International Entomology Congress held in Moscow. His research interests center on serological diagnosis of parasitic diseases (ascariasis, echinococcosis, and filariasis).

He has published a number of papers in the field of microbiology, and he is a co-author of a book *Tropical Diseases* published by the Microbiology Department of the University of Medicine, in Pecs, Hungary.

Klara Czako-Ver received her Ph.D. in Biological Sciences from Eotvos Lorand University, Hungary. She is presently a faculty member of Janus Pannonius University, Hungary, and teaches microbiology.

Her present research interest is in isolation of chromium sensitive and tolerant *Schizosaccharomyces pombe mutants*, cytological, biophysical, and genetical investigations.

Her publications have mostly dealt with the effects of toxic metals on the microbial life.

Acknowledgment

We owe a great debt of gratitude to all the people who participated in making this book possible. We thank those who have supported us, inspired our thoughts, and given us ideas. Without their time, effort, and helpful suggestions, we could not have forged ahead so easily.

Our sincere appreciation goes to Dick Fencher, Florida State Department of Environmental Protection (DEP) and James Tucci, environmental consultant for their friendship, excellent ideas, helpful suggestions, and support both before and after completing this text.

Our thanks go to Salman Elawad, biology professor, for his support, and his patience in proofreading and correcting the text.

We are honored to thank Attila Borhidi, Head of the Hungarian Academy of Science. He has been responsible for the development of our scientific relations with the outstanding staff of the Janus Pannonius University, Hungary. His inspiration and motivation with his international scientific aspects has made this book more valuable.

Many thanks to Laszlo Szabo, Head of the Biology Department of Janus Pannonius University, Hungary, for his helpful comments and support.

We are pleased to express our gratitude to our students and friends, Steven Holowach and Joice Wood, for their review of the text, and Kenneth Carron, for the preparation of the figures and tables.

Our appreciation to Barna Csuros, retired library director for the literature he sent for our review and for his always available helping hand.

It has been a pleasure to thank those who have contributed in their special way to the completion of this project. Thanks are also going to Ken McCombs, acquiring editor, for his friendship, understanding, and consideration over the longer than expected period of time that it has taken to complete this project (caused by Maria's sudden sickness). Not to be forgotten in this adventure is Susan Alfieri, production manager, who personifies warmth, kindness, and care. Their patience and steady support in bringing out this text are appreciated.

We gratefully acknowledge also the support of the outstanding editorial and production staff of CRC/Lewis Publishers.

Warm words of thanks to our sons, Geza and Zoltan, and to our grandchildren, Aaron, Andrew, Daniel, Jordan, and Sebastian for their love, encouragement, and for their cheerful spirit.

To all of you, thank you!

Dedication

Our work is dedicated with all of our love to our sons Geza and Zoltan.

M. C. and C. C.

I am honored to dedicate this book to the memory of my parents,
Maria and Geza Csuros.

C.C.

Contents

Chapter 1
Scope and History of Microbiology ... 1
1.1 Microbes and Microbiology ... 1
1.1.1 General Concepts of Microbiology ... 1
1.1.2 Size of Microorganisms ... 1
1.1.3 Unicellular and Noncellular Organisms ... 1
1.1.4 Basic Characteristics of Living Systems ... 2
1.1.5 Investigation of Microorganisms ... 2
1.2 Brief History of Microbiology ... 2
1.2.1 The Microscope and Microbes ... 2
1.2.2 Spontaneous Generation vs. Biogenesis Theory ... 2
1.2.3 Aseptic Techniques ... 5
1.3 The Golden Age of Microbiology ... 5
1.3.1 Fermentation and Pasteurization ... 6
1.3.2 The Germ Theory of Disease ... 6
1.3.3 Vaccination ... 7
1.4 Chemotherapy ... 8
1.4.1 The First Synthetic Drugs ... 9
1.4.2 Antibiotics ... 9
1.5 Modern Developments in Microbiology ... 10
1.6 Naming and Classifying of Microorganisms ... 10
1.6.1 Naming of Microorganisms ... 10
1.7 The Diversity of Microorganisms ... 11
1.7.1 Bacteria ... 11
1.7.2 Fungi ... 11
1.7.3 Protozoa ... 12
1.7.4 Algae ... 12
1.7.5 Viruses ... 12
1.7.6 Multicellular Parasites ... 12

Chapter 2
Procaryotic Cells ... 15
2.1 Introduction of Procaryotic and Eucaryotic Cells ... 15
2.2 The Procaryotic Cell ... 15
2.2.1 Morphology of the Bacterial Cell ... 15
2.2.1.1 Spiral Bacteria ... 16
2.2.1.2 Bacilli ... 17
2.2.1.3 Cocci ... 17
2.3 Structures External to the Cell Walls ... 18
2.3.1 Glycocalyx ... 18
2.3.2 Flagella ... 19
2.3.3 Pili ... 19
2.4 The Cell Wall ... 19
2.4.1 Gram Positive Bacteria ... 20
2.4.2 Gram Negative Bacteria ... 20

2.4.3 Atypical Cell Wall ... 20
2.4.4 Damage to the Cell Wall ... 21
2.5 Structures Internal to the Cell Wall ... 21
2.5.1 Plasma or Inner Membrane ... 21
2.5.2 Movement of Material Across Membranes ... 22
2.5.2.1 Diffusion ... 22
2.5.2.2 Transport by Carriers ... 23
2.5.2.3 Endocytosis and Exocytosis ... 23
2.5.3 Cytoplasm ... 23
2.5.4 Nuclear Area ... 24
2.5.5 Ribosomes ... 24
2.5.6 Inclusion ... 25
2.5.6.1 Metachromatic Granules ... 25
2.5.6.2 Polysaccharide Granules ... 26
2.5.6.3 Lipid Inclusions ... 26
2.5.6.4 Sulfur Granules ... 26
2.5.6.5 Gas Vacuoles ... 26
2.5.7 Endospores ... 26
2.5.7.1 Sporulation ... 26
2.5.7.2 Germination ... 26

Chapter 3

Microbial Metabolism ... 29
3.1 Meaning of Metabolism ... 29
3.1.1 The First and Second Law of Thermodynamics ... 29
3.1.1.1 The First Law of Thermodynamics ... 30
3.1.1.2 The Second Law of Thermodynamics ... 30
3.1.2 Energy Transport through ADP and ATP ... 30
3.2 Metabolic Pathways ... 30
3.3 Enzymes ... 31
3.3.1 Mechanism of Enzyme Action ... 32
3.3.2 Naming the Enzymes ... 32
3.3.3 Temperature and pH Effect on Enzymes ... 32
3.3.4 Inhibition of Enzymes ... 33
3.3.4.1 Competitive Inhibition of Enzymes ... 33
3.3.4.2 Noncompetitive Inhibition of Enzymes ... 33
3.4 Nutritional Classification of Organisms ... 34
3.4.1 Energy Production ... 34
3.4.1.1 Carbon and Energy Sources ... 35
3.4.1.2 Nitrogen Sources ... 35
3.4.1.3 Other Chemical Requirements ... 35
3.5 Energy Production of Organisms ... 35
3.5.1 Oxidation–Reduction ... 36
3.5.2 Phosphorylation as ATP Formation in a Biological System ... 36
3.5.2.1 Substrate Level Phosphorylation ... 36
3.5.2.2 Oxidative Phosphorylation ... 36
3.5.2.3 Photophosphorylation ... 36
3.5.3 Formation of ATP during Respiration ... 36
3.5.3.1 Aerobic Respiration ... 37
3.5.3.2 Anaerobic Respiration ... 37
3.5.3.3 Fermentation ... 37
3.5.4 Formation of ATP during Fermentation ... 37

3.6 Carbohydrate Catabolism 37
3.6.1 Glycolysis 38
3.6.2 Alternative Pathway of Glycolysis by Bacteria 38
3.6.3 Transition Reaction 38
3.6.4 Krebs Cycle or Citric Acid Cycle 39
3.6.5 Respiratory Chain or Electron Transport System or Cytochrome System 41
3.6.6 Aerobic Respiration 42
3.6.7 Fermentation 43

Chapter 4
Microbial Growth and Its Control 45
4.1 Microbial Growth 45
4.1.1 The Lag Phase 45
4.1.2 The Log Phase 45
4.1.3 The Stationary Phase 45
4.1.4 The Death Phase 45
4.2 Physical Requirements 46
4.2.1 Temperature 46
4.2.1.1 Cold Loving Microbes: Psychrophiles 46
4.2.1.2 Moderate Temperature Loving Microbes: Mesophiles 46
4.2.1.3 Heat Loving Microbes: Thermophiles 47
4.2.2 pH and Buffers 47
4.2.3 Osmotic Pressure 48
4.3 Chemical Requirements 48
4.3.1 Carbon 48
4.3.2 Nitrogen, Sulfur, and Phosphorus 48
4.3.3 Trace Elements 48
4.3.4 Oxygen 49
4.3.5 Organic Growth Factors 49
4.4 Control of Microbial Growth 50
4.4.1 Terms Related to Destruction of Organisms 51
4.4.1.1 Sterilization 51
4.4.1.2 Disinfection 51
4.4.1.3 Pasteurization 51
4.4.1.4 Germicide 51
4.4.1.5 Antiseptics 51
4.4.1.6 Asepsis 51
4.4.1.7 Degerming 51
4.4.1.8 Sanitization 51
4.5 Physical Methods of Microbial Control 52
4.5.1 High Heat 52
4.5.1.1 Moist Heat Sterilization 52
4.5.1.2 Dry Heat Sterilization 52
4.5.2 Filtration 52
4.5.3 Low Temperature 52
4.5.4 Desiccation 53
4.5.5 Radiation 53
4.5.5.1 Ionizing Radiation 53
4.5.5.2 Nonionizing Radiation 53
4.6 Chemical Methods of Microbial Control 53
4.6.1 Surfactants or Surface Acting Agents 53
4.6.1.1 Soap 53

4.6.1.2 Acid–Anionic Surface Acting Sterilizers 53
4.6.1.3 Quaternary Ammonium Compounds (Quats) 54
4.6.2 Organic Acids 54
4.6.3 Aldehydes 54
4.6.4 Ethylene Oxide 54
4.6.5 Oxidizing Agents 54
4.6.6 Phenol (Carbolic Acid) 55
4.6.7 Halogens 55
4.6.8 Alcohols 55
4.6.9 Heavy Metals 55

Chapter 5
Microorganisms in the Environment 57
5.1 Biochemical Cycles 57
5.1.1 The Water Cycle or Hydrologic Cycle 57
5.1.2 The Carbon Cycle 58
5.1.2.1 The Greenhouse Effect 59
5.1.3 The Nitrogen Cycle 59
5.1.3.1 Nitrogen-Fixing Bacteria 60
5.1.3.2 Nitrifying Bacteria 61
5.1.3.3 Denitrifying Bacteria 62
5.1.4 The Sulfur Cycle 62
5.1.4.1 Sulfate Reducing Bacteria 62
5.1.4.2 Sulfur Reducing Bacteria 62
5.1.4.3 Sulfur Oxidizing Bacteria 63
5.1.5 The Phosphorus Cycle 63
5.2 Microorganisms Found in Air 63
5.2.1 Sources of Indoor Pollution 64
5.2.1.1 Outdoor Environment 64
5.2.1.2 Indoor Contamination 64
5.2.2 Factors Affecting Indoor Microbial Levels 65
5.2.3 Controlling Microorganisms in Air 65
5.2.3.1 Chemical Agents 65
5.2.3.2 Ultraviolet Radiation 65
5.2.3.3 Air Filtration 65
5.3 Microorganisms in Soil 66
5.3.1 Components of Soil 66
5.3.1.1 Inorganic Components 66
5.3.1.2 Organic Matter 66
5.3.1.3 Root Systems 67
5.3.1.4 Living Organisms 67
5.3.2 Microorganisms in Soil 67
5.3.2.1 Bacteria 67
5.3.2.2 Actinomycetes 68
5.3.2.3 Fungi 68
5.3.2.4 Cyanobacteria, Algae, Protista and Viruses 69
5.3.3 Abiotic Factors Influence Microorganisms in the Soil 69
5.3.3.1 Water and Oxygen 69
5.3.3.2 Soil pH 69
5.3.3.3 Temperature 69

5.3.4 Importance of Decomposers in the Soil ... 69
5.3.5 Soil Pathogens ... 70
5.4 Aquatic Microbiology ... 70
5.4.1 Freshwater Environment ... 70
5.4.1.1 Ground water ... 70
5.4.1.2 Surface Water ... 70
5.4.2 Factors Affecting Microorganisms in Aquatic Environments ... 71
5.4.2.1 Temperature ... 71
5.4.2.2 pH ... 71
5.4.2.3 Nutrients ... 71
5.4.2.4 Oxygen ... 71
5.4.2.5 Depth ... 71
5.4.3 Marine Environments ... 71
5.4.3.1 Temperature ... 71
5.4.3.2 pH ... 72
5.4.3.3 Salinity ... 72
5.4.3.4 Hydrostatic Pressure ... 72
5.4.3.5 Sunlight and Oxygen Concentration ... 72
5.4.3.6 Nutrient Concentrations ... 72
5.4.4 Water Pollution ... 72
5.4.4.1 Organic Wastes ... 73
5.4.4.2 Industrial Wastes ... 73
5.4.4.3 Synthetic Chemicals ... 73
5.4.4.4 Radioactive Substances ... 73
5.4.4.5 Agricultural, Mining, and Construction Activities ... 73
5.4.4.6 Heat ... 73
5.4.5 Pathogens in Water ... 73
5.5 Water Purification ... 74
5.5.1 Flocculation ... 74
5.5.2 Filtration ... 74
5.5.3 Chlorination ... 74
5.6 Sewage Treatment ... 75
5.6.1 Primary Treatment ... 75
5.6.2 Secondary Treatment ... 75
5.6.2.1 Trickling Filter Systems ... 76
5.6.2.2 Activated Sludge Systems ... 76
5.6.3 Tertiary Treatment ... 76
5.6.4 Septic Tanks ... 77

Chapter 6

Bioremediation of Organic Contaminants ... 79
6.1 Bioremediation ... 79
6.1.1 The Objective of Bioremediation ... 79
6.1.2 Advantage of Bioremediation ... 79
6.1.3 Disadvantage of Bioremediation ... 80
6.1.4 Development of Bioremediation ... 80
6.1.5 Biotransformation by Subsurface Microorganisms ... 80
6.2 Application of Bioremediation ... 80
6.2.1 Wood Preservative Industries ... 81
6.2.2 Soil and Hazardous Waste Bioremediation ... 81

6.2.2.1 Bioreactors ... 82
6.2.2.2 Solid Phase Bioremediation ... 82
6.2.2.3 Soil Heaping ... 82
6.2.2.4 Composting ... 82
6.2.3 Degradation of Synthetic Chemicals in Soil ... 82
6.2.4 Biological Control of Groundwater Pollution ... 83
6.2.5 Modern Microbiological Concepts and Pollution Control ... 84

Chapter 7

Microbiological Quality of Environmental Samples ... 85
7.1 Monitoring Microbiological Quality ... 85
7.1.1 Waterborne Disease Outbreaks ... 85
7.1.1.1 Swimming Associated Outbreaks ... 86
7.1.2 Recovery of Pathogens from Environmental Samples ... 87
7.1.3 Indicator Bacteria ... 87
7.1.3.1 Coliform Group ... 87
7.1.3.2 Fecal Coliform Bacteria ... 87
7.1.4 Standards on Microbiological Quality ... 88
7.2 Introduction to Microbiological Parameters to the Sanitary Quality of Environmental Samples ... 88
7.2.1 Heterotrophic Plate Count (HPC) ... 88
7.2.2 Coliform Bacteria Group ... 88
7.2.2.1 Total Coliform Group ... 88
7.2.2.2 Fecal Coliform, *Escherichia coli (E. coli)* ... 88
7.2.2.3 Fecal streptococcus ... 89
7.2.3.4 Klebsiella ... 89
7.3 Pathogenic Microorganisms ... 89
7.3.1 Pathogenic Bacteria ... 89
7.3.1.1 Salmonella ... 90
7.3.1.2 *Salmonella typhi* ... 90
7.3.1.3 Shigella ... 90
7.3.1.4 *Vibrio cholera* ... 90
7.3.1.5 Enteropathogenic *Escherichia coli* ... 90
7.3.1.6 Yersinia ... 90
7.3.1.7 Campylobacter ... 91
7.3.1.8 *Legionella pneumophila* ... 91
7.3.1.9 *Clostridium* ... 91
7.3.2 Viruses ... 92
7.3.2.1 Hepatitis A Virus (HAV) ... 93
7.3.2.2 Norwalk Agent ... 93
7.3.3 Pathogenic Protozoa ... 93
7.3.3.1 *Giardia lamblia* ... 93
7.3.3.2 *Entamoeba histolytica* ... 94
7.4 Iron and Sulfur Bacteria ... 94
7.5 Actinomycetes ... 96
7.6 Fungi ... 96
7.6.1 *Aspergillus* ... 96
7.6.2 *Candida albicans* ... 96
7.7 Regulations for Drinking Water Quality ... 96
7.7.1 Monitoring Agencies ... 97
7.7.1.1 World Health Organization (WHO) ... 97

7.7.1.2 European Economic Community (EEC) 97
7.7.1.3 Environmental Protection Agency (EPA) 97
7.7.2 Monitoring Requirements 98
7.7.3 Minimum Coliform Monitoring Requirements 98
7.7.4 Standards 99
7.7.5 Disinfection Requirements for All Public Water Supplies 99
7.7.6 Disinfection By-Product Regulations 100
7.8 Groundwater 100
7.8.1 Classification of Groudwater 100
7.8.2 Groundwater Standards 101
7.9 Surface Waters 101
7.9.1 Surface Water Classification 101
7.9.2 Surface Water Treatment Rule (SWTR) 102
7.9.3 Surface Water Quality Standards 103
7.10 Wastewater 103
7.10.1 Industrial Wastewater 103
7.10.2 Permit for Industrial and Domestic Wastewater Effluents 104
7.10.3 Minimum Treatment Standard of Domestic Wastewater Effluents 104
7.10.3.1 Effluents for Direct Surface Water Disposal or Surface Water Disposal via Ocean Outfall 104
7.10.3.2 Discharging to Open Ocean Water 104
7.10.3.3 Effluent Disinfection 104
7.10.3.3.1 Basic Disinfection 104
7.10.3.3.2 High-Level Disinfection 105
7.10.3.3.3 Intermediate Disinfection 105
7.10.3.3.4 Low-Level Disinfection 105
7.11 Soil and Sediment 105
7.12 Air 106

Chapter 8

Safety in Environmental Microbiology Laboratory 107
8.1 Laboratory Facilities 107
8.2 Laboratory Safety Considerations 108
8.2.1 General Laboratory Safety Rules 108
8.2.2 Standard Safety Practices in a Microbiological Laboratory 109
8.2.3 General Handling and Storage of Chemicals and Gases 111
8.2.4 Electrical Precautions 111
8.3 Summarized Safety Check List for Environmental Microbiology Laboratories 111
8.3.1 Administrative Considerations 111
8.3.2 Personal Conduct 114
8.3.3 Laboratory Equipment 114
8.3.4 Disinfection/Sterilization 115
8.3.5 Biohazard Control 115
8.3.6 Handling and Storage of Chemicals and Gases 115
8.3.7 Emergency Precautions 115

Chapter 9

Laboratory Quality Assurance and Quality Control 117
9.1 Introduction 117
9.1.1 Quality Assurance (QA) 117
9.1.2 Quality Control (QC) 117

9.1.3 Quality Assessment 117
9.2 Requirements for Facilities and Personnel 117
9.2.1 Ventilation 117
9.2.2 Laboratory Bench Areas 117
9.2.3 Walls and Floors 118
9.2.4 Laboratory Cleanliness 118
9.2.5 Air Monitoring 118
9.2.5.1 RODAC Plates 118
9.2.5.2 Air Density Plates 118
9.2.6 Personnel 119
9.3 Quality Control for Laboratory Equipment and Instrumentation 119
9.3.1 Thermometers and Temperature-Recording Instruments 119
9.3.2 Balances 119
9.3.3 pH Meter 120
9.3.4 Water Deionization Unit 120
9.3.5 UV Sterilizer 121
9.3.6 Membrane Filter Apparatus 123
9.3.7 Centrifuge 123
9.3.8 Hot-Air Oven 125
9.3.9 Autoclave 125
9.3.10 Refrigerators and Freezers 126
9.3.11 Incubators 126
9.3.12 Water Bath 126
9.3.13 Safety Hood 126
9.3.14 Microscope 127
9.3.15 Spectrophotometer 127
9.4 Quality Control of Laboratory Supplies 127
9.4.1 Glassware 127
9.4.2 Glass, Plastic, and Metal Utensils for Media Preparation 128
9.4.3 Culture Dishes 128
9.4.4 Sterility Check on Glassware 128
9.4.5 Chemicals and Reagents 128
9.4.6 Dyes and Stains 129
9.4.7 Membrane Filters and Pads 129
9.4.8 Culture Media 129
9.4.8.1 Quality Control Criteria for Prepared Media 130
9.5 Quality Control of Laboratory Pure Water 130
9.5.1 Checks and Monitoring Criteria 130
9.5.2 Test for Bacterial Quality 130
9.5.2.1 Reagents 131
9.5.2.2 Sample Preparation 132
9.5.2.3 Preparation of Bacterial Suspension 132
9.5.2.4 Test 132
9.5.2.5 Calculation 132
9.5.2.6 Interpretation of the Results 133
9.6 Analytical Quality Control Procedures 133
9.6.1 Quality Control in Routine Analysis 133
9.6.1.1 Negative (Sterile) Control 133
9.6.1.2 Positive Control 134
9.6.1.3 Duplicate Analysis 134
9.6.2 Measurement of Method Precision 134

9.6.3 Reference Sample ... 134
9.6.4 Performance Sample ... 134
9.6.5 Membrane Filter Method (MF) Verification ... 135
9.6.5.1 Total Coliform ... 135
9.6.5.2 Fecal Coliforms ... 136
9.6.5.3 Fecal Streptococci ... 136
9.7 Records and Data Reporting ... 136
9.8 Interlaboratory Quality Control ... 136

Chapter 10

Collecting and Handling Environmental Samples for Microbiological Examination ... 139
10.1 Sampling ... 139
10.1.1 Sampling Program ... 139
10.1.2 Type of Samples ... 139
10.1.3 Sample Containers ... 139
10.1.4 Dechlorinating Agent ... 140
10.1.5 Chelating Agents ... 140
10.1.6 Sampling Procedures ... 140
10.2 Sample Collection from Different Sources ... 141
10.2.1 Collecting Potable Water Samples ... 141
10.2.1.1 Sampling from Distribution System ... 141
10.2.1.2 Sampling from Wells ... 142
10.2.2 Collect Samples from River, Stream, Lake, Spring, or Shallow Well ... 143
10.2.3 Sample Collection from Bathing Beaches ... 143
10.2.4 Marine and Estuarine Sampling ... 146
10.2.5 Sample Collection from Domestic and Industrial Discharges ... 146
10.2.6 Collecting Samples from Sediments and Sludges ... 146
10.2.6.1 Bottom Sediments ... 146
10.2.6.2 Sludges ... 147
10.2.7 Soil Sampling ... 147
10.3 Sample Identification ... 148
10.4 Sample Transportation, Preservation and Holding Time ... 149
10.4.1 Sample Transportation and Preservation ... 149
10.4.2 Laboratory Custody Procedure ... 150
10.4.3 Holding Time ... 150
10.4.4 Discard Samples ... 151

Chapter 11

Laboratory Equipment and Supplies in the Environmental Microbiology Laboratory ... 153
11.1 Laboratory Equipment ... 153
11.1.1 Incubators ... 153
11.1.2 Hot-Air Sterilizing Oven ... 153
11.1.3 Autoclaves ... 153
11.1.4 pH Meter ... 154
11.1.5 Balances ... 155
11.1.6 Optical Counting Equipment ... 155
11.1.7 Refrigerator ... 156
11.1.8 Membrane Filtration Equipment ... 156
11.1.9 Line Vacuum or Electric Vacuum Pump ... 157
11.1.10 Inoculating Needles or Loops ... 157

11.1.11 Microscope ... 157
11.1.11.1 Parts of the Microscope ... 157
11.1.11.2 Light Microscope ... 158
11.1.11.3 Focusing the Microscope ... 160
11.1.11.3.1 Low Power (4× or 10×) ... 160
11.1.11.3.2 Higher Power ... 160
11.1.11.4 Rules for Using Microscopes ... 161
11.1.11.5 Dark Field Microscopy ... 161
11.1.11.6 Phase-Contrast Microscopy ... 161
11.1.11.7 Fluorescent Microscopy ... 162
11.1.11.8 Electron Microscopy ... 162
11.2 Laboratory Glassware ... 162
11.2.1 Petri Dishes ... 162
11.2.1.1 Petri Dishes with Tight Fitting Lids ... 162
11.2.1.2 Petri Dishes with Loose Fitting Lids ... 162
11.2.2 Pipets ... 162
11.2.3 Graduated Cylinders ... 166
11.2.4 Vacuum Filter Flask ... 166
11.2.5 Safety Trap Flask ... 167
11.2.6 Dilution (Milk Dilution) Bottles ... 167
11.2.7 Fermentation Tubes and Vials ... 168
11.2.8 Thermometers ... 168
11.2.9 Cleaning Laboratory Glassware ... 169
11.2.10 Glassware Sterilization ... 169
11.3 Chemicals and Reagents ... 169
11.3.1 Chemicals and Reagents ... 169
11.3.2 Laboratory Pure Water ... 169

Chapter 12

Culture Media ... 171
12.1 Culture Media and Culture ... 171
12.1.1 Chemically Defined Media ... 171
12.1.2 Complex Media ... 171
12.1.3 Selective and Differential Media ... 172
12.1.3.1 Selective Media ... 172
12.1.3.2 Differential Media ... 172
12.1.4 Storage of Dehydrated Culture Media ... 172
12.1.5 Preparation of Media ... 172
12.1.5.1 Preparation and Sterilization of Media ... 173
12.1.6 Sterilization of the Media ... 174
12.1.7 pH Check of the Media ... 175
12.1.8 Storage of Culture Media ... 175
12.1.9 Sterile Media from Commercial Sources ... 176
12.2 Bacterial Growth in Media ... 176
12.2.1 Bacterial Growth in Liquid (Broth) Media ... 176
12.2.1.1 Observing Growth Patterns on Broth Media ... 176
12.2.2 Bacterial Growth in Agar Slant Cultures ... 178
12.2.2.1 Preparation of Agar Slants ... 178
12.2.3 Bacterial Selection by Sugar Fermentation ... 178
12.2.4 Obtaining Pure Culture: Streak Plate Method ... 180
12.2.4.1 How to Prepare Agar Plates ... 181

12.2.4.2 How to Inoculate Agar Plates ... 182
12.2.4.3 How to Incubate Agar Plates ... 182

Chapter 13

Direct Measurement of Bacterial Growth ... 183
13.1 Dilutions ... 183
13.1.1 Single Dilution ... 183
13.1.2 Serial Dilution ... 183
13.1.3 Prompt Use of Dilutions ... 184
13.1.4 Dilution Water ... 184
13.1.4.1 Phosphate Buffered Dilution Water ... 184
13.1.4.1.1 Stock Phosphate Buffer Solution ... 184
13.1.4.1.2 Working Phosphate Buffer Solution ... 185
13.2 Plate Counts ... 185
13.2.1 Pour Plate Method ... 185
13.2.2 Spread Plate Method ... 186
13.2.3 Streak Plate Method ... 186
13.3 Membrane Filter (MF) Technique ... 186
13.3.1 Advantage of the Membrane Filter Technique ... 187
13.3.2 Limitations of the Membrane Filter Technique ... 187
13.3.3 Outline of the Membrane Filter Technique ... 187
13.4 Most Probable Number (MPN) Method ... 188
13.4.1 Presumptive Test ... 188
13.4.2 Confirmed Test ... 188
13.4.3 Completed Test ... 188
13.4.4 Calculation and Reporting of MPN Values ... 189
13.5 Staining Procedures ... 189
13.5.1 Preparation of Bacterial Smears ... 189
13.5.2 Gram Stain ... 189
13.6 Direct Microscopic Counts ... 191

Chapter 14

Estimation of Bacterial Numbers by Indirect Methods ... 193
14.1 Turbidity ... 193
14.2 Biochemical Reactions and Enzymatic Tests ... 194
14.2.1 Catalase Test ... 194
14.2.2 Clumping Factor Test (Slide Coagulase Test) ... 194
14.2.3 Nitrate Reduction Test ... 195
14.2.4 Oxidase Test (Kovacs Method) ... 196
14.2.5 Indole Test ... 197
14.2.5.1 The Rapid Indole Test ... 197
14.2.5.2 Regular Indole Test ... 197
14.2.6 Rapid Urease Test ... 198
14.2.7 IMViC Tests ... 198
14.2.8 Methyl Red and Voges-Proskauer Test ... 198
14.2.8.1 Methyl Red (MR) Test ... 199
14.2.8.2 Voges-Proskauer (V-P) Reaction ... 199
14.2.9 Citrate Utilization Test ... 199
14.2.10 Decarboxylase Test ... 200

14.2.11 Motility Test ... 200
14.3 Rapid Identification Systems (Multitest Systems) ... 200

Chapter 15

Methods for Analyzing Microbiological Quality of the Environment ... 201
15.1 Methods and Techniques ... 201
15.1.1 Methods for the Total Coliform Group ... 201
15.1.2 Methods for Fecal Coliforms ... 201
15.1.3 Heterotrophic Plate Count ... 202
15.1.4 Method for Fecal Streptococcus ... 203
15.1.5 Detection of Stressed Organisms ... 203
15.1.5.1 Ambient Temperature Effect ... 203
15.1.5.2 Chlorinated Effluents and Toxic Wastes ... 203
15.2 Rapid Detection Methods ... 203
15.2.1 7 Hour Fecal Coliform Test ... 204
15.3 Methods for Microbiological Examination of Recreational Waters ... 204
15.3.1 Swimming Pools ... 204
15.3.1.1 Disinfected Indoor Pools ... 204
15.3.1.2 Disinfected Outdoor Pools ... 204
15.3.1.3 Untreated Pools ... 204
15.3.2 Whirlpools ... 205
15.3.3 Natural Bathing Beaches ... 205
15.4 Detection of Pathogenic Bacteria ... 205
15.5 Detection of Soil Microorganisms ... 206
15.5.1 Sample Collection ... 206
15.5.2 Sample Handling and Storage ... 206
15.5.3 Moisture Determination ... 206
15.5.4 Plate Count ... 207
15.5.5 Most Probable Number (MPN) Method ... 207

Chapter 16

Heterotrophic Plate Count ... 209
16.1 Introduction ... 209
16.2 Pour Plate Method ... 209
16.2.1 Equipment and Material ... 209
16.2.2 Media Preparation ... 210
16.2.3 Preparation of Agar ... 210
16.2.4 Dilution Preparation ... 210
16.2.5 Plate Preparation ... 211
16.2.6 Counting ... 213
16.2.6.1 Select Plates Having 30 to 300 Colonies ... 213
16.2.6.2 Plates Having More Than 300 Colonies ... 213
16.2.6.3 Plates Have No Colonies ... 214
16.2.6.4 Number of Colonies per Plate Greatly Exceeding 300 ... 214
16.2.6.5 Spreading Colonies Are on the Plate(s) ... 214
16.2.6.6 Plates Are Uncountable ... 214
16.2.6.7 Examples for Counting and Reporting ... 214
16.3 Spread Plate Method ... 215
16.3.1 Equipment and Material ... 215
16.3.2 Media and Agar Preparation, Sample Dilution ... 215
16.3.3 Preparation of Agar Plates ... 215

16.3.4 Procedure 215
16.3.5 Counting and Reporting 216
16.4 Plate Count from Soils and Sediments 216
16.4.1 Moisture Determination 216
16.4.2 Preparation of Dilutions 216
16.4.3 Calculation and Reporting 217
16.4.4 Pour Plates 217
16.4.5 Spread Count 217

Chapter 17

Determination of Total Coliform 219
17.1 Membrane Filter Technique 219
17.1.1 Introduction 219
17.1.2 Application 219
17.1.3 Equipment and Glassware 220
17.1.4 Culture Media 221
17.1.4.1 Preparation of the M-Endo Broth Media 221
17.1.4.2 Preparation of the M-Endo Agar Medium 222
17.1.4.3 Preparation of Lauryl Tryptose Broth 222
17.1.4.4 Preparation of Brilliant Green Bile Broth 222
17.1.4.5 Preparation of Dilution Water 222
17.1.4.6 Preparation of Magnesium Chloride Solution 222
17.1.5 Set Up Filtration Apparatus 222
17.1.6 Suggested Sample Volumes for Membrane Filter Total Coliform 222
17.2 Membrane Filtration (MF) Procedure 223
17.2.1 Procedure Using Broth Media 223
17.2.2 Procedure Using Agar Media 223
17.2.3 Counting and Recording Colonies 223
17.2.3.1 General Rules for Counting and Reporting 224
17.2.3.2 Special Rules in Counting and Reporting for Potable Waters 225
17.2.4 Verification 225
17.2.4.1 Verification Procedures 225
17.3 Delayed Incubation by the MF Method 226
17.3.1 Application 227
17.3.2 Equipment and Glassware 227
17.3.3 Culture Media 227
17.3.3.1 Preparation of M-Endo Preservative Medium 227
17.3.4 Procedure, Counting, and Recording 227
17.4 Most Probable Number (MPN) Method 227
17.4.1 Introduction 227
17.4.2 Application 228
17.4.3 Equipment and Glassware 228
17.4.4 Culture Media 228
17.4.4.1 Lauryl Tryptose Broth 228
17.4.4.2 Brilliant Green Lactose Bile Broth (BGLB) 228
17.4.4.3 Eosine Methylene Blue Agar (EMB Agar) 228
17.4.5 Dilution Water 229
17.4.6 Procedure 230
17.4.6.1 Presumptive Test 230
17.4.6.2 Confirmed Test 230
17.4.6.3 Completion Test 230
17.4.7 Calculation 231

17.4.7.1 Calculation of Reported Value When 10, 1.0, and 0.1 ml Portions Are Used ... 231
17.4.7.2 Calculation of Reported Values When the Series of Decimal Dilutions Is Other Than 10, 1.0, and 0.1 ml ... 231
17.4.7.3 More Than Three Sample Volumes Are Inoculated ... 231
17.5 Determination of *Klebsiella* ... 233
17.5.1 Introduction ... 233
17.5.2 Membrane Filter Procedure ... 234
17.5.2.1 Apparatus ... 234
17.5.2.2 Culture Medium ... 234
17.5.2.2.1 Modified M-FC Agar (M-FCIC) Agar ... 234
17.5.2.2.2 M-Kleb Agar ... 234
17.5.2.3 Procedure ... 235
17.5.2.4 Counting Colonies ... 235
17.5.2.5 Verification ... 235
17.6 Application for Soil, Sediment, and Sludge Samples ... 235
17.7 Analytical Quality Control ... 236

Chapter 18

Determination of Fecal Coliform ... 237
18.1 Definition of the Fecal Coliform Group ... 237
18.2 Membrane Filter Method ... 237
18.2.1 Application ... 237
18.2.2 Equipment and Glassware ... 237
18.2.3 Culture Media ... 237
18.2.3.1 Preparation of M-FC Broth ... 238
18.2.3.1.1 Rosolic Acid Solution ... 238
18.2.3.1.2 0.2 *N* NaOH ... 238
18.2.3.2 Preparation of M-FC Agar ... 238
18.2.3.3 Preparation of Lauryl Tryptose Broth ... 238
18.2.3.4 Preparation of EC Medium ... 238
18.2.3.5 Preparation of Sterile Dilution Water ... 238
18.2.4 Procedure ... 238
18.2.4.1 Set Up Filtration Apparatus ... 238
18.2.4.2 Suggested Sample Volumes for the Membrane Filter Fecal Coliform Test ... 238
18.2.4.3 Procedure Using Broth Media ... 239
18.2.4.4 Procedure Using Agar Media ... 239
18.2.5 Counting and Recording Colonies ... 239
18.2.5.1 Countable Membranes with 20 to 60 Colonies ... 239
18.2.5.2 Countable Membrane Filters with Less Than 20 Blue Colonies ... 239
18.2.5.3 Membranes with No Colonies ... 239
18.2.5.4 Countable Membranes with More Than 60 Colonies ... 240
18.2.5.5 Uncountable Membranes with More Than 60 Colonies ... 240
18.2.6 Verification ... 240
18.2.6.1 Procedure ... 240
18.3 Delayed Incubation MF Method ... 240
18.3.1 Application ... 240
18.3.2 Equipment and Glassware ... 240
18.3.3 Culture Media ... 241
18.3.3.1 Preparation of the M-VFC Medium ... 241
18.3.4 Counting and Recording Colonies ... 242
18.4 The Most Probable Number (MPN) Method ... 242

18.4.1 Application ... 242
18.4.2 Equipment and Glassware ... 242
18.4.3 Media ... 242
18.4.3.1 Lauryl Tryptose Broth ... 242
18.4.3.2 EC Medium ... 242
18.4.3.3 Dilution Water ... 242
18.4.4 Procedure ... 242
18.4.5 Calculation and Reporting ... 244
18.5 The MUG Test ... 244
18.6 Application for Soil, Sediment, and Sludge Samples ... 244
18.7 Analytical Quality Control (QC) Procedure ... 244

Chapter 19

Determination of Fecal Streptococcus ... 245
19.1 Introduction ... 245
19.1.1 The Fecal Coliform and Fecal Streptococci, FC/FS Ratio ... 245
19.2 Enterococci Portion of the Fecal Streptococcus Group ... 246
19.3 Membrane Filter (MF) Technique ... 246
19.3.1 Application ... 246
19.3.2 Equipment and Glassware ... 246
19.3.3 Culture Media ... 247
19.3.3.1 Preparation of KF Streptococcus Agar ... 247
19.3.3.1.1 1 Percent Solution of 2,3,5-triphenyl-tetrazolium-chloride (TTC) ... 247
19.3.3.1.2 10 Percent Sodium Carbonate, Na_2CO_3 ... 247
19.3.3.2 Preparation of Brain Heart Infusion Broth (BHI) ... 247
19.3.3.3 Brain Heart Infusion Agar (BHI Agar) ... 247
19.3.3.4 Brain Heart Infusion Broth with 40 Percent Bile ... 248
19.3.3.4.1 Preparation of 10 percent Oxgall Solution ... 248
19.3.3.5 Dilution Water ... 248
19.3.4 Procedures ... 248
19.3.4.1 Sample Volume ... 248
19.3.5 Counting and Recording Colonies ... 248
19.3.6 Verification ... 248
19.4 Delayed MF Procedure ... 249
19.5 Most Probable Number (MPN) Method ... 249
19.5.1 Culture Media ... 249
19.5.1.1 Azide Dextrose Broth ... 249
19.5.1.2 Pfizer Selective Enterococcus, (PSE) Agar ... 250
19.5.2 Procedure ... 251
19.5.2.1 Presumptive Test ... 251
19.5.2.2 Confirmed Test ... 251
19.5.3 Calculation ... 251
19.6 Pour Plate Method ... 251
19.6.1 Media ... 252
19.6.2 Procedure ... 252
19.6.3 Counting and Reporting ... 252
19.6.3.1 Plates with 30 to 300 Colonies ... 252
19.6.3.2 All Plates Greater than 300 Colonies ... 253
19.6.3.3 All Plates with Fewer than 30 Colonies ... 253
19.6.3.4 Plate with No Colonies ... 253

19.6.3.5 All Plates Are Crowded 253
19.7 Procedures for Soils, Sediments, and Sludges 254
19.8 Analytical Quality Control (QC) Procedures 254

Chapter 20
Enterobacteriaceae 255
20.1 Introduction 255
20.1.1 Coliform Organisms 255
20.1.2 Differentiation of Enterobacteriaceae 255
20.2 ENTEROTUBE II 256
20.2.1 Operation of ENTEROTUBE II 256
20.2.2 To Read ENTEROTUBE II 257
20.2.3 To Indicate Positive Reactions 257
20.2.4 Ordering Information 257
20.3 BBL OXI/FERM Tube II 260
20.4 API 20E System 260
20.4.1 Operation of API 20E 263

Chapter 21
Iron and Sulfur Bacteria 265
21.1 Introduction to Iron and Sulfur Bacteria 265
21.2 Iron Bacteria 265
21.2.1 Identification of Iron Bacteria 266
21.3 Sulfur Bacteria 267
21.3.1 Common Forms of Sulfur Bacteria 267
21.3.1.1 Sulfate-Reducing Bacteria 268
21.3.1.2 Photosynthetic Green and Purple Sulfur Bacteria 268
21.3.1.3 Colorless Filamentous Sulfur Bacteria 268
21.3.1.4 Aerobic Sulfur Oxidizers 268
21.3.2 Identification of Sulfur Bacteria 269
21.3.2.1 Green and Purple Sulfur Bacteria 269
21.3.2.2 Colorless Filamentous Sulfur Bacteria 269
21.3.2.3 Colorless Nonfilamentous Sulfur Bacteria 270
21.3.2.4 Colorless Small Sulfur Bacteria and Sulfate-Reducing Bacteria 270
21.4 Enumeration, Enrichment, and Isolation of Iron and Sulfur Bacteria 270

Chapter 22
Detection of Actinomycetes 271
22.1 General Introduction 271
22.1.1 Actinomycetes 271
22.1.2 Streptomyces 271
22.2 Determination of Actinomycetes Density 272
22.2.1 Plating Method 272
22.2.2 Preparation and Dilution of Samples 272
22.2.3 Preparation of Soil Samples 272
22.2.4 Medium 273
22.2.5 Procedure 273
22.2.6 Counting 273
22.2.7 Calculation and Reporting 274
22.2.7.1 Alternative Calculations 275

Appendix A

Exponential Notation 277
A.1 General Discussion 277
A.1.1 Numbers Greater Than 10 in Exponential Notation 277
A.1.2 Numbers Less Than 1 in Exponential Notation 278
A.1.3 Adding and Subtracting Numbers in Exponential Notation 278
A.1.4 Multiplying and Dividing Numbers in Exponential Notation 278

Appendix B

International System of Units (Metric System) 279
B.1 Derived SI Units 279
B.2 Derived SI Units with Special Names 279
B.3 Prefixes 280
B.4 Useful Conversion Factors 280

Appendix C

Units and Conversion Factors 281
C.1 Conversion to Metric Measures 281
C.2 Conversion from Metric Measures 281

Appendix D

Biochemical Oxygen Demand (BOD) 283
D.1 Scope and Application 283
D.1.1 Carbonaceous vs. Nitrogenous BOD 283
D.1.2 Dilution Requirements 283
D.1.3 Summary of the Method 284
D.1.4 Sampling and Storage 284
D.1.5 Sample Collection for the Winkler Method 284
D.1.6 Interferences 285
D.2 Apparatus and Material 285
D.3 Reagents 285
D.4 Procedure 286
D.4.1 DO Meter Calibration 286
D.4.2 Preanalysis Checking 287
D.4.3 Analysis 287
D.5 Calculation 288
D.5.1 BOD Concentration 288
D.5.2 Correction Factor for Seed Control 288
D.5.3 Dilution Factor 289
D.5.4 Alternative Calculation 289
D.6 Quality Control 289
D.6.1 Method Blank 289
D.6.2 Seed Control 289
D.6.3 Reference Standard (Glucose–Glutamic Acid Standard) 289
D.6.4 DO Meter Performance Check 290
D.6.5 Accuracy 290
D.6.6 Precision 290
D.7 Safety 290

Appendix E

Determination of Solids ... 291
E.1 General Discussion ... 291
E.1.1 Sample Collection and Handling ... 291
E.1.2 Sample Pretreatment ... 291
E.2 Determination of Total Solids (TS) Dried at 103 to 105°C ... 291
E.2.1 Principle of the Method ... 291
E.2.2 Interferences ... 291
E.2.3 Apparatus and Material ... 292
E.2.4 Procedure ... 292
E.2.5 Calculation ... 292
E.2.6 Quality Control ... 292
E.2.6.1 Method Blank ... 292
E.3 Determination of Total Dissolved Solids (TDS) Dried at 180°C ... 292
E.3.1 Principle of the Method ... 292
E.3.2 Interferences ... 293
E.3.3 Apparatus and Material ... 293
E.3.4 Reagents ... 293
E.3.4.1 Reference Standards, 500 mg per l ... 293
E.3.5 Procedure ... 293
E.3.5.1 Preparation of Evaporation Dishes ... 293
E.3.5.2 Sample Analysis ... 294
E.3.6 Calculation ... 294
E.3.7 Quality Control ... 295
E.3.7.1 Method Blank ... 295
E.3.7.2 Reference Standard ... 295
E.3.7.3 Duplicates ... 295
E.4 Determination of the Total Suspended Solids (TSS) Dried at 103 to 105°C ... 295
E.4.1 Principle of the Method ... 295
E.4.2 Interferences ... 295
E.4.3 Apparatus and Material ... 295
E.4.4 Procedure ... 295
E.4.4.1 Filter Preparation ... 295
E.4.4.2 Sample Analysis ... 296
E.4.5 Calculation ... 296
E.5 Fixed and Volatile Solids Ignited at 500°C ... 296
E.5.1 Principle of the Method ... 296
E.5.2 Interferences ... 297
E.5.3 Apparatus and Material ... 297
E.5.4 Procedures ... 297
E.5.5 Calculation ... 297
E.6 Total, Fixed, and Volatile Solids in Solid and Semisolid Samples ... 297
E.6.1 Principle of the Method ... 297
E.6.2 Interferences ... 297
E.6.3 Apparatus and Material ... 298
E.6.4 Procedure ... 298
E.6.4.1 Preparation of Evaporation Dishes ... 298
E.6.4.2 Sample Analysis ... 298
E.6.5 Calculation ... 298
E.6.6 Quality Control ... 298

E.7 Settleable Solids 298
E.7.1 Principle of the Method 298
E.7.2 Apparatus 298
E.7.3 Procedure 299

Appendix F

Determination of pH 301
F.1 General Discussion 301
F.2 Sample Collection and Holding Time 301
F.3 Potentiometric Determination of pH 301
F.3.1 Principle of the Method 301
F.3.2 Interferences 302
F.3.3 Apparatus and Materials 302
F.3.4 Reagents 302
F.3.5 Meter Calibration 303
F.3.6 Measurement of Aqueous Samples and Wastes Containing Greater Than 20 Percent Water 303
F.3.7 Measurement of Solid Samples 303
F.3.8 General Rules 304
F.4 Quality Control (QC) 304
F.4.1 Duplicate Analysis 304
F.4.2 Initial Calibration Verification (ICV) 304
F.4.3 Continuing Calibration Verification (CCV) 305
F.4.4 Calculate Accuracy 305

Appendix G

Bacteriophages 307
G.1 Coliform Counts 307
G.1.1 Total Coliforms 307
G.1.2 Fecal Coliforms 307

Bibliography 309

Index 311

List of Tables and Figures

TABLES

5.1 Distribution of microorganisms in numbers per gram of typical garden soil at various depths . . . 68
5.2 Human pathogens transmitted in water . . . 74
5.3 Effects of water pollution . . . 75
6.1 Organic compounds known to be biotransformed by surface microorganisms . . . 81
6.2 Favorable and unfavorable chemical and hydrological site conditions for in situ bioremediation . . . 81
6.3 Biodegradability of volatile organic pollutants . . . 83
7.1 Water-borne disease outbreaks reported during 1971–1988 . . . 85
7.2 Bacterial diseases generally transmitted by contaminated drinking water . . . 86
7.3 Viral and parasitic diseases generally transmitted by drinking water . . . 86
7.4 Recreational water-associated disease outbreaks in the United States during 1986 and 1988 . . . 87
7.5 SDWA drinking water standard for microbiological quality . . . 98
7.6 Drinking water standards for disinfection by-products . . . 100
7.7 Primary and secondary drinking water standards . . . 101
7.8 Bacteriological criteria for surface water quality classification . . . 103
8.1 Contents of a laboratory first aid kit . . . 108
8.2 Material safety data sheet (MSDS) . . . 112
8.3 Lab guard safety label system . . . 113
8.4 Lab guard safety label system: acids and bases . . . 114
9.1 Incubator temperature control log (temperature 37°C ± 2°C) . . . 120
9.2 Incubator temperature control log (temperature 44°C ± 2°C) . . . 121
9.3 Refrigerator temperature control log (temperature 4°C ± 2°C) . . . 122
9.4 Hot air oven temperature control log (temperature 160°C–170°C) . . . 123
9.5 Calibration chart for thermometer . . . 124
9.6 Balance check . . . 124
9.7 Calibration stock and standard solution log form . . . 125
9.8 Time and temperature for autoclave sterilization . . . 126
9.9 Holding times for prepared media . . . 131
9.10 Quality check of laboratory pure water . . . 131
9.11 Reagent addition for water quality test . . . 133
9.12 Calculation of precision criteria . . . 135
9.13 Daily checks on precision of duplicate counts . . . 135
10.1 Sampling frequency for drinking waters based on population . . . 141
10.2 Chain of custody . . . 148
10.3 Sample log sheet . . . 149
10.4 Sample label . . . 150
11.1 Summary of various types of microscopes . . . 163
12.1 Holding times for prepared media . . . 175
15.1 Methods for microbiological tests . . . 202
17.1 MPN table with 95 percent confidence limits . . . 229
17.2 MPN indices and 95 percent confidence limits . . . 232

20.1 ENTEROTUBE II color reactions 260
20.2 Summary of test results, API 20E System 262

FIGURES

1.1 Anton van Leeuvenhoek 3
1.2 Leeuvenhoek's simple microscope 3
1.3 Redi's experiments 4
1.4 Louis Pasteur is shown in his laboratory 5
1.5 Lord Joseph Lister applied the aseptic technique 7
1.6 Edward Jenner introduced the modern method of vaccination to prevent smallpox 8
1.7 Edward Jenner is shown vaccinating a child against smallpox 8
1.8 Paul Ehrlich is shown in his laboratory 9
1.9 Alexander Fleming discovered penicillin 10
1.10 The diversity of the microbial world 13
2.1 Ultrastructure of a bacterial cell 16
2.2 The basic shapes of bacteria 16
2.3 Spiral bacteria 17
2.4 Diagram of a fictionalized bacterial cell 17
2.5 Three variations of cocci 18
2.6 Details of various types of flagellation 19
2.7 Comparison of structures and the chemical composition of the Gram posit21 and Gram negative cell wall 21
2.8 Simple diffusion 22
2.9 Effect of osmotic pressure on cells 23
2.10 Endocytosis and exocytosis 24
2.11 Bacterial chromosomes and plasmids 25
2.12 Stages in endospore formation 27
3.1 Relationship between anabolism, catabolism, ADP, and ATP 29
3.2 Reactions without enzymes and with enzymes 31
3.3 Mechanisms of enzymatic activity 32
3.4 Factors influencing enzymatic activity 33
3.5 Enzyme inhibition 33
3.6 Energy source of organisms 34
3.7 The intermediate of glycolysis 39
3.8 Formation of acetyl CoA from pyruvic acid 40
3.9 The Krebs cycle or the citric acid cycle 40
3.10 Respiratory chain or electron transport system or cytochrome system 41
3.11 Overview of cellular respiration 42
4.1 Bacterial growth curve 46
4.2 Comparison of the growth rates of psychrophile, mesophile, and thermophile organisms 47
4.3 Different organism sensitivity to oxygen 49
4.4 Ignaz Semmelweis and Joseph Lister 50
5.1 The hydrologic cycle 57
5.2 The carbon and oxygen cycles 58
5.3 The nitrogen cycle 59
5.4 Nodules on the roots of a bean plant 61
5.5 Microcolony of the nitrifying bacteria 61

5.6 A relative proportion of various kinds of organisms found in soil 68
5.7 Trickling filters used in secondary sewage treatment 76
5.8 Two sheathed bacteria used in trickling filters 77
5.9 An overview of a sewage treatment plant showing primary, secondary, and tertiary treatment facilities 78
7.1 Gram stain of smear made from *vibrio cholera* 91
7.2 Colonies of *Legionella pneumophila* 92
7.3 *Giardia lamblia* trophozoite and cysts 94
7.4 *Entamoeba histolytica* trophozoite containing ingested red blood cells 95
7.5 *Entamoeba histolytica* trophozoite 95
8.1 Eye protective device 110
8.2 Eye wash fountain 110
8.3 Hazard warning signs 110
9.1 Label formats used for reagents and solvents 128
10.1 Suggested sample containers 140
10.2 Weighed bottle frame and sample bottle 142
10.3 Demonstration of technique used in grab sampling 143
10.4 Kemmerer depth sampler 144
10.5 Sampling a lake or impoundment 144
10.6 Sampling a large stream 145
10.7 Sampling a water supply reservoir 145
10.8 Eckman bottom grab sampler 146
10.9 Typical location of subsamples 147
11.1 Typical laboratory incubator 154
11.2 Membrane filtration funnels 156
11.3 Sterilizing the wire inoculating loop 157
11.4 Parts of a compound light microscope 159
11.5 The same organism seen under different microscopes 161
11.6 Tight fitting petri dish lids 163
11.7 Loose fitting petri dish lids 164
11.8 Technique for using a volumetric pipet 164
11.9 Different types of pipets 165
11.10 Useful technique for reading the meniscus 166
11.11 Avoiding parallax error in reading the meniscus 167
11.12 Vacuum filtration 168
11.13 Milk dilution bottle 168
12.1 Gauze covered nonabsorbable cotton plug and aluminum foil cover for flasks used in media preparation 174
12.2 Use of a membrane filter system to sterilize solutions 174
12.3 Inoculating procedures 177
12.4 Cultural characteristics of broth cultures 178
12.5 Inoculate agar slant by streaking back and forth across the surface 179
12.6 Cultural characteristics of agar slant culture 179
12.7 Broth culture tubes with Durham tubes 180
12.8 Using aseptic technique 181
12.9 Rotate the petri plate so that the medium covers the bottom 181
12.10 Proper handling of the cover of the petri dish 182
12.11 Steps in the preparation of a streak plate 182
13.1 Serial dilution technique 184
13.2 Preparation of a smear from a solid culture 190
13.3 Heat fixation of an air dried specimen 190
13.4 Steps in the Gram-stain procedure 191

13.5 Gram-staining procedure ... 192
14.1 Catalase test ... 195
14.2 Kovacs oxidase test ... 196
14.3 Spot indole test ... 197
16.1 Melting media in tubes ... 210
16.2 Testing agar temperature ... 211
16.3 Typical dilution series for heterotrophic plate count ... 211
16.4 Pipetting sample into the petri dish ... 212
16.5 Using aseptic technique ... 213
16.6 Rotate the petri plate ... 213
17.1 Membrane filtration setup ... 221
17.2 Inoculation in membrane filter technique ... 224
17.3 Verification outline for total coliform the MF technique ... 226
17.4 Flow chart for total coliform by MPN method ... 232
18.1 Verification of fecal coliform colonies on the membrane filter ... 241
18.2 Flow chart for the fecal coliform MPN tests ... 243
19.1 Isolation and identification of fecal streptococci ... 250
20.1 To use ENTEROTUBE II for the identification of Gram-negative bacteria (Enterobacteriaceae) ... 258
20.2 Biochemical reaction of the compartments of ENTEROTUBE II ... 259
20.3 ENTEROTUBE II pad for interpretation, indication and reporting result ... 261
20.4 To read the ENTEROTUBE II System ... 262
20.5 API 20E System, microtube consists of tube and a cupule section ... 263
21.1 Iron bacteria, *Gallionella ferruginea* ... 266
21.2 Filamentous iron bacteria, *Spherotylus natans* ... 267
21.3 Filamentous iron bacteria, *Crenothrix polyspora* ... 268
21.4 Filamentous sulfur bacteria, *Thiodendron mucosum* (left) and *Beggiatoa alba* (right) ... 269
21.5 Nonfilamentous and photosynthetic sulfur bacteria ... 270

1 Scope and History of Microbiology

1.1 MICROBES AND MICROBIOLOGY

1.1.1 General Concepts of Microbiology

The science of microbiology is study of microorganisms and their activities. You may ask, exactly what is a microorganism and how do microorganisms differ from other organisms?

The word microorganism originates from the Greek word micro, meaning small. Microorganisms are very small life forms—so small that individual microorganisms usually can only be seen with magnification. Microbiology is also concerned with the forms, structure, reproduction, physiology, metabolism, and identification of microbes. To study microorganisms, we must be familiar with their distribution in nature, their relationship to each other and other living organisms, their beneficial and detrimental effects on man, and the physical and chemical changes they make in their environment.

The major groups of microorganisms are bacteria, algae, fungi, viruses, and protozoa. All are widely distributed in nature.

1.1.2 Size of Microorganisms

Microbes range in size from small viruses 20 nm in diameter to large protozoans 5 mm or more in diameter (see Appendix B for a review of the metric system).

Resolution or resolving power refers to the ability of lenses to reveal fine details or two points distinctly separated. The resolution of the naked eye is 0.1 mm. This means that the naked eye can distinguish between two subjects that are not smaller than 0.1 mm. Most microorganisms are smaller than 0.1 mm, so we need a microscope to increase the resolving power to make the microorganisms visible.

1.1.3 Unicellular and Noncellular Organisms

For the most part, microbiology deals with unicellular microscopic organisms. In unicellular organisms, all the life processes are performed in a single cell. Regardless of the complexity of an organism, the cell is the true, complete unit of life. According to cell theory, the smallest unit of life is one cell, nothing smaller or simpler is alive.

All living cells are basically similar. They are composed of protoplasm, a colloidal organic complex, consisting largely of proteins, lipids, and nucleic acids. All are bordered by lining cell membrane or cell membrane and cell wall, and all contain a nucleus or an equivalent nuclear substance with genetic material.

In the study of microbiology, we also encounter organisms that may represent the borderline of life. These organisms are viruses and viruses do not fit the basic characteristics of any living system.

1.1.4 Basic Characteristics of Living Systems

Any living system, unicellular or multicellular, must have the following basic characteristics:

1. The ability to reproduce new cells either for growth, repair or maintenance, or production of a new individual.
2. The ability to ingest food and metabolize it for energy and growth.
3. The ability to excrete waste products.
4. The ability to respond to changes in the environment, called irritability.
5. The ability to mutate.

1.1.5 Investigation of Microorganisms

Microorganisms have provided specific systems for the investigation of the physiologic, genetic, and biochemical reactions that are basis of life. They can be conveniently grown in test tubes or in petri dishes. They grow rapidly and reproduce at an unusually high rate.

Some bacteria go through about 100 generations in one 24 h period. In microbiology, we can study organisms in great detail and observe their life processes while they are actively metabolizing, growing, reproducing, aging, and dying. By modifying their environment, we can alter metabolic activities, regulate growth, or destroy the organism. Microbiologists have been remarkably successful in explaining the useful microorganisms and in combating the harmful ones.

1.2 BRIEF HISTORY OF MICROBIOLOGY

1.2.1 The Microscope and Microbes

Considering the small size of the microorganism, it is no wonder that the existence of microorganisms was not recognized until a few centuries ago. The science of microbiology was born when men learned to grind lenses from pieces of glass and to combine the lenses to produce magnifications large enough to see microbes.

In about 1665, an English scientist named *Robert Hook* prepared a compound microscope and used it to observe thin slices of cork which was composed of walls of dead plant cells. Hook called the small boxes cells, because they reminded him of monks' cells. This observation and the discovery of building blocks in plants of a cell structure marked the beginning of cell theory.

It was the Dutchman *Anton van Leeuwenhoek*, an amateur scientist and cloth merchant, who first made and used lenses to observe our environment and living organisms; see Figure 1.1. The lenses Leeuwenhoek made were of excellent quality; some gave magnification up to 200 to 300 times and were remarkably free from distortion. Making these lenses and looking through them were the passions of Leeuwenhoek's life. Starting in the 1670s, he wrote numerous letters to the Royal Society of London describing the *animalcules* he saw through his simple, single lens microscope.

Leeuwenhoek made detailed drawings of the microbes he saw under the microscope in rain water, peppercorn infusion, and in material taken from teeth scrapings. They were identified as representations of bacteria in spherical, rod, and spiral forms, protozoa, algae, yeast, and fungi. He pursued his studies until his death in 1723 at the age of 91. His picture and his simple microscope are seen in Figures 1.1 and 1.2.

Leeuwenhoek refused to sell his microscopes to others and so he failed to foster the development of microbiology as much as he might have.

1.2.2 Spontaneous Generation vs. Biogenesis Theory

After Leeuwenhoek discovered the invisible world of microorganisms, the scientific community became interested in the origins of these tiny living things. Until the second half of the nineteenth

FIGURE 1.1 Anton van Leeuwenhoek (1632–1723) was the first to describe microbes (bacteria and protozoa) under a microscope. He is known as the father of microbiology and he was the first to examine our environment.

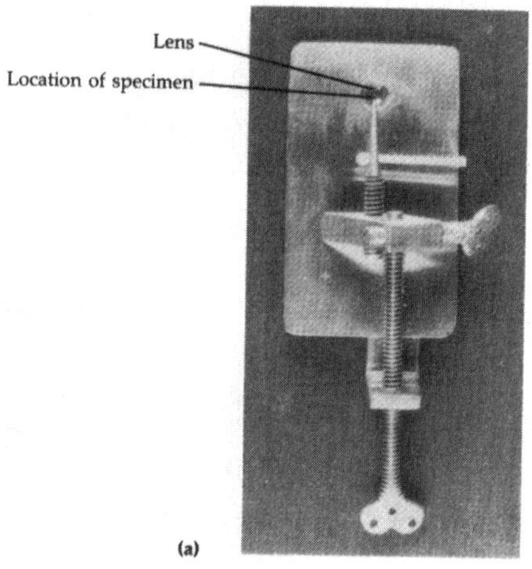

FIGURE 1.2 a. Leeuwenhoek's simple microscope.

century, it was generally believed that some forms of life could arise spontaneously from nonliving matter. This process was known as spontaneous generation.

People thought that toads, snakes, and mice could be born of moist soil, that flies emerged from manure, and maggots could arise from decaying corpses. Common people embraced the idea, for even they could see slime breeding toads, and meat generating worm-like maggots.

In 1749, British clergyman *John Needham* put forth the notion that microorganisms arise by spontaneous generation in flasks of molten gravy. Needham experimented with meat exposed to hot ashes, observed the appearance of organisms not present at the start of the experiment, and concluded that the bacteria originated from the meat.

Francesco Redi disputed the theory of spontaneous generation. Redi performed a series of tests in which he covered jars of meat with fine lace preventing the entry of flies and showed that the meat would not produce maggots when protected from flies. Years and years passed with arguments and experiments against the spontaneous generation theory without this proof being accepted. Redi's experiment is shown in Figure 1.3.

In 1858, the German scientist *Rudolf Virchow* challenged the spontaneous generation theory with the concept of biogenesis. He said, "Omnis cellula ex cellulae," that living cells originated from another preexisting living cell.

Finally, in 1861, *Louis Pasteur* (1822–1895) performed experiments that ended the argument for all time, see Figure 1.4. "There is no condition known today in which you can affirm that microscopic beings come into the world without germs, without parents like themselves. They who alleged it have been the sport of illusions of ill-made experiments, vitiated by errors which they have not known how to avoid." Pasteur demonstrated that microorganisms are indeed in the air and can contaminate seemingly sterile solutions, but air itself does not create microbes. He began by filling several short-necked flasks with beef broth and then boiling their contents. Some were then left open and allowed to cool. In a few days, these flasks were found to be contaminated with microbes. The other flasks, sealed after boiling, remained free of microorganisms. From these results, Pasteur reasoned that microbes in the air were the agents responsible for contaminating nonliving matter.

Pasteur next placed broth in open-ended long-necked flasks and bent the necks into S-shaped curves. The contents of these flasks were then boiled and cooled. Even after months, the broth in the flasks did not decay and showed no signs of life. Pasteur's unique design allowed air to pass into the

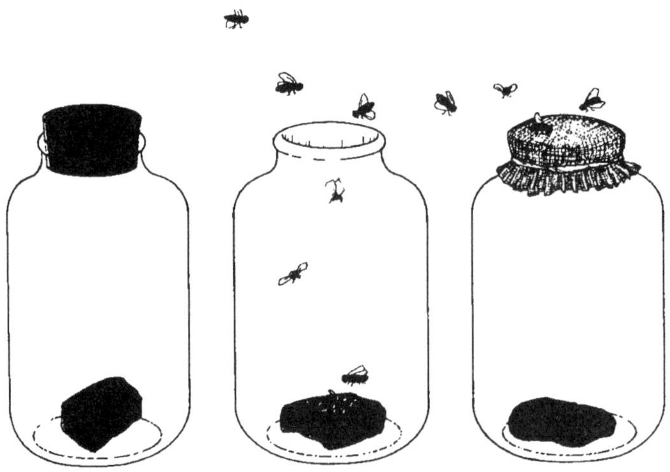

FIGURE 1.3 Redi's experiments refuting the spontaneous generation of maggots in meat. When meat is exposed in an open jar, flies lay their eggs on it, and the eggs hatch into maggots. In a sealed jar, no maggots appear. If the jar is covered with gauze, maggots hatch from eggs that flies lay on top of the gauze, but still no maggots appear on the meat.

FIGURE 1.4 Louis Pasteur is shown in his laboratory.

flask, but the curved neck trapped any airborne microorganisms that might contaminate the broth. Pasteur showed that microorganisms can be present in nonliving matter—on solids, in liquids, and in the air. Furthermore, he demonstrated conclusively that microbial life can be destroyed by heat and that methods can be devised to block the access of airborne microorganisms to a nutrient environment.

1.2.3 Aseptic Techniques

Pasteur's discoveries form the basis of aseptic techniques (techniques to prevent contamination by unwanted microorganisms). The techniques are now standard practice in laboratory and many medical procedures. Modern aseptic techniques are among the first and most important things that a beginner microbiologist must learn. Pasteur's work provided evidence that microorganisms cannot originate from mystical forces present in nonliving materials. Rather, any appearance of spontaneous life in nonliving solutions can be attributed to microorganisms that were already present in the air or in the fluids themselves. Pasteur was the first to relate the decay of milk by bacteria and he further concluded that it was the same within the human body.

1.3 THE GOLDEN AGE OF MICROBIOLOGY

The science of microbiology blossomed during a period of about 60 years, referred to as the Golden Age of Microbiology. The period began in 1857 with the work of Pasteur and continued until the advent of World War I. During these years, numerous branches of microbiology were established and the foundation was laid for the maturing process that led to modern microbiology. Discoveries during these years included both the agents of many diseases and the role of immunity in the prevention and cure of diseases. Microbiologists studied the chemical activities of organisms, improved techniques for performing microscopy and culturing microorganisms, and developed vaccines and surgical techniques.

1.3.1 Fermentation and Pasteurization

In the early days of microbiology, Pasteur found a practical method of preventing the sickness or spoilage of beer and wine. At this time, many scientists believed that air had the power to convert sugars into alcohols. Pasteur found that in the absence of air, microorganisms (yeasts) are responsible for this conversion. It is called fermentation. However, in the presence of air bacteria changes alcohol to acetic acid, causing souring and spoilage of alcohol.

Pasteur used mild heating, which was sufficient to kill the organisms that caused the particular spoilage problem without damaging the taste of the product. This process is called pasteurization. The same principle was later applied to milk to kill the harmful bacteria in it. In the classic pasteurization treatment of milk, the milk was exposed to a temperature of 63°C for 30 min. Most pasteurization done today uses a higher temperature, at least 72°C for about 15 s. Milk can also be treated so that it can be stored without refrigeration. It is heated to approximately 138°C for a few seconds and then cooled rapidly to minimize the scalded taste.

1.3.2 The Germ Theory of Disease

Before Pasteur, effective treatment of many diseases was discovered by trial and error, but the cause of the diseases was unknown. The idea that microorganisms might cause diseases was called the germ theory of disease. The germ theory was a difficult concept for many to accept because for centuries people believed that disease was a punishment for an individual's crime or misdeeds. When the inhabitants of an entire village became ill, foul odors from swamps were often blamed for the sickness. Most people born in Pasteur's time found it inconceivable that invisible microbes could travel through the air to infect plants and animals or remain on clothes and bedding to be transmitted from one person to another.

In the 1840s, the Hungarian physician *Ignaz Semmelweis* (1818–1865) demonstrated that physicians who went from one obstetric patient to another without disinfecting their hands transmitted puerperal (childbirth) fever. Deaths owing to infection were reduced in the cases handled according to his instructions, in which chances for infection were minimized by using a chlorine solution to rinse the hands after the usual handwashing process. As part of his crusade he published "The cause, concept, and prophylaxis of childbed fever" in 1861.

In the 1860s, *Joseph Lister*, an English surgeon, applied the germ theory to medical issues; see Figure 1.5. Disinfectants were unknown at that time. Lister had heard of Pasteur's work connecting microbes to disease, and he become interested in the work of Semmelweis. Lister began soaking surgical dressings in a mild solution of carbolic acid (aqueous solution of phenol). The practice so reduced the incidence of infections and deaths that other surgeons quickly adopted it. Lister's aseptic technique was one of the earliest medical attempts to control the infections caused by microorganisms.

The first proof that bacteria actually caused disease was given by *Robert Koch* in 1876. Koch, a brilliant German physician, discovered the rod shaped bacteria now known as *Bacillus anthraces* in the blood of cattle that had died of anthrax.

Koch demonstrated that microorganisms cause specific diseases; bacteria are present in the blood of infected animals, but not present in healthy animals. In short, Koch showed that a specific infectious disease is caused by a specific microorganism that can be isolated and cultured in an artificial medium. Koch later used the same methods to isolate the causative agent of tuberculosis (*Mycobacterium tuberculosis*).

Today we refer to Koch's experimental conclusions as *Koch's Postulates:*

1. The same pathogen must be present in every case of the disease.
2. The pathogen must be isolated from the diseased host and grown in pure culture.
3. The pathogen from a pure culture must cause the disease when inoculated into a healthy susceptible laboratory animal.
4. The pathogen must again be isolated from the inoculated animal and must be identical to the original organism.

FIGURE 1.5 A photograph of Lord Joseph Lister who applied the aseptic technique.

Koch's Postulates are useful in determining the causative agent of most bacterial diseases. It can also be used today with some exceptions.

Koch also developed the cultivation technique that we use today, called the pure culture technique. He demonstrated his pure culture technique in 1881 to the International Congress meeting at Lister's laboratory in London.

Koch, Pasteur, Semmelweis, and Lister contributed the so-called "Germ Theory of Disease," they proved that bacteria is the actual causative agent of special diseases.

1.3.3 Vaccination

The next step in the history of microbiology is to protect or prevent the human body from contracting the diseases caused by microorganisms.

Edward Jenner, a young British physician, was the first to try to find protection from smallpox; see Figure 1.6. Smallpox epidemics were greatly feared. The disease periodically swept through Europe killing thousands, and smallpox wiped out 90 percent of American Indians on the East Coast when European settlers brought the infection to the New World. The disease usually started with the appearance of small red spots (pox) all over the body. The pox became hard pimples that broke down into blisters filled with pus. Jenner observed that individuals who had been sick from cowpox (a much milder disease) could not get smallpox. He wanted to put this observation to a test. Jenner collected scrapings from cowpox blisters. Then he made inoculations with the cowpox materials by scratching a patient's arm with a pox-infected needle. In a few days, the patient became mildly sick and after recovering would never contract either cowpox or smallpox again.

The process was called vaccination from the Latin word vacca for cow and the vaccination of a child against smallpox is shown in Figure 1.7. The protection from disease provided by vaccination is called immunity. We now know that cowpox and smallpox are caused by viruses. Years later, around 1880, *Pasteur* discovered why vaccinations work. He discovered that microorganisms with

FIGURE 1.6 Edward Jenner introduced the modern method of vaccination to prevent smallpox.

FIGURE 1.7 Edward Jenner is shown vaccinating a child against smallpox.

decreased virulence still retained the ability to induce immunity against subsequent infections by virulent counterparts. Pasteur called the cultures of the avirulent microorganisms used for preventive inoculation a vaccine.

1.4 CHEMOTHERAPY

The treatment of disease by using chemical substances is called chemotherapy. The term also commonly refers to chemical treatment of noninfectious diseases such as cancer.

Chemotherapeutic agents prepared from chemicals in the laboratory are called synthetic drugs. Agents produced naturally by bacteria or fungi are called antibiotics.

1.4.1 The First Synthetic Drugs

Paul Ehrlich, a German physician, launched a search for a magic bullet that could kill or destroy a pathogen without harming the infected host; see Figure 1.8. In 1910, he found a chemotherapeutic agent called salvarsan, an arsenic derivate that was effective against syphilis. By the late 1930s, researchers had developed several other synthetic drugs that could kill microorganisms. The sulfa drugs were one important group discovered at the same time.

1.4.2 Antibiotics

Alexander Fleming, a Scottish physician and bacteriologist, almost tossed out some culture plates that had been contaminated by mold, when he observed a clear area around the mold, where the bacterial culture had stopped growing; see Figure 1.9. The mold was later identified as *Penicillium notatum* and, in 1928, Fleming called this active inhibitor of bacterial growth penicillin. Penicillin is an antibiotic produced by a fungus. Finally, penicillin was produced and clinically tested in 1940. Since then, many other antibiotics have been found.

Unfortunately, antibiotics and other chemotherapeutic drugs are not without problems. Many antimicrobial chemicals are too toxic to humans for practical use, or they kill the pathogen microbes, but they also damage the infected host. The inadequate targeting of drugs and the lack of antiviral drugs are only two limitations of modern chemotherapy. An equally important problem is the emergence and spread of new varieties of microorganisms that are resistant to antibiotics. Sophisticated

FIGURE 1.8 Paul Ehrlich is shown in his laboratory. He was searching for a substance he called a magic bullet that could destroy a pathogenic organism without harming the infected host. He developed an arsenic compound, the salvarsan, that was effective against *Treponema pallidum*.

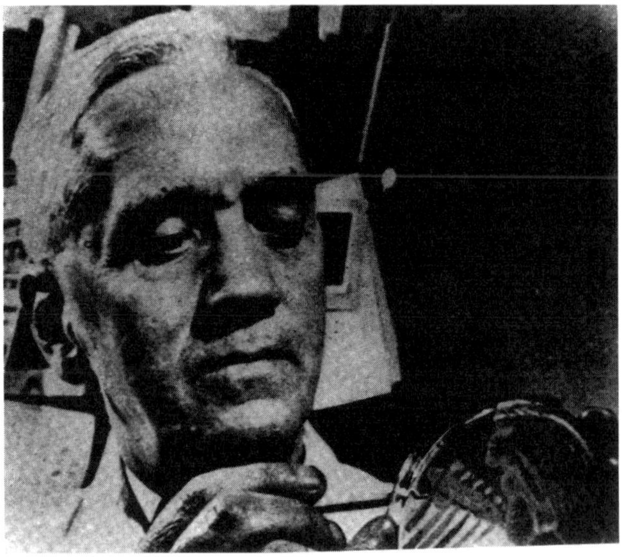

FIGURE 1.9 A photograph of Alexander Fleming who discovered penicillin.

research techniques and correlated studies are required to solve these problems. Microbiologists, pharmacologists, and clinicians are now working on various techniques to target antimicrobial drugs.

1.5 MODERN DEVELOPMENTS IN MICROBIOLOGY

After the groundwork of microbiology, new branches were developed including immunology and virology. Most recently, the development of a set of new methods called the recombinant DNA technique is developing research and practical applications in all areas of microbiology. DNA stands for deoxyribonucleic acid, a molecule that contains the genetic code (gene).

1.6 NAMING AND CLASSIFYING OF MICROORGANISMS

1.6.1 NAMING OF MICROORGANISMS

The system of naming (nomenclature) that we use was established in 1735 by *Carolus Linnaeus* (Linne). Scientific names are Latinized because Latin was the language related to sciences.

Scientific nomenclature assigns each organism two names: the genus is the first name and is always capitalized, and it precedes the species (specific epithet) name that is not capitalized. By custom, after the scientific name has been mentioned once, it can be abbreviated with the initial of the genus followed by the name of the species. For example, *Escherichia coli* can be written as *E. coli*. Scientific names can, among other things, describe the organism, honor a researcher, or identify the habitat of the species. For example, a bacterium commonly found on the skin of humans is *Staphylococcus aureus*. *Staphylo* describes the clustered arrangement of the cells; *coccus* indicates their spherical shapes; *aureus* is latin for golden, the color of the bacterial colonies.

The genus of the bacterium *Escherichia coli* is named for a scientist, *Theodore Escherich*, whereas its specific epithet, *coli*, indicates to us that *E. coli* live in the colon, or large intestine.

Scientists group all organisms as follows: similar species are included in the same genus, similar genera are placed in a family, related families are in an order, related orders are in a class, related classes are in a phylum, and related phyla constitute a kingdom.

In 1969, *Robert H. Whittaker* proposed the five kingdom system of biological classification, which considered procaryotes to be the ancestors to all eukaryotes.

1. Kingdom *Prokaryote*—bacteria or Monera.
2. Kingdom *Protista*—slime molds, protozoa, and some algae.
3. Kingdom *Fungi*—unicellular yeasts, multicellular molds, and mushrooms.
4. Kingdom *Plantae*—some algae and all plants.
5. Kingdom *Animalia*—sponges, worms, insects, and vertebrates.

1.7 THE DIVERSITY OF MICROORGANISMS

1.7.1 Bacteria

Bacteria are very small, relatively simple organisms, whose genetic material is not enclosed in a special nuclear membrane-bounded organelle called a nucleus.

For this reason, bacteria are called procaryotes. Procaryotes have no nucleus, but have genetic material located in the cytoplasm.

The other group of living organism cells are the *eucaryotic* cells containing genetic material enclosed by a membrane-forming nucleus.

Bacteria cells appear in variations of three different shapes:

1. Rodlike (*bacillus*).
2. Spherical or ovoid (*coccus*).
3. Spiral (*corkscrew* or *curved*).

Individual bacteria may form:

1. Pairs.
2. Chains.
3. Clusters, or other grouping.

Such formations are usually characteristic of a particular genus or species.
General features of bacteria that can be observed with the light microscope include:

1. Almost all bacteria have a semirigid cell wall.
2. If bacteria are motile, their motility is usually achieved by flagella.
3. Bacteria are unicellular. Although they are frequently found in characteristic groups, each cell carries out all the functions of the organism.
4. Most bacteria multiply by binary fission, a process by which a single cell divides into two identical daughter cells.

1.7.2 Fungi

Like bacteria, fungi are also extremely diverse. Fungi are eucaryotes—organisms whose cells have a distinct nucleus which contains the cell genetic material and which is surrounded by a special envelope called the nucleus membrane. Fungi are heterotroph organisms requiring organic compounds for energy and carbon sources. Fungi are aerobes or facultative anaerobes. No anaerobic fungi are known. The majority of fungi are saprophytes in soil and water, where they primarily decompose organic material to obtain their nutrients. Fungi are chemoheterotrophs and similar to bacteria, they contribute significantly to the decomposition of matter and recycling of nutrients. By using extracellular enzymes, such as cellulase and pectinase, fungi are primarily decomposers of the hard parts of plants that cannot be digested by most animals.

Of the more than 100,000 species of fungi, only about 100 are pathogenic for humans and other animals. On the other hand, thousands of fungi are pathogenic to plants. Originally, every ecologically important plant is attended by one or more fungi.

Fungi may be unicellular, or multicellular. The unicellular forms of fungi, yeasts, are oval microorganisms that are larger than bacteria.

The most typical fungi are called filamentous fungi or molds and form tube like filaments called hyphae. Hyphae are composed of many cells and can form integrated masses called mycelia. Mycelia are the visible structures seen when molds grow on bread and other substrates.

Large multicellular fungi, such as mushrooms, may look somewhat like plants.

1.7.3 Protozoa

Protozoa are unicellular, eucaryotic microbes and belong to the Kingdom Protista. They usually lack a cell wall. Protozoa have a heterotrophic metabolism, meaning that they obtain cellular energy from organic substances such as proteins. They live either as free entities or as parasites (organisms that derive nutrients from living organisms).

Motility is a major characteristics traditionally used in classifying protozoa.

1. Amoebas move by using extensions of their cytoplasm called pseudopods (false feet).
2. Flagellate move by means of one or more whiplike flagella.
3. Ciliata have numerous hairlike projections, called cilia.

1.7.4 Algae

Algae are photosynthetic eucaryotes with a wide variety of shape and both sexual and asexual reproductive forms. The algae are usually unicellular. Algae are classified according to the biochemical characteristic of their pigments, storage products, cell walls, and by the kind of flagellas they have.

They are abundant in fresh and salty water, soil, plant, and other media. As photosynthesizers, algae need light and air for food production and growth, but do not generally require organic compounds from the environment. By producing oxygen, which is utilized by other organisms including animals, algae play an important role in the balance of nature.

1.7.5 Viruses

Viruses are very different from the other microorganisms. They are very small, most can only be seen with an electron microscope, and they are not cellular. Structurally, the viruses are very simple and contain only one type of nucleic acid, either DNA or RNA. All living cells have RNA and DNA, can carry out chemical reactions, and can reproduce themself. Viruses, however have no cell machinery for metabolism and can reproduce only inside the cells of other organisms. Thus, viruses are parasites of other forms of life.

There are three groups of viruses. Animal viruses infect and replicate within animal cells, plant viruses infect and replicate within plant cells, and bacteriophages infect and replicate within bacterial cells.

Viruses that infect bacterial cells cannot infect and replicate within cells of other organisms. It is possible to utilize the specificity to identify bacteria, a procedure called phage typing.

1.7.6 Multicellular Parasites

Multicellular parasites are also a part of microbiology. They have medical importance. The two major groups of parasites are the *flatworm* and the *roundworm*, collectively called *helminths*. During

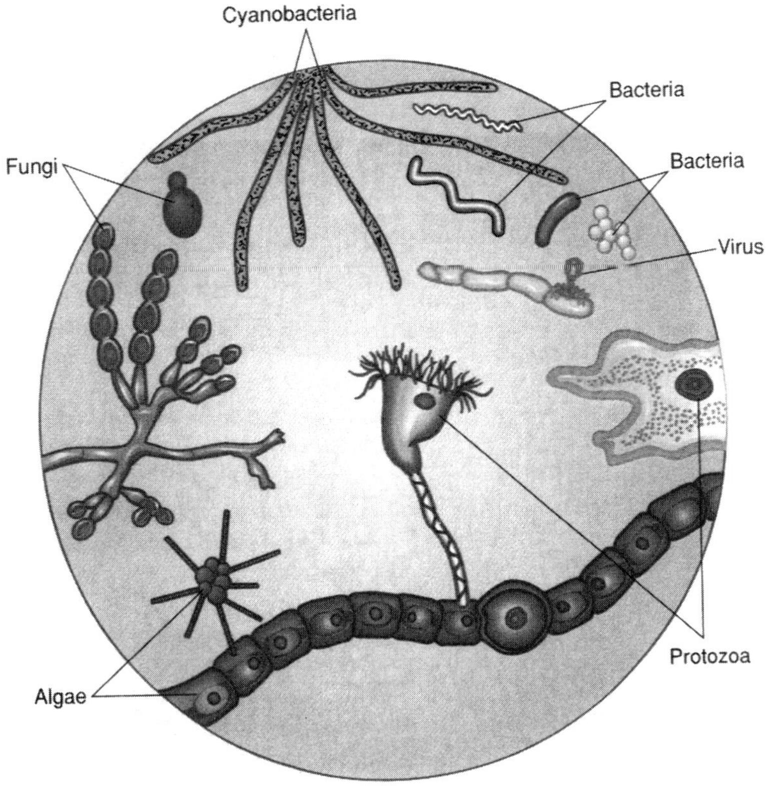

FIGURE 1.10 The diversity of the microbial world.

reproduction, the eggs and the developing helminths are microscopic in size. Many techniques, including the use of a microscope, must be used in the laboratory identification of these organisms.

The diversity of microorganisms is shown in Figure 1.10.

2 Procaryotic Cells

2.1 INTRODUCTION OF PROCARYOTIC AND EUCARYOTIC CELLS

Despite the complexity and variety of life, all living cells can be divided into two groups, procaryotic and eucaryotic. Procaryotic is derived from the Greek for pronucleus; eucaryotic is also derived from the Greek for true nucleus. In the microbe world, bacteria and cyanobacteria (formerly known as blue-green algae) are assigned to their own kingdom, that is, the kingdom procaryote. Other microbes (protozoans, yeasts, molds, algae, plants, and animals) are entirely composed of eucaryotic cells. As noncellular elements, with some cell-like properties, viruses do not fit into any organization scheme of living cells. They are unable to perform the usual chemical activity of living cells. Procaryotes and eucaryotes are chemically similar; both contain nucleic acids, proteins, lipids, and carbohydrates. Both use the same kind of chemical reactions to metabolize food, build proteins, and store energy. It is primarily the structure of the cell walls, cell membranes, and various internal components that distinguish procaryotes and eucaryotes.

2.2 THE PROCARYOTIC CELL

As mentioned previously, bacteria are composed of procaryotic cells. The thousands of species of bacteria are differentiated by many factors, including morphology (shape), chemical composition, nutritional requirements, and biochemical activities.

Bacteria are unicellular organisms and can be found in characteristics groupings, but each cell carries out all the functions of the organism. All bacteria (with few exception) have a semirigid cell wall. If the bacteria are motile, movement is achieved by flagella. The structure of the bacterial cell is shown in Figure 2.1.

2.2.1 Morphology of the Bacterial Cell

Morphology means the external appearance, including the size, shape, and arrangement of the bacterial cell. Sizes fall within a range of 0.2 to 2.0 μm in diameter with three basic shapes: the spherical, called coccus; the rod, called bacillus; and the curved or corkscrew, called spiral. The basic shapes of bacteria are shown in Figure 2.2.

The Chief Characteristic Differences between Procaryotic and Eucaryotic Cells

Procaryotic Cell	Eucaryotic Cell
They lack a nucleus.	They have nucleus.
Genetic material is not enclosed within the nucleus.	Genetic material is enclosed within the nucleus.
They lack organelles.	They have organelles.
DNA is not associated with the histone protein.	DNA is associated with the chromosomal protein, histone.
Cell wall contains peptidoglycan (a complex carbohydrate).	They have no cell wall; other eucaryotes contain cellulose (as in plants).
Procaryotes divide by binary fission.	They have a mitotic apparatus that participates in a type of nuclear division called mitosis.

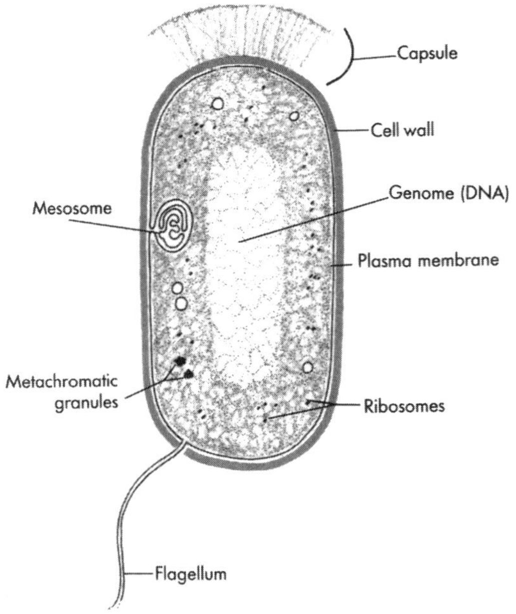

FIGURE 2.1 Ultrastructure of a bacterial cell.

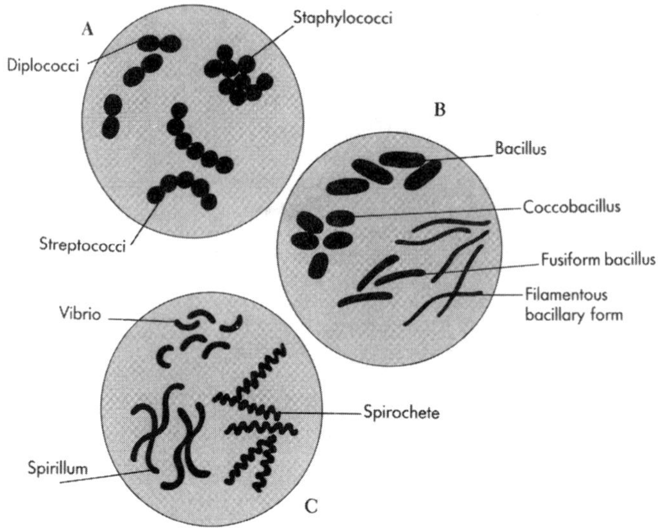

FIGURE 2.2 The basic shapes of bacteria: A. The coccus is the spherical one, a pair of cocci is diplococci, a chain is the streptococci, and a grapelike cluster is the staphylococci. B. The bacillus is rod shaped. C. The vibrio, spirillum, and spirocheta are spiral shaped.

2.2.1.1 Spiral Bacteria

Spiral bacteria have one or more twists and are never straight:

1. *Vibrios* are curved rods that look like commas.
2. *Spirillas* look like a corkscrew and have flagella.
3. *Spirochetes* move by axial filament.

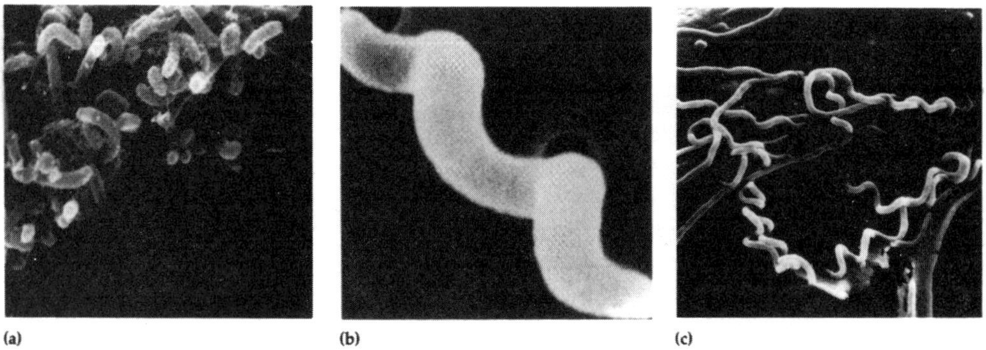

FIGURE 2.3 Spiral bacteria: a. Vibrios. b. Spirillum. c. Spirochetes.

The shape of bacteria is determined by heredity, called monomorphic bacteria. However, a number of environmental conditions can alter the shape of the organisms, causing a different shape, called pleomorphic bacteria. Unfortunately, if this happens identification becomes even more difficult. Spiral bacteria are shown in Figure 2.3.

2.2.1.2 Bacilli

Bacilli are rod shaped bacteria, and when divided can be:

1. Single.
2. Diplobacilli, pairs.
3. Streptobacilli, chains.
4. Coccobacilli, oval shaped rods.

The diagram of a fictionalized bacterial cell is shown in Figure 2.4.

2.2.1.3 Cocci

Cocci means "berries." They are not necessarily perfectly round, but may be somewhat elongated, oval, or flattened on one side. When cocci divide to reproduce, the cell can be:

1. Single cell.
2. Diplococci, pairs.

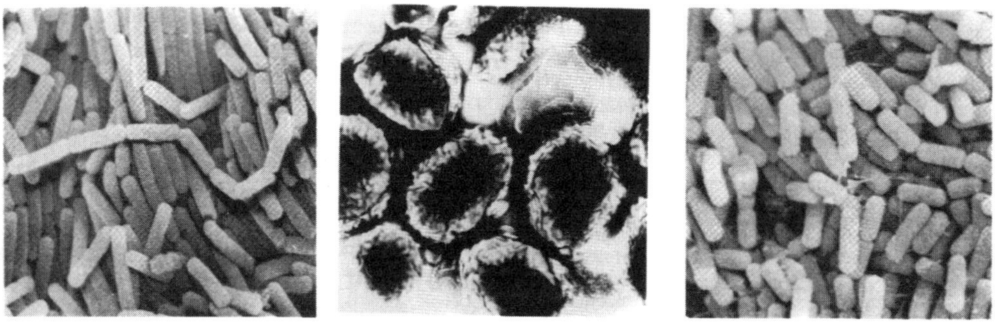

FIGURE 2.4 Diagram of a fictionalized bacterial cell.

3. *Streptococci*, chains.
4. *Staphylococci*, cocci divide at random planes and form grape-like clusters or broadsheets.
5. *Tetrad*, four cells group.
6. *Sarcinae*, cube-like groups of eight.

The variation of the cocci is shown in Figure 2.5.

2.3 STRUCTURES EXTERNAL TO THE CELL WALLS

The substance found external to the cell wall, from the thickest capsules to the thinnest slime layers is the glycocalyx, but also external to the cell wall are the flagella, axial filaments, fimbriae, and pili.

2.3.1 Glycocalyx

Glycocalyx is a gelatinous or gummy layer around the bacterial cell. It is a polymer, composed of polysaccharide, polypeptide, or both. If this viscous material is organized and attached to the cell wall, it forms a capsule that is highly antigenic and enhances the virulence of an organism. A capsule does not stain with ordinary bacteriologic dyes, but appears as a clear halo around the bacterium.

If unorganized, it will be loosely attached to the cell wall, or if it is easily deformed, it is described as a slime layer.

Generally the important mechanism of glycocalyx is to increase bacterial virulence, the degree of pathogenicity, and to protect bacterial cells from phagocytosis of the host cell. For example, *Streptococcus pneumoniae, Klebsiella pneumoniae, and Bacillus anthracis* are important encapsulated pathogenic organisms.

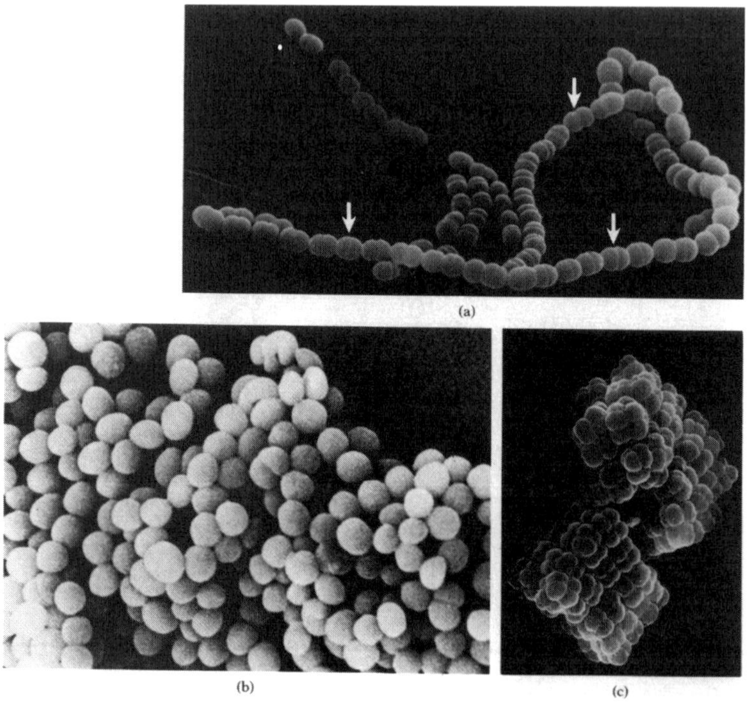

FIGURE 2.5 Three variations of Cocci: a. Streptococcus. b. Staphyloccus. c. Cluster of sarcinae.

2.3.2 Flagella

If the bacteria can move, we call it motile. It can run or swarm on the culture medium. Many bacteria are capable of this and this is usually owing to the presence of a special organelle of motility, the flagellum.

Bacterial flagella are long, thin appendages, are free at one end, and are anchored to the cell at the other end. They are so thin, that they may only be seen under the microscope with special staining methods. Flagella are arranged differently on different organisms.

Monotrichous—bacteria may have one polar filament.
Amphitrichous—have single flagellum at each end of the cell.
Lophotrichous—two or more flagella at one or both poles of the bacteria.
Peritrichous—flagella over the entire cell.
Axial filaments—inner flagella, which arise at the poles, within the cell wall, and spiral around the cell. The rotation of the filaments causes the cell to rotate much like a corkscrew; found in spirochetes. Eucaryotic cells do not contain axial filaments.

Various types of flagellation are shown in Figure 2.6.

2.3.3 Pili

Pili are also referred to as attachment organelles. They are the filamentous surface projections like flagella found in certain Gram-negative bacteria. They are shorter and finer than flagella and are not used for motility.

Pili are of two different types. Bacteria have common pili that can adhere to mucous membranes and other cell surfaces in a susceptible host, and increase the virulence of the bacteria. Sex pili help to join bacterial cells in order to exchange DNA from one cell to another during bacterial conjugation.

2.4 THE CELL WALL

The complex, semirigid structure that is responsible for the characteristic shape of the cell and protects it from damage is the cell wall. The cell envelope refers to the cell or plasma membrane plus all external structures, such as the cell wall and capsule.

The major function of the cell wall is to prevent the bacterial cell from rupturing when the osmotic pressure is different from outside to inside the cell. Clinically, the cell wall is an important target of antibiotics action.

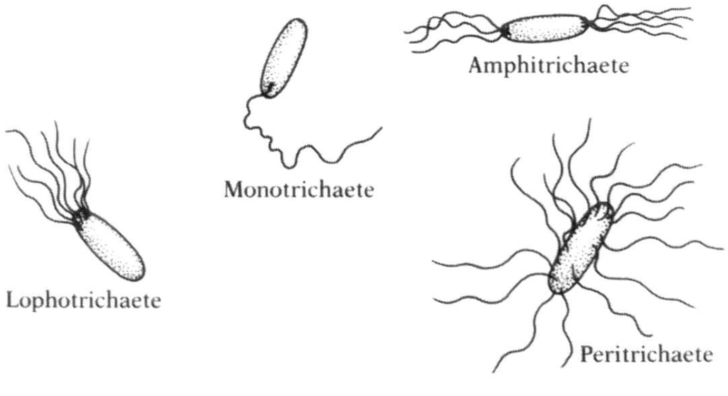

FIGURE 2.6 Details of various types of flagellation.

The cell wall is chemically composed of a macromolecule called the peptidoglycan (murein) layer.

We can distinguish two groups of bacteria with a special type of staining method, called the Gram stain. This method of staining was introduced in 1884 by *Hans Christian Joachim Gram* (1853–1938), a Danish physician. It is an important method of differential staining that remains essentially unaltered in its use today. It divides bacteria into two groups: those that are Gram positive and those that are Gram negative. This method depends on the fact that bacteria cell walls can be of two different types, showing two different colors with the stain, either purple or red. Gram staining method is discussed further in Chapter 13.5.

2.4.1 Gram Positive Bacteria

Gram positive (Gr^+) bacteria cell walls consist of several peptidoglycan layers with teichoic acids that might aid the transport of ions into and out of the cell. Teichoic acid is negatively charged and promotes the movement of cations (positive ions) into and out of the cell.

Most Gr^+ bacteria produce exotoxin during metabolism and exotoxin is deposited into the bloodstream of the host. Exotoxins are protein and act as an antigen. They can travel from the focus point of infection to distant parts of the body and may cause a fever.

2.4.2 Gram Negative Bacteria

Gram negative (Gr^-) bacteria also contain a peptidoglycan layer, but in very small amounts, located in periplasmic space and Gram negative bacteria contain no teichoic acid.

The peptidoglycan layer is surrounded by an outer membrane, found only in Gr^- bacteria. This is built up by a phospholipids bilayer which consists of lipoprotein and lipopolysaccharide. The polysaccharide of the external cell membrane of Gr^- bacteria is an excellent antigen used for identification. It distinguishes the species of Gr^- bacteria as the Salmonella species in an immunological serological way (also called the typisation of bacteria). The lipid portion of the external cell membrane of Gr^- bacteria is endotoxin. Endotoxin can be liberated and deposited into the host's bloodstream when the Gr^- bacteria die. The Gr^- bacteria cell wall undergoes lysis, causing fever and intravascular hemolysis.

The outer membrane in Gr^- bacteria is an important barrier and forms a protective layer against antibiotics, dyes, bile salts, and heavy metals. The outer membrane of Gr^- bacteria is permeable because of porin channels.

Figure 2.7 compares the structures and chemical composition of the Gr^+ and Gr^- bacteria cell wall.

2.4.3 Atypical Cell Wall

Among procaryotes there are cells that have no natural walls. *Mycoplasma*, the smallest known bacteria, are capable of autonomous growth and are of special evolutionary interest because of their extremely simple cell structure. Mycoplasma cells are usually small, and they are highly pleomorphic, a consequence of their lack of rigidity. Their plasma membranes have sterols which are thought to help protect them from osmotic lysis. *Mycoplasma pneumonia* is the causative agent of atypical or walking pneumonia.

L-form bacteria, named after Lister, have a defective cell wall. In certain species of bacteria, such as Proteus, Bacteroides, Pseudomonades, and some coliform, a normal cell may swell in response to special chemicals and disintegrate into numerous particles known as L-forms. Certain chemicals and antibiotics induce the production of L-form organisms.

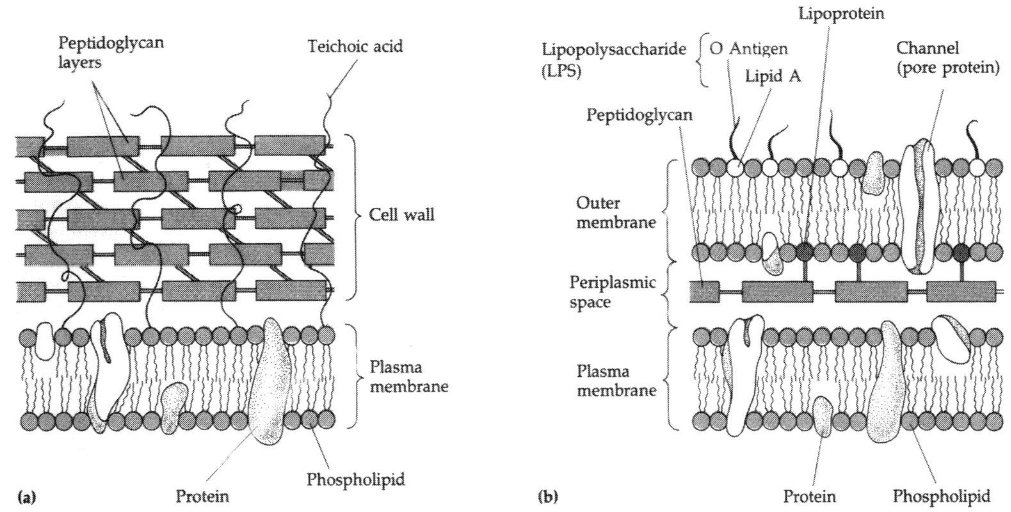

FIGURE 2.7 Comparison of structures and chemical composition of: a. Gram-positive cell wall. b. Gram-negative cell wall.

2.4.4 Damage to the Cell Wall

Some bacteria cell walls can be damaged and loosened by antibiotics or lysozyme enzyme activity. The remaining cellular structures without cell walls are called protoplast in Gr^+ bacteria and spheroplast in Gr^- bacteria. Both are very vulnerable in different osmotic solutions and cannot tolerate the osmotic pressure causing osmolysis or plasmolysis.

2.5 STRUCTURES INTERNAL TO THE CELL WALL

2.5.1 Plasma or Inner Membrane

In both Gr^+ and Gr^- bacteria, the plasma membrane lies below the cell wall enclosing the cytoplasm of the cell. The plasma membrane consists of the phospholipid bilayer, containing a charged polar head, composed of the phosphate (PO_4^{3-}) group and glycerol. It is water soluble, called *hydrophilic* (water loving). The nonpolar tail is insoluble in water, called *hydrophobic* (water fearing).

The membrane is an important barrier through which materials enter and exit the cell. In this function, the plasma membrane is selectively semipermeable. This indicates that certain molecules and ions pass through the membrane, but others cannot. The movement through the membrane depends on its size, charge, and lipid solubility. The plasma membrane in a procaryotic cell takes over the function of mitochondria and it is the site of energy production. There are pigments, called *chromatophores*, and enzymes in the cell membrane of the photosynthetic and cyanobacteria organisms involved in capturing light energy for the synthesis of sugars.

In some locations, the cell membrane extends inwardly into coiled passages, or sacs in the cytoplasm, called *mesosomes*. They increase the internal surface area available for membrane activities. Mesosomes participate in cell wall synthesis and they may function during bacterial division, called binary fission, forming a transverse septum to separate one cell into two. Mesosomes are prominent in Gr^+ bacteria.

2.5.2 Movement of Material Across Membranes

2.5.2.1 Diffusion

Diffusion is the driving force of the passive random transport of particles from an area of higher concentration to an area of lower concentration. For diffusion to occur, there must be a difference in concentration levels. Simple diffusion is shown in Figure 2.8. The rate of diffusion is affected by the concentration gradient, the size of particles, the temperature, and the medium where diffusion is occurring. Diffusion can take place in the air or in the water.

Some material can cross the cell membrane as diffusion. If in an aqueous solution the solvent (as water) moves through the cell membrane, the phenomenon is called *osmosis*; if the solvent moves through the cell membrane, it is called *dialysis*.

Osmotic pressure is the pressure needed to stop the flow of water across the selectively permeable membrane. In other words, osmotic pressure is the pressure required to prevent the movement of pure water into a solution containing some solute. A cell may be subjected to three kinds of osmotic solutions: isotonic, hypotonic, and hypertonic as shown in Figure 2.9.

Isotonic (iso means equal) solution occurs when the overall concentration of solutes are the same on both sides of the membrane. Cell contents are in equilibrium with the solution outside the cell wall. In that case, the same concentration of solutions is separated by the cell membrane, the net movement of the water molecules is zero, and a dynamic equilibrium exists between the two solutions.

Hypotonic (hypo means less) solution occurs when a solution outside the cell, the solute concentration, is lower than the concentration inside the cell. A cell is in a hypotonic solution when water molecules tend to move from the more concentrated area to the less concentrated one. They diffuse inward across the cell membrane and exert an osmotic pressure on the other side of the membrane. So much water can enter the cell that it swells and bursts, causing osmolysis in the bacterial cell.

Hypertonic (hyper means above) solution occurs when the solution surrounding the cell has a greater solute concentration. It has a higher osmotic pressure than the pressure inside the cell and water tends to flow out from the cell through the cell membrane. The cell may become dehydrated, shrink, and die causing bacterial cell plasmolysis.

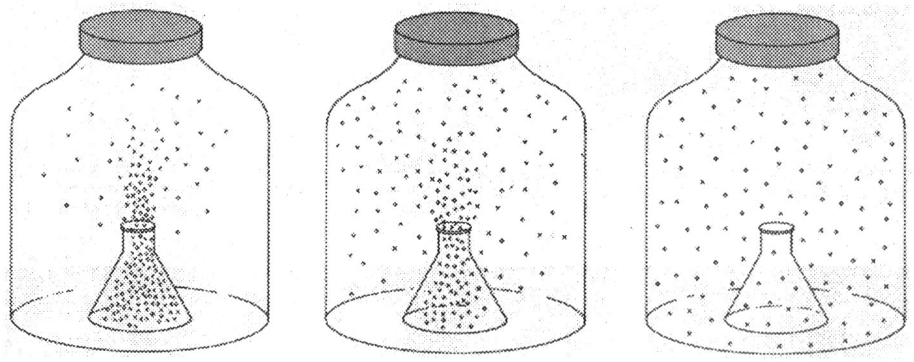

FIGURE 2.8 Simple diffusion: the random movement of particles causes the particles to spread out (diffuse) from an area of high concentration to areas of lower concentration until eventually they are equally distributed throughout the available space.

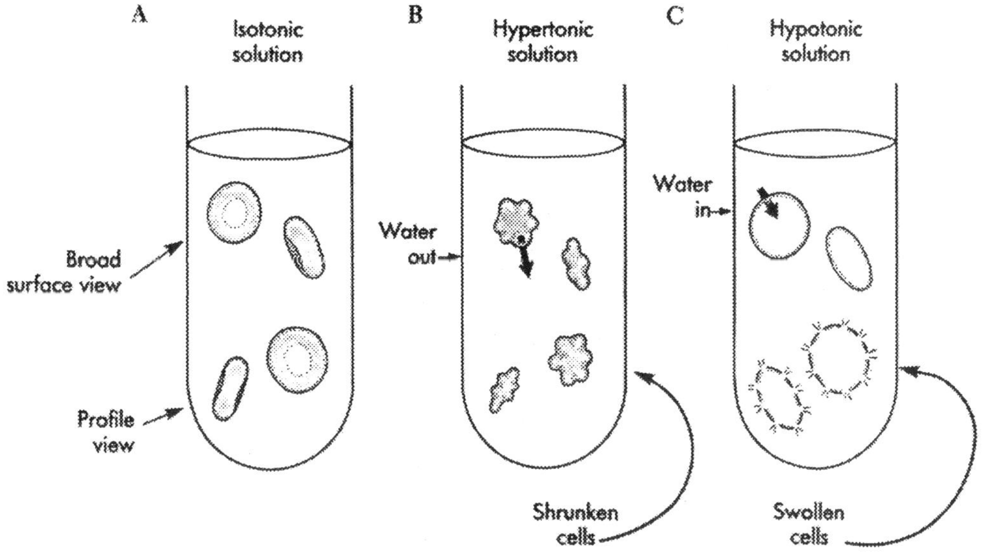

FIGURE 2.9 Effect of osmotic pressure on cells: A. In isotonic solution cell does not change size or shape. B. Cell in hypertonic solution shrinks and water moves out from the cell by osmosis. C. Cell in hypertonic solution gains water, swells, and bursts.

2.5.2.2 Transport by Carriers

As in all cells, the microorganism also provides the necessary nutrients that must be taken into the cell from the surroundings. Survival also requires that cells transport products, such as waste materials into the environment from the cell. Whatever the direction, transport occurs across the cell membrane—the structure specialized for this role.

The general type of transport by carriers is passive transport which follows physical laws and generally does not require direct energy input from the cell as the material moves from an area of high concentration to a less concentrated area. This kind of transportation only needs the transporter, not energy.

If the transportation is from an area of lower concentration to an area of higher concentration against the concentration gradient, energy needs to be input. We call this situation active transport. Active transport requires carrier proteins in the membrane of living cells and the expenditure of energy.

2.5.2.3 Endocytosis and Exocytosis

Endocytosis and exocytosis occur only in eucaryotic cells and not in procaryotic cells.

Endocytosis is the movement of material into the cell, known as phagocytosis, when the eucaryotic cell can engulf a procaryotic cell.

Exocytosis is the movement material from the cell and the release of the material into the surrounding matter. Endocytosis and exocytosis are shown in Figure 2.10.

2.5.3 Cytoplasm

For a procaryotic cell, the term cytoplasm refers to the internal matrix of the cell contained inside the plasma membrane. Cytoplasm consists of about 80 percent water and contains primarily proteins (enzymes), carbohydrates, lipids, and inorganic ions. Cytoplasm is a thick, aqueous,

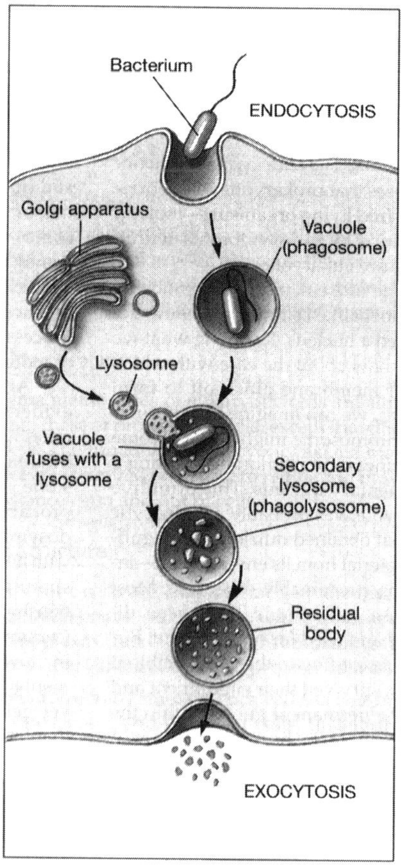

FIGURE 2.10 Endocytosis is the process of taking material into the cell. This process is called phagocytosis. Exocytosis is the process of releasing material from the cell.

semitransparent, and elastic matter. Cytoplasm is the stage where chemical reactions take place. Procaryotic cytoplasm lacks certain features of eucaryotic cytoplasm.

2.5.4 Nuclear Area

The nuclear area or the nucleoid in procaryotic cell is one of the key features that differentiate procaryotic cells from eucaryotic cells. This is owing to the absence of a nucleus bounded by a nuclear membrane. Instead of a nucleus, bacteria have a nuclear region or nucleoid area only that contains a single, long circular DNA, the bacterial chromosomes.

Bacteria often contain in addition to the bacterial chromosomes, small circular DNA molecules called *plasmids*. See Figure 2.11. These molecules are extrachromosomal genetic elements. They are not connected to the main bacterial chromosomes and they replicate independently of chromosomal DNA. Plasmids may carry genes for such activities as antibiotic resistance, tolerance to toxic metals, production of toxins, and synthesis of enzymes.

2.5.5 Ribosomes

Ribosomes are small organelles where protein synthesis occurs and are composed of protein and a type of RNA, called ribosomal RNA. Procaryotic ribosomes differ from eucaryotic ribosomes in the

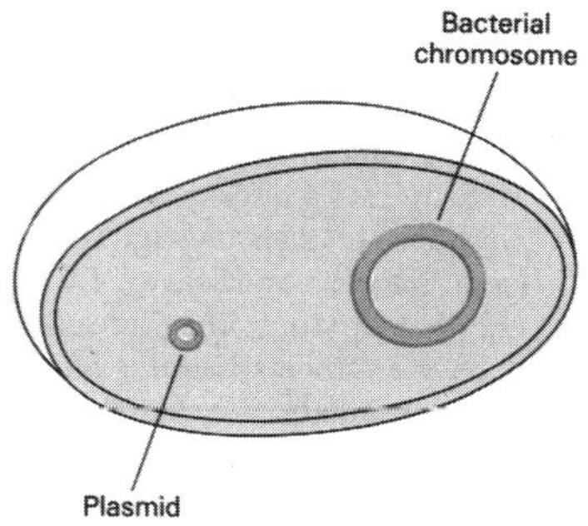

FIGURE 2.11 A bacterial chromosome and plasmid.

number of proteins and RNA molecules they contain. Procaryotic ribosomes are called 70S ribosomes, and those of eucaryotic ribosomes are known as 80S ribosomes. The letter S refers to *Svedberg unit*, which indicates the relative rate of sedimentation during ultra-high-speed centrifugation. Sedimentation rate is a function of the size, weight, and shape of the particles.

The function of ribosomes is the production of protein. Protein anabolism, or protein synthesis, is a process in which amino acids are bound together in a sequence determined by the hereditary information in the cell. The central theme of protein synthesis holds the segments of DNA on the chromosome, known as genes, that provide a code for production of fragments of RNA. The genetic code of DNA is expressed in RNA by a process called *transcription*.

One type of RNA then functions as a messenger by carrying the code to other areas of cytoplasm where amino acids are fitted together in a precise sequence to form the protein. This sequencing process, called *translation*, reflects the genetic code in the DNA.

The overall process is summarized as follows:

$$\text{DNA} \xrightarrow{\text{transcription}} \text{RNA} \xrightarrow{\text{translation}} \text{protein}$$

Several antibiotics act as inhibition of protein synthesis in the bacterial cells by interfering with bacterial metabolism. These include streptomycin, neomycin, and tetramycin.

2.5.6 Inclusion

Within the cytoplasm there are several kinds of reserved deposits, known as inclusions. Some inclusions are common in several bacteria, whereas others are limited to a small number of species, and therefore, serve as a basis for identification.

2.5.6.1 Metachromatic Granules

Metachromatic granules stain red with certain blue dyes such as methylene blue and they name these inclusions as *volutin*. Metachromatic granules contain inorganic polyphosphate that can be used in the synthesis of ATP and is found mostly in algae, fungi, and protozoa.

These granules are characteristic of *Corynebacterium diphtheriae*, the causative agent of diphtheria.

2.5.6.2 Polysaccharide Granules

Polysaccharide granules consist of glycogen and starch. Their presence can be demonstrated when iodine is applied to the cells. In the presence of iodine, glycogen granules appear as reddish brown, and starch granules are color blue.

2.5.6.3 Lipid Inclusions

Lipid inclusions are in *Mycobacterium*, *Azotobacter*, *Spirillum*, and other genera.

2.5.6.4 Sulfur Granules

Sulfur granules are common in sulfur bacteria, such as *Thiobacillus*, derive energy by oxidizing sulfur and sulfur containing compounds. These sulfur granules serve as an energy reserve.

2.5.6.5 Gas Vacuoles

Gas vacuoles are hollow cavities found in many aquatic procaryotes. They maintain buoyancy so that the cells can remain in the water, and receive sufficient amount of oxygen, light, and nutrients.

2.5.7 Endospores

2.5.7.1 Sporulation

When essential nutrients are depleted or when water is unavailable, certain Gr^+ bacteria form resting cells called spores. Clostridium and Bacillus species are the best known spore formers. The spore forming bacteria grow, mature, and reproduce for several hours as vegetative cells. Spore formation then begins and is called sporulation.

The bacterial chromosome replicates, a small amount of cytoplasm gathers with it, and the cell membrane grows in to seal off the developing spore. Thick layers of peptidoglycan form and a series of coats are synthesized to protect the contents further. The cell wall of the vegetative cell then disintegrates and the spore is freed. The function is called sporulation and the products are in the dormant stage, helping it to survive.

This is an important clinical view point, because the spores are probably the most resistant living things known. They are resistant to any process that normally kills vegetative cells, such as heating, freezing, desiccation, chemicals, and radiation. The structural changes that occur during the conversion of a vegetative cell to a spore can be studied readily with the electron microscope. Under certain conditions, mainly as a result of nutrient exhaustion, instead of dividing, the cell undergoes the complex series of events leading to spore formation; these are illustrated in Figure 2.12.

2.5.7.2 Germination

Germination is the process when the endospore returns to its vegetative stage.

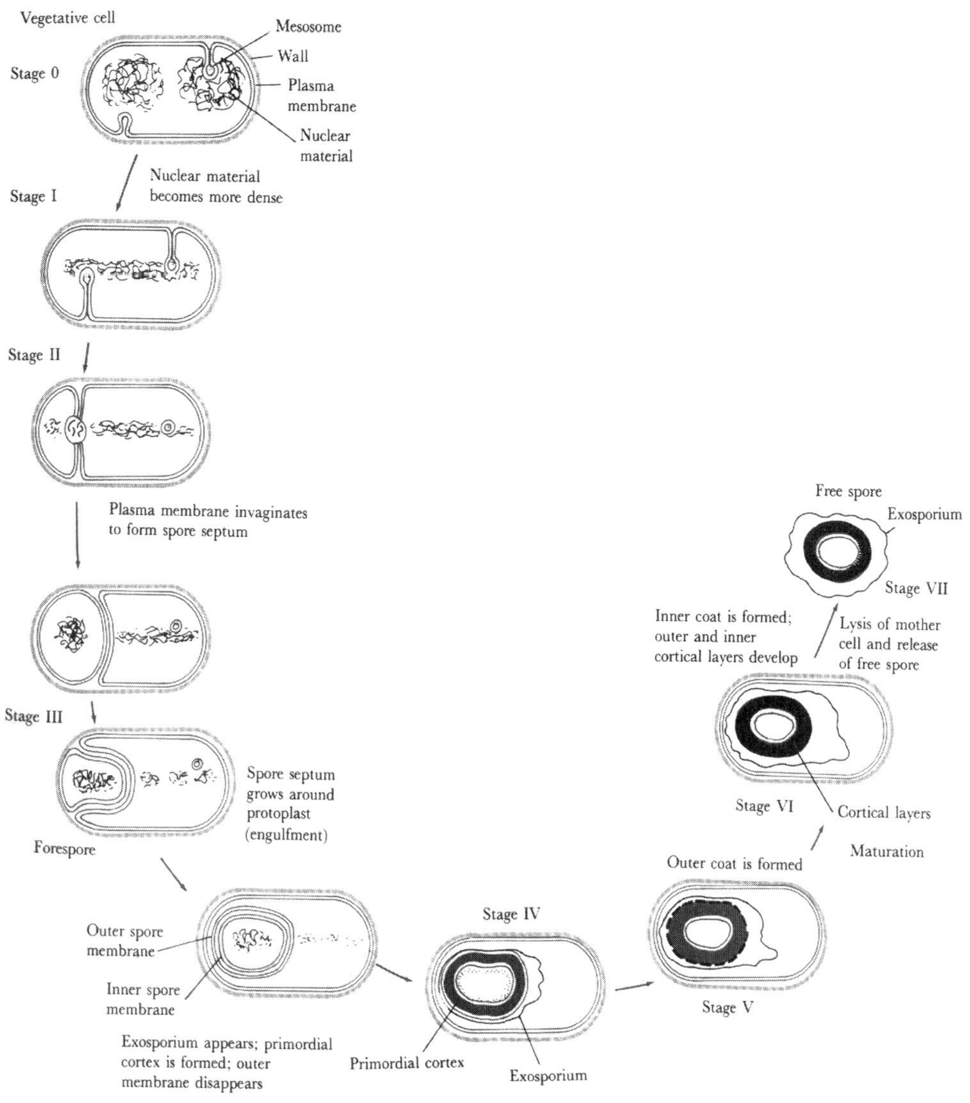

FIGURE 2.12 The stages in endospore formation.

3 Microbial Metabolism

3.1 MEANING OF METABOLISM

The life support activities of organisms with even the simplest structure involve a large number of complex biochemical reactions, which we called metabolism.

There was a time, not long ago when people did not know what caused fruit juice to become wine or caused milk to sour. *Pasteur* was the first to prove in 1857 that the alteration was owing to microorganisms. Later, he identified microbes from samples of fermented juice and sour milk. He established a new theory, and he was the first to study these chemical processes in living organisms. We learned from him that there are some representative chemical reactions that either produce or liberate energy. We call these reactions metabolism.

If the reaction requires energy, it is called *anabolic reaction*; if the reaction releases energy, it is called *catabolic reaction*. The relationship between these two reactions is shown in Figure 3.1.

Generally, all chemical reactions within a cell are called metabolism and consist of two types of reactions: a synthetic reaction that needs energy, an anabolic reaction, and a degradative reaction that releases energy, a catabolic reaction.

metabolism	=	anabolism (synthesis)	and	catabolism (degradation)
		needs energy		releases energy

3.1.1 The First and Second Law of Thermodynamics

Thermodynamics is the study of energy relationship and exchange.

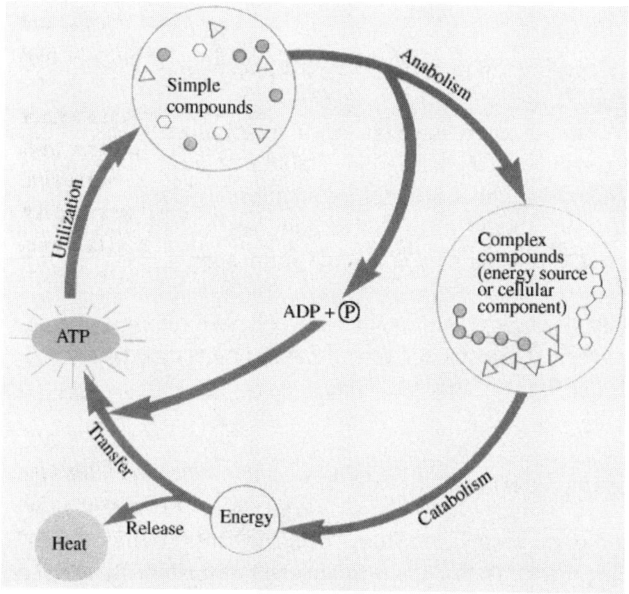

FIGURE 3.1 The relationship between anabolism, catabolism, ADP, and ATP.

3.1.1.1 The First Law of Thermodynamics

The first law of thermodynamics says that energy can neither be created nor destroyed, however, energy can be transformed or converted from one form to another form.

3.1.1.2 The Second Law of Thermodynamics

The second law of thermodynamics says that one usable form of energy cannot be completely converted into usable form. During these transformations, some usable energy is degraded into a less usable form, usually heat, that disperses into the environment. As a result, the amount of usable energy (energy that can do work) in the universe decreases over time and the less usable energy increases. According to the two thermodynamic laws, as far as we know, the energy present in the universe when it formed billions of years ago, equals the amount of energy that we have at present in the universe.

3.1.2 Energy Transport Through ADP and ATP

Each organism needs two sources: energy sources and carbon sources. Energy sources can be the sun's ray energy for photosynthetic organisms or can be organic or inorganic compounds for chemotropic organisms.

In a biological system, energy must be available nonstop, and the required energy is temporarily stored in the energy storing molecule, *adenosine triphosphate*, ATP. ATP is the energy currency in cells because it has high energy phosphate bonds.

ATP breaks down to *adenosine diphosphate*, ADP and P and energy, and rebuilds again from the same compounds.

The rebuilding process, however, requires an input of energy, and this energy comes from the metabolism of glucose products in mitochondria. This rejoining of ADP and P provides a constant supply of ATP in all cells.

$$\text{ADP} + \text{P} + \text{E} \longrightarrow \text{ATP} \tag{3.1}$$

where ADP equals adenosine diphosphate, P equals phosphorus, E equals energy, and ATP equals adenosine triphosphate.

ATP is continually made and remade in cells. For this process, the requirement is 7 kcal energy input, the amount that can be freed during the catabolic reaction. This means ATP has high energy phosphate bonds. When these bonds are broken, an unusually large amount of energy is released.

An average male needs about 8 kg of ATP an hour, yet the body has on hand only about 50 grams at any time. The answer to this paradox is that the entire supply of ATP is recycled about once each minute.

ATP is the organism's energy source. When the cell needs energy, it must be available from previously stored ATP energy. It is a nonstop process. ATP builds up on ADP and ATP is used as energy sources and returns as ADP (see Figure 3.1).

3.2 METABOLIC PATHWAYS

Metabolism is the sum of all chemical reactions occurring inside a living cell. The reactions are stepwise processes that begin with a particular reactant and terminate with an end product. Actually there are many steps between the reactant and the end product that are called metabolic pathways. Generally, reactants yield the products.

In the metabolic pathway, one reaction leads to the next reaction, which leads to the next, and so forth as a chain reaction.

$$A \xrightarrow{E_1} B \xrightarrow{E_2} C \xrightarrow{E_3} D \xrightarrow{E_4} E \xrightarrow{E_5} F \xrightarrow{E_6} G \xrightarrow{E_7} H \xrightarrow{E_n} I$$

The capital letters represent the reactants and products. A is the reactant and B is the product, now B is the reactant and C is the product, and so on.

The capital E represents the enzyme and indicates that each step of the metabolic pathway is supported by a different enzyme.

The reactants in an enzymatic reaction are called substrates for the enzyme. In the first reaction A is the substrate for E_1 and B is the product. Now B becomes the substrate for E_2 and C is the product. The process continues until the final product forms.

Metabolic pathways contain chemical reactions created by making and breaking chemical bonds within the cell. Chemical bonds keep atoms together to form molecules. The reactants must collide with each other according to the collision theory. The collision frequency depends on temperature and on the available energy level. To complete the reaction, the reaction needs special speed, energy, and temperature.

3.3 ENZYMES

A substance whose presence changes the rate of a chemical reaction without itself undergoing permanent change in its composition is called a catalyst. Catalysts that participate in biochemical reactions are called biocatalysts or enzymes. All chemical reactions in the cell require enzymatic actions. Enzymes in a biological system speed up chemical reactions by increasing the frequency of collision, lowering the quantity of the activation energy, and drastically reducing the time needed for the reaction. Activation energy is the energy required to start a chemical reaction as shown in Figure 3.2.

Enzymes in a living organism are specific and each enzyme can metabolize only one reaction. Enzymes are large, huge molecule complexes with a molecular weight of more than 10,000 and have 3-dimensional shapes.

An enzyme is composed of two parts: a protein part, called *apoenzyme*, and a nonprotein part, called *co-enzyme* or co-factor. Co-enzymes are vitamins, co-factors are inorganic ions such as Fe, Cu, Mg, Mn, Zn, Ca and Co.

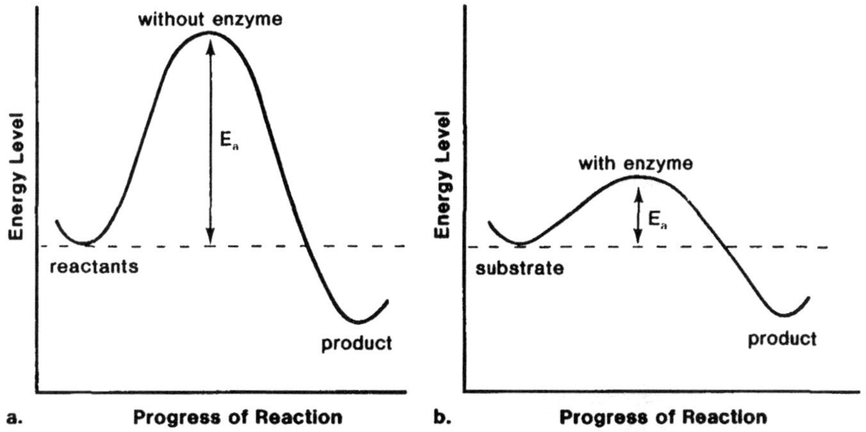

FIGURE 3.2 The progress of a reaction without an enzyme and the progress of a reaction with an enzyme shows the necessary energy of activation.

The co-enzyme or co-factor with the apoenzyme is called *holoenzyme* or active enzyme.

$$\text{apoenzyme} + \text{co-enzyme} \longrightarrow \text{holoenzyme}$$

3.3.1 Mechanism of Enzyme Action

The mechanisms of the enzyme activity are summarized in Figure 3.3:

1. The surface of the enzyme has a specific site called an active site.
2. The enzyme reacts with the substrate at the active site forming a substrate–enzyme complex.
3. During the substrate formation, the reaction takes place.
4. The enzyme–substrate separates into the product and the enzyme.
5. The enzyme is unaltered and can start the reaction again with another substrate.

3.3.2 Naming the Enzymes

Enzymes are named by the reaction they can complete with the suffix of -ase. All enzymes can be grouped into six classes according to the type of chemical reaction they catalyze. Enzymes within each of the major classes are named according to the more specific types of reaction they assist.

The main six classes are:

Transferase—transfers functional groups such as the amino group, the acetyl group, and the phosphate group.
Hydrolase—the addition of water to a substrate.
Oxidoreductase—participates in oxidation–reduction reactions in which oxygen or hydrogen are gained or lost.
Lyase—the removal of groups of atoms without hydrolysis.
Isomerase—the rearrangement of atoms within a molecule.
Ligase—joining two molecules together by using the energy usually derived from the breakdown of ATP.

3.3.3 Temperature and pH Effect on Enzymes

Enzyme activity can be affected by temperature and pH. As temperature rises, the movement of enzyme and substrate molecules increases, and the amount of product increases. However, further increases in temperature denature the protein part of the enzyme and will stop the reaction.

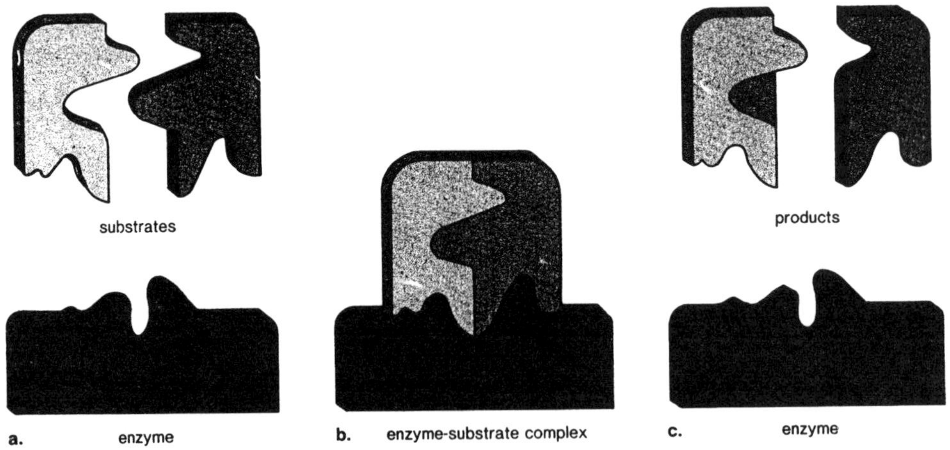

FIGURE 3.3 The mechanism of enzymatic action.

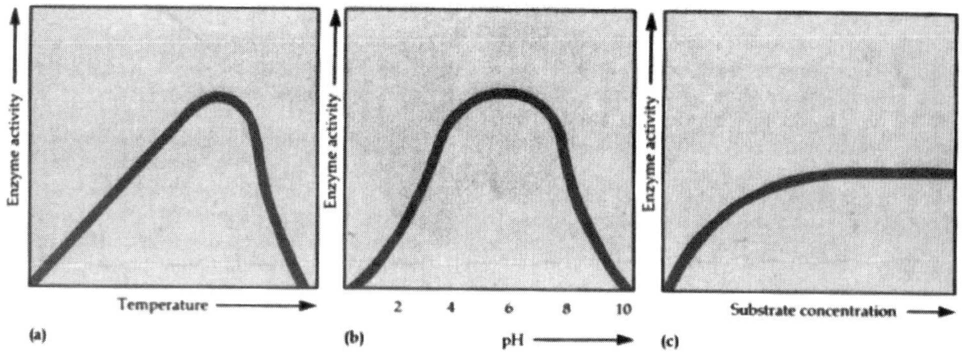

FIGURE 3.4 Factors influencing enzymatic activity.

Most enzymes, for the proper function, need an optimum pH level at which their activity is characteristically at its maximum. The factors that affect enzyme activity are shown in Figure 3.4.

3.3.4 Inhibition of Enzymes

One way to control the growth of bacteria is to control enzymes. Certain poisons such as cyanide, arsenic, and mercury combine with enzymes and prevent the enzymes from functioning. As a result, cells stop functioning and die.

Enzymes can be inactivated by competitive and noncompetitive inhibitions.

3.3.4.1 Competitive Inhibition of Enzymes

A competitive inhibitor is a chemical compound that competes with the normal substrate for the active site of the enzyme (see Figure 3.5b). It has the same shape and chemical structure as the normal substrate that is needed for the metabolic reaction. If it is mistakenly incorporated in the metabolic pathway, it will break down the chain reaction. For example, sulfur drugs have competitive inhibition in bacterial metabolism and can mistakenly form *substrate–enzyme* complexes. The cell will not be able to complete the reaction and will die. We use competitive inhibition when we apply chemicals to control pathogenic organisms.

3.3.4.2 Noncompetitive Inhibition of Enzymes

Noncompetitive inhibition of enzymes occurs when a molecule combines with the enzyme at a site other than the active site. The binding of the molecule causes the enzyme to assume a shape that prevents it

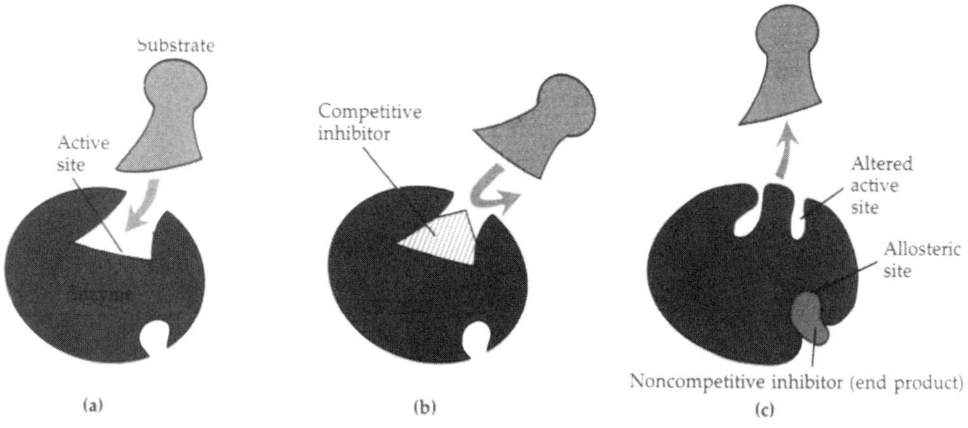

FIGURE 3.5 Enzyme inhibition.

from binding with its normal substrate. Noncompetitive, reversible inhibition is the normal way by which metabolic pathways are regulated in cells (see Figure 3.5c).

Another type of noncompetitive inhibition can operate on enzymes that require metallic ions for their activity. Certain chemicals can bind or tie up metallic ion activators and, thus, prevent enzymatic reaction. For example, cyanide can bind to iron, and fluoride can bind to calcium or magnesium.

3.4 NUTRITIONAL CLASSIFICATION OF ORGANISMS

3.4.1 ENERGY PRODUCTION

Nutrient molecules like all molecules have energy associated with the electrons that form bonds between their atoms. Energy is stored in the chemical bonds between nutrient molecules. Energy sources of organisms are summarized in Figure 3.6.

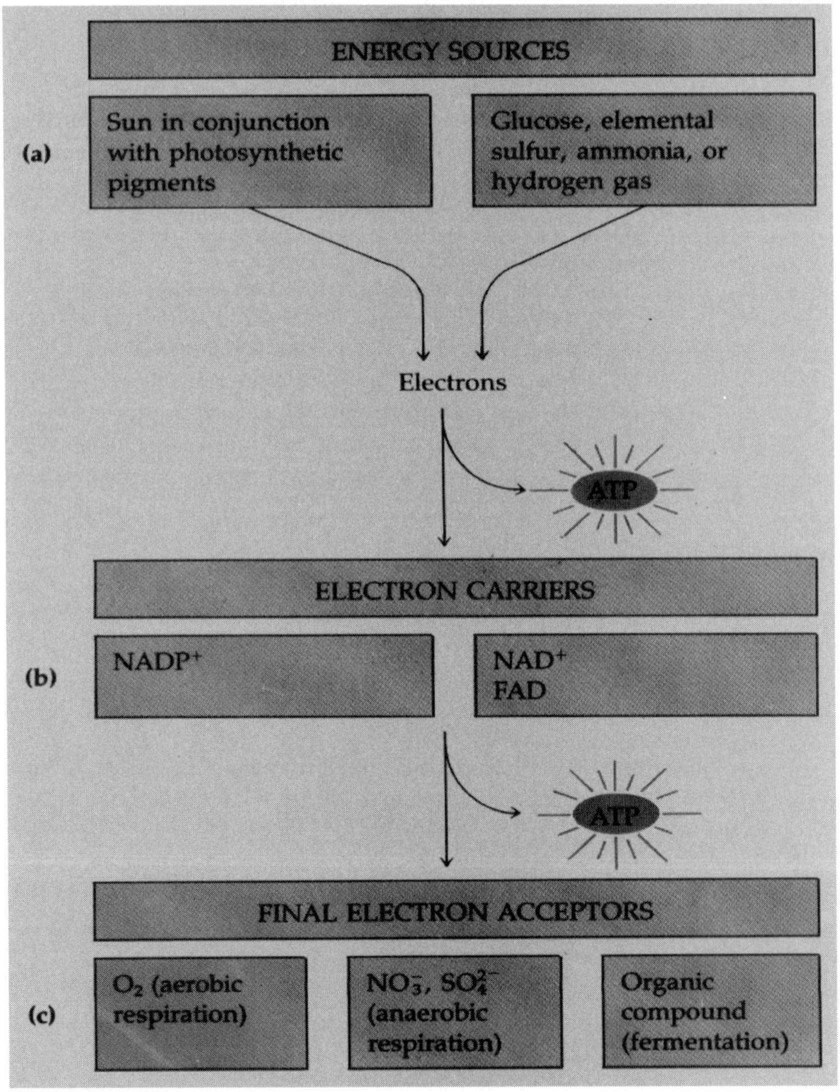

FIGURE 3.6 Energy source for organisms.

3.4.1.1 Carbon and Energy Sources

All biological systems, from microorganisms to multicellular eucaryotes have a set of nutritional requirements with regard to carbon and energy sources. Depending on the origin of carbon and energy, we can group all organisms into four groups:

Photoautotrophs—Some organisms can employ radiant energy as their energy source and carbon dioxide (CO_2) as their carbon source. These organisms are designated as photoautotrophs. They include photosynthetic bacteria (green and purple sulfur bacteria and cyanobacteria), algae, and green plants.

Photoheterotrophs—They use light as a source of energy, but cannot convert CO_2 to sugar; rather, they use organic compounds, such as alcohols, fatty acids, other organic acids, and carbohydrates as a source of carbon. Among the photoheterotrophs are the green and the purple nonsulfur bacteria.

Chemotrophs—Another form of life is incapable of utilizing radiant (solar) energy. Their primary source of energy comes from the oxidation–reduction reaction of organic or inorganic compounds. They are called chemotrophs.

Chemoautotrophs—Chemotroph organisms that use carbon dioxide (CO_2) as their primary source of carbon are called chemoautotrophs. Inorganic sources of energy for these organisms include sulfur for *Thiobacillus*, hydrogen sulfide (H_2S) for *Beggiatoa,* ammonia (NH_3) for *Nitrosomonas*, nitrite ions (NO_{2-}) for *Nitrobacter*, and ferrous ion (Fe^{2+}) for *Thiobacillus ferrooxidans*.

Chemoheterotrophs—The energy source and the carbon source are usually the same organic compound: glucose, for example. They are further classified according to their source of organic molecules: saprophytes live on dead organic matter and parasites derive nutrients from a living host.

According to the energy and carbon sources living organisms are classified as follows:

Nutritional Type	Energy Source	Carbon Source	Examples
Photoautotroph	Light	CO_2	Photosynthetic bacteria, algae, and plants
Photoheterotroph	Light	Organic compounds	Purple nonsulfur and green nonsulfur bacteria
Chemoautotroph	Inorganic compounds	CO_2	Hydrogen, sulfur, iron, nitrifying bacteria
Chemoheterotroph	Organic compounds	Organic compounds	Most bacteria, fungi, protozoan, and animals

3.4.1.2 Nitrogen Sources

Besides the energy and carbon sources, all organisms require nitrogen sources. Plants utilize nitrogen in the form of inorganic salts, such as potassium nitrate. Animals require organic nitrogen such as proteins. Bacteria are extremely versatile, some types use atmospheric nitrogen, some thrive on inorganic nitrogen compounds, and others derive nitrogen from protein or from organic nitrogen compounds.

3.4.1.3 Other Chemical Requirements

Other chemical compounds such as sulfur, phosphorus, metallic elements, electrolytes, and vitamins are also important nutritional components of all living organisms.

3.5 ENERGY PRODUCTION OF ORGANISMS

Energy is stored in the chemical bonds between atoms. Several types of chemical reactions are involved in energy production. Energy production means liberation of existing energy. This happens during a very sophisticated chemical chain reaction called metabolic pathway and it strictly obeys the Second Law of Thermodynamics as stated previously in Section 3.1.1.2.

3.5.1 Oxidation–Reduction

The most common energy production (liberation) reaction in living systems is oxidation–reduction. Oxidation is the addition of oxygen to or the removal of electron(s) from an atom or molecule. When a molecule loses electron(s), it is oxidized, and when it picks up the lost electron(s), it is reduced. Whenever a substance is oxidized, another substance is reduced. The pairing of these reactions is called the redox reaction. In many cellular oxidations, electrons and protons are removed simultaneously (by removing hydrogen atoms) because a hydrogen atom is made up of one proton and one electron. Most biological oxidations involve the loss of hydrogen atoms. These reactions are called dehydrogenation.

Oxidation always liberates energy. Reduction uses energy to build up a new bond between the hydrogen atoms and the recipient molecule. So, when a compound loses hydrogen atoms, it liberates energy and becomes oxidized.

3.5.2 Phosphorylation as ATP Formation in a Biological System

The energy storage molecule in all living organisms is adenosine triphosphate, ATP. When adenosine diphosphate, ADP, becomes adenosine triphosphate, ATP, 7 kcal of energy are stored. The reaction is reversible and can go from left to right (anabolism) or from right to left (catabolism).

$$\text{ADP} + \text{P} + 7\,\text{kcal} \longrightarrow \text{ATP} \tag{3.2}$$

When a compound such as ADP gains phosphorus (P), the reaction is called phosphorylation. The energy involved is 7 kcal. Addition of phosphorus to an organic compound is also a phosphorylation reaction. Organisms use phosphorylation three ways.

3.5.2.1 Substrate Level Phosphorylation

ATP is generated by direct transfer of a high energy phosphorus group from a phosphorylated compound to ADP. The following example shows only the carbon (C) skeleton and the high energy phosphorus (P) of a typical substrate.

$$\text{C}-\text{C}-\text{C}-\text{P} + \text{ADP} = \text{C}-\text{C}-\text{C} + \text{ATP} \tag{3.3}$$

3.5.2.2 Oxidative Phosphorylation

An oxidative phosphorylation occurs when electrons are removed from an organic compound, usually by a co-enzyme, such as *nicotinamide–adenine–dinucleotide* (NAD), and *flavin–adenine–dinucleotide* (FAD). An organic compound is oxidized when a co-enzyme is reduced and energy is removed and used by the reduction reaction. The reduced co-enzyme will finally generate ATP in the respiratory chain. The series of electron carriers used in oxidative phosphorylation is called the *electron transport chain* (3.6.5).

3.5.2.3 Photophosphorylation

Photophosphorylation occurs only in photosynthetic cells. The light energy is converted to the chemical energy of ATP.

3.5.3 Formation of ATP During Respiration

The metabolic pathway occurs by aerobic respiration, anaerobic respiration, and fermentation.

3.5.3.1 Aerobic Respiration

Aerobic respiration is a type of metabolism, involving oxidation–reduction reactions, and the final electron acceptor is molecular oxygen, so that the process is air dependent.

3.5.3.2 Anaerobic Respiration

In some other organisms the final electron recipient is not oxygen, rather it is some other inorganic compound such as nitrate (NO_3^-), sulfate (SO_4^{2-}), or carbonate (CO_3^{2-}).

3.5.3.3 Fermentation

When the electron recipient is an organic compound, such as lactic acid or alcohol, it is called fermentation. Both aerobic and anaerobic respiration produces 38 ATP and the metabolic pathway uses glycolysis (see Section 3.6.1), the Krebs cycle (see Section 3.6.4) and the electron transport chain (see Section 3.6.5). The fermentation metabolic pathway uses only glycolysis and produces only two ATP.

3.5.4 Formation of ATP During Fermentation

Fermentation is an anaerobic catabolic process, and does not require oxygen. During fermentation energy is released from an organic compound and the final electron recipient is also an organic compound. During the process a small amount of ATP is liberated (two ATP).

The reaction pathway uses only glycolysis and does not use the Krebs cycle or the electron transport system.

3.6 CARBOHYDRATE CATABOLISM

Microorganisms oxidize carbohydrates to provide most of the cell's energy. Carbohydrate catabolism is the breakdown of carbohydrate molecules to produce energy during the oxidation reaction. Therefore, it is of great importance in cell metabolism.

Glucose is the most common carbohydrate energy source used by cells. To produce energy from glucose, microorganisms use two general processes: respiration and fermentation.

The first step in both respiration and fermentation is the oxidation of glucose to pyruvic acid. This is most commonly accomplished by a process called glycolysis. *Glycolysis* is the oxidation of glucose to pyruvic acid.

During aerobic respiration, glycolysis is followed by the *Krebs cycle* and the *electron transport system*. The Krebs cycle is the oxidation of a derivative of pyruvic acid: acetyl co-enzyme A to CO_2 with the production of some *adenosine triphosphate* (ATP), *nicotineamide–adenine–dinucleotide* (NAD) reduced form (NADH) and another reduced electron carrier, *flavin–adenine–dinucleotide reduced form* ($FADH_2$).

ATP is a nucleotide that is of fundamental importance as a carrier of chemical energy in all living organisms. It conists of adenine linked to *D-ribose* (e.g., adenosine); the D-ribose component bears three phosphate groups, linearly linked together by covalent bond. These bonds can undergo hydrolysis to yield either a molecule of *adenosine–diphosphate* (ADP) and inorganic phosphate or a molecule of *adenosine monophosphate* (AMP) and pyrophosphate. Both these reactions yield a large amount of energy that is used to bring about biological processes as muscle contraction, the active transport of ions and molecules across cell membranes, and the synthesis of biomolecules. The reactions bringing about these processes often involve the enzyme-catalyzed transfer of the phosphate group to intermolecular substrates. Most ATP-mediated reactions require Mg^{2+} ions as cofactors.

ATP is regenerated by the rephosphorilation of AMP and ADP using the chemical energy obtained from the oxidation of food. This takes place during glucolysis and the Krebs cycle but, most significantly, is also a result of the oxidation–reduction reactions of the electron transport chain, that ultimately reduces molecular oxygen to water (oxidative phosphorylation).

NAD is a co-enzyme, derived from the B vitamin nicotinic acid that participates in many biological dehydrogenation reactions. It normally carries a positive charge and can accept one hydrogen atom and two electrons to become the reduced form NADH. NADH is generated during the oxidation of food; it then gives up two electrons (and single proton) to the electron transport chain, thereby reverting to NAD^+ and generating three molecules of ATP per molecule of NADH.

FAD (flavin–adenine–dinucleotide) is also a co-enzyme. It comprises a phosphorylated vitamin B_2 (riboflavin) molecule linked to the AMP. It functions as a hydrogen acceptor in dehydrogenation reactions, being reduced to $FADH_2$. This in turn is oxidized to FAD by the electron transport chain, thereby generating ATP (two molecules of ATP per molecule of $FADH_2$).

In the electron transport system, NADH and $FADH_2$ are oxidized, contributing the electrons they have carried from the substrate to a cascade of oxidation reactions. Energy from these reactions is used to generate a considerable amount of ATP.

In fermentation, the first step is also glycolysis. However once glycolysis has taken place the pyruvic acid is converted to one or more different products that might include alcohol or lactic acid. Much less ATP is produced.

3.6.1 Glycolysis

Glycolysis, the oxidation of glucose to pyruvic acid, is the first stage in carbohydrate catabolism. It occurs within the cytoplasm in both procaryotic and eucaryotic cells. Glycolysis is also called the *Embden–Meyerhof Pathway*. Glycolysis means splitting of glucose. The six carbon glucose molecule splits into two three carbon sugars. The three carbon sugar is oxidized to pyruvic acid and the removed hydrogen will reduce temporarily two nicotinamide adenine dinucleotide, NAD co-enzyme to $NADH_2$ and two ATP is formed by substrate level phosphorylation.

The end products of glycolysis are two $NADH_2$, two ATP, and two pyruvic acids.

The glycolysis pathway is shown in Figure 3.7. The glucolysis pathway does not require oxygen, and it can occur under aerobic or anaerobic conditions.

3.6.2 Alternate Pathway of Glycolysis by Bacteria

Many bacteria have, in addition to glycolysis, alternate pathways for the oxidation of sugars.

One of these pathways is the pentose phosphate pathway that provides for the breakdown of five carbon sugars, such as pentoses.

Another pathway is the *Entner Doudoroff Pathway* (EDP). The bacteria that possess this pathway are some species in the genus *Rhizobium* and several species in the genus *Pseudomonas*. Test for the bacteria using EDP pathway is an important laboratory procedure as an oxidase test (see Chapter 14.2.4) for identification of *Pseudomonas* in clinical laboratories.

3.6.3 Transition Reaction

After glucose has been oxidized to pyruvic acid during the glycolysis, pyruvic acid enters the mitochondria for respiration which is a further oxidation of pyruvic acid if molecular oxygen is available. Without molecular oxygen only fermentation occurs. In aerobic respiration, the final electron acceptor is oxygen; in anaerobic respiration, the final electron acceptor is an inorganic molecule, rarely an organic compound.

Before pyruvic acid enters the mitochondria, it must undergo more oxidation, forming *acetyl co-enzyme A* which can enter into the Krebs cycle as shown in Figure 3.8.

We named the reaction between glycolysis and the Krebs cycle transition reaction, which connects glycolysis to the Krebs cycle.

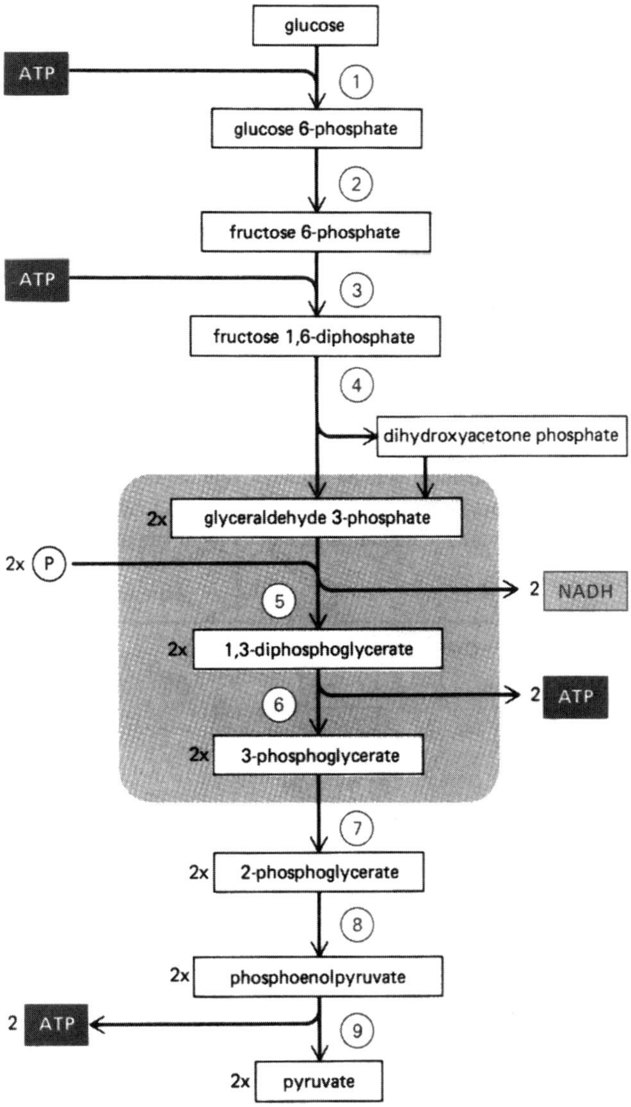

FIGURE 3.7 Intermediates of glycolysis. Each reaction above is catalyzed by a different enzyme. In the series of reactions designated as Step 4, a six-carbon sugar is cleaved to give two three-carbon sugars, so that the number of molecules at every step after this is doubled. Steps 5 and 6 in the gray box are the reactions responsible for the net synthesis of the ATP and NADH molecules.

During the transition reaction the cell produces two molecules of $NADH_2$ and two molecules of co-enzyme A.

3.6.4 Krebs Cycle or Citric Acid Cycle

The Krebs cycle is a series of biochemical reactions in which the large amount of potential chemical energy stored in acetyl Co-A is released step-by-step. In this cycle, a series of oxidation and reduction transfer the potential energy in the form of electrons to the electron carrier co-enzymes, chiefly NAD^+. The pyruvic acid derivatives are oxidized; the co-enzymes are reduced.

The cycle is also called the citric acid cycle or oxidative decarboxylation. During a series of biochemical reactions, all hydrogen atoms are removed from the organic compound and, temporarily,

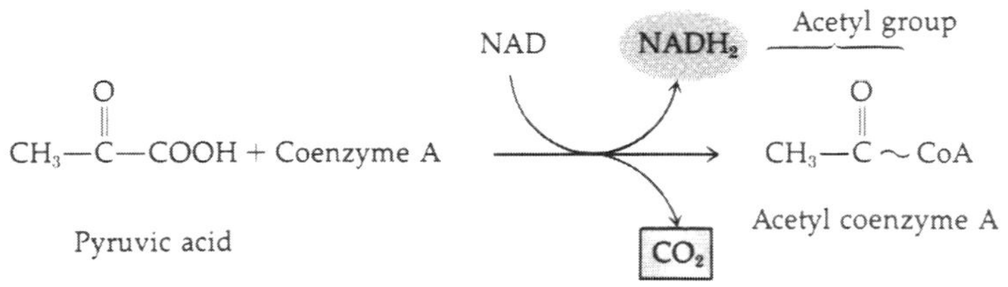

FIGURE 3.8 Formation of acetyl CoA from pyruvic acid.

the co-enzyme is reduced by the removed H atoms as NAD and FAD, forming $NADH_2$ and $FADH_2$, respectively.

The end products of the organic compound during the Krebs Cycle are six molecules of $NADH_2$, two molecules of $FADH_2$, and two molecules of ATP.

The reactions during the Krebs cycle are illustrated in Figure 3.9.

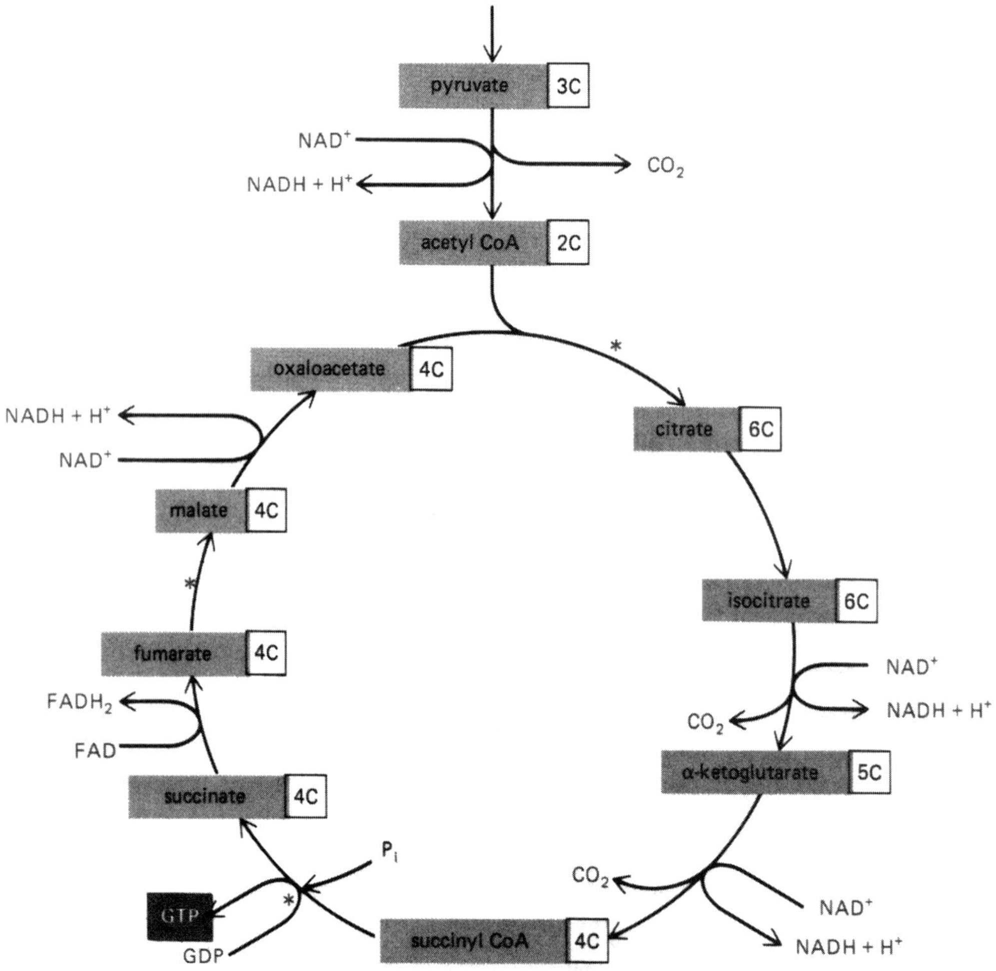

FIGURE 3.9 The citric acid cycle. In mitochondria and in aerobic bacteria, the acetyl groups produced from pyruvate are further oxidized. The carbon atoms of the acetyl groups are converted to CO_2, while the hydrogen atoms are transferred to the carrier molecules NAD^+ and FAD. Additional oxygen and hydrogen atoms enter the cycle in the form of water at the steps marked with an asterisk (∗).

3.6.5 Respiratory Chain or Electron Transport System or Cytochrome System

During glycolysis, transition reactions, and the Krebs cycle, all of the hydrogen atoms are removed. The co-enzymes NAD and FAD are reduced, having only temporarily, the hydrogen atoms and energy.

All of the reduced co-enzymes, as produced, move to the last stage of cellular respiration which we call the respiratory chain, the electron transport system, or the cytochrome system. The electron transport chain consists of a sequence of carrier molcculcs that arc capablc of oxidation and reduction. The electron flow down the chain and are accompanied at several points by the active transport of protons. The result is a buildup of protons on one side of the membrane. This buildup of protons provides energy to produce ATP. One molecule of $NADH_2$ has the energy to produce three ATP; one molecule of $FADH_2$ has the energy to produce two ATP (see Figure 3.10).

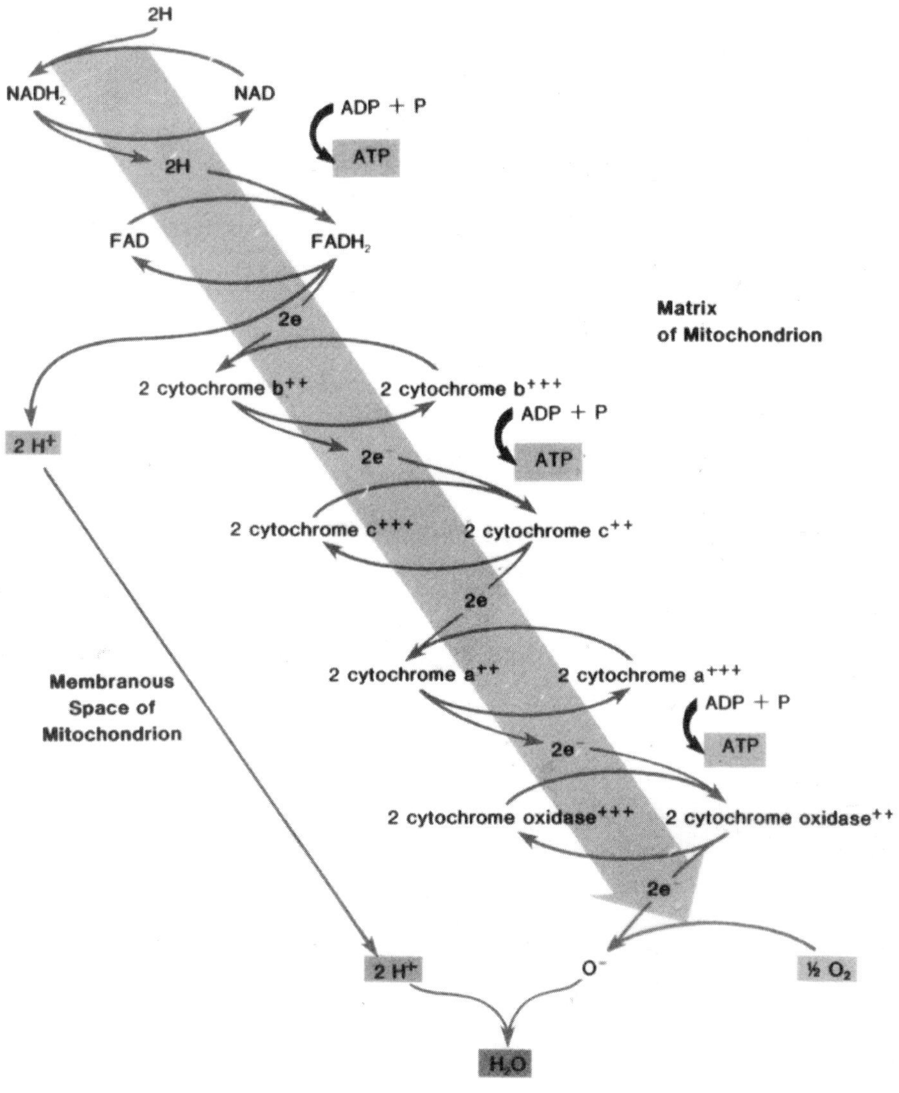

FIGURE 3.10 Respiratory chain.

3.6.6 Aerobic Respiration

The various electron transfers in the electron transport chain generate about 34 molecules of ATP from each molecule of glucose oxidized, approximately 3 from each of the molecules of NADH (total of 30) and approximately 2 from each of the 2 molecules of $FADH_2$ (total of 4). These 34 ATP molecules are added to those generated by oxidation in glycolysis (2) and the Krebs cycle (2), all together 8 ATP molecules are generated. One ATP is equivalent to 7 kcal. Therefore, during respiration, 266 kcal of effective biological energy are generated. This represents 38 percent of the total energy in one molecule of glucose. The remaining energy is the missing energy used as biological work or as heat energy.

The respiratory chain is shown in Figure 3.10.

> During aerobic respiration, a total of 38 molecules of ATP can be generated from 1 molecule of glucose.

The overall reaction for aerobic respiration in procaryotes is:

$$C_6H_{12}O_6 + 6\,O_2 + 38\,ADP + 38\,P \longrightarrow 6\,CO_2 + 6\,H_2O + 38\,ATP \qquad (3.4)$$

Overview of cellular respiration is shown in Figure 3.11.

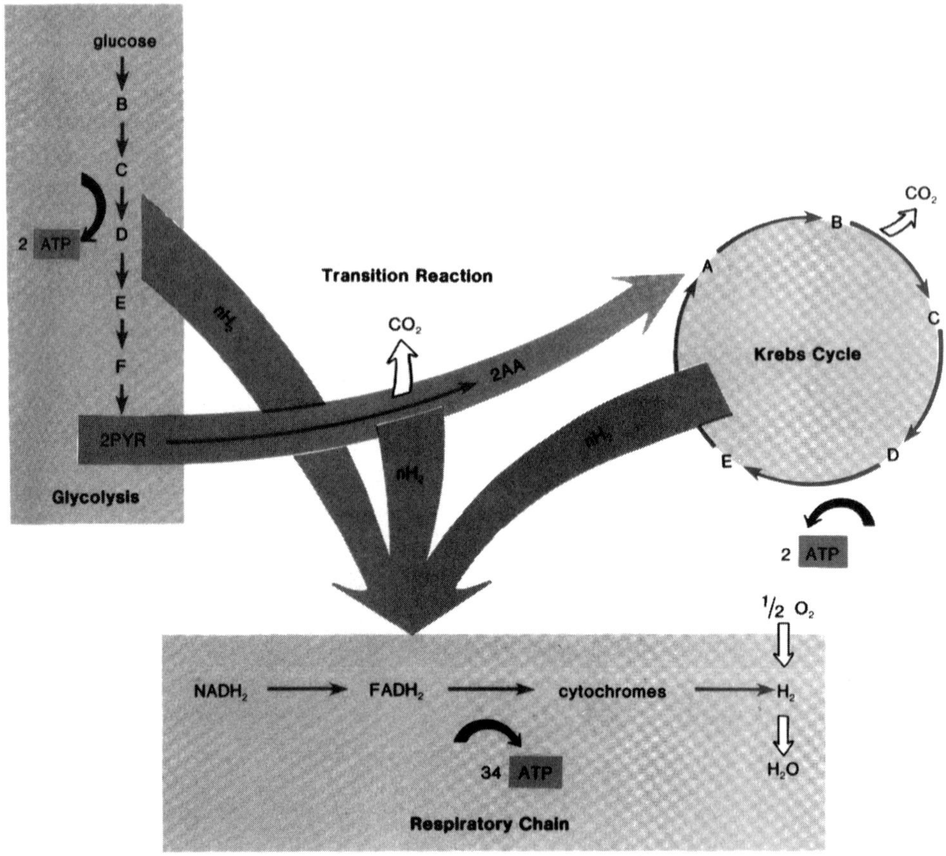

FIGURE 3.11 Overview of cellular respiration.

3.6.7 Fermentation

If molecular oxygen is not available after glycolysis, the reaction cannot go to respiration through the transition reaction, Krebs cycle, and the respiratory chain.

The end product of glycolysis, pyruvic acid, without oxygen, follows the fermentation pathway and produces different kinds of organic compounds, such as lactic acid, ethanol, propionic acid, butanol, acetic acid, formic acid, and so on, and generates two ATP molecules.

Fermentation produces different organic compounds. For example, ethyl alcohol and carbon dioxide are produced in yeast cells and it produces lactic acid in animal cells. Lactic acid causes the typical chest pain sensation (angina) or other muscle pain; it indicates that there is not enough oxygen for normal cellular respiration and the ATP production is drastically reduced from 38 ATP to only 2 ATP. It means much less energy is available for normal heart and muscular functions.

During fermentation the net ATP production is only two molecules.

Other fermentation processes are:

- Production of alcoholic beverages.
- Any spoilage of foods by microorganisms.
- Any energy releasing metabolic processes that take place only under anaerobic conditions.
- All metabolic processes that release energy from an organic compound do not require oxygen and use an organic molecule as a final electron acceptor.

4 Microbial Growth and Its Control

4.1 MICROBIAL GROWTH

When we talk about microbial growth, we are really referring to the number of cells, not the size of the cells. Microbial populations can become very large in a very short time and unchecked microbial growth can cause serious disease or food spoilage. By understanding microbial growth, we can determine how to control the growth.

Bacterial growth refers to an increase in bacterial cell number and it normally happens by binary fission. During active bacterial growth, the size of the microbial population is continuously doubling. One cell divides to form two, each of these cells divides so that four cells form, and so forth in a geometric progression. The time required to achieve a doubling of the population size, known as the generation time or doubling time, is the unit of measure of the microbial growth rate. When a bacteria is inoculated into a new culture medium, it exhibits a characteristic pattern or change in cell numbers. This pattern is a growth curve. The normal growth curve of bacteria has four phases, the lag phase, the log phase or exponential growth phase, the stationary phase, and the death or logarithmic decline phase.

4.1.1 The Lag Phase

The lag phase can last for an hour or several days and it is practically an adaptation period. There is no increase of cell number, but the organism is not in a dormant state during which the bacteria is preparing for division.

4.1.2 The Log Phase

This phase begins when bacteria start to divide and enter a period of growth or the exponential growth phase. Cellular reproduction is most active during this period and shows a minimum constant generation time.

The log phase is also when cells are most active metabolically and most vulnerable to chemical or antibiotic treatment and radiation. These treatments interfere with some important steps in the growth process and are, therefore, most harmful to the cell during this phase.

4.1.3 The Stationary Phase

This phase often occurs when the maximum population density that can be supported by the available resources is reached. There is no further net increase in bacterial cell numbers. During this phase, the growth rate is equal to the death rate and cell number, therefore, remains constant and the population stabilizes.

4.1.4 The Death Phase

This phase occurs when the number of living cells decreases because the rate of cell death exceeds the rate of new cell formation. This phase continues until the population is diminished to a tiny fraction of

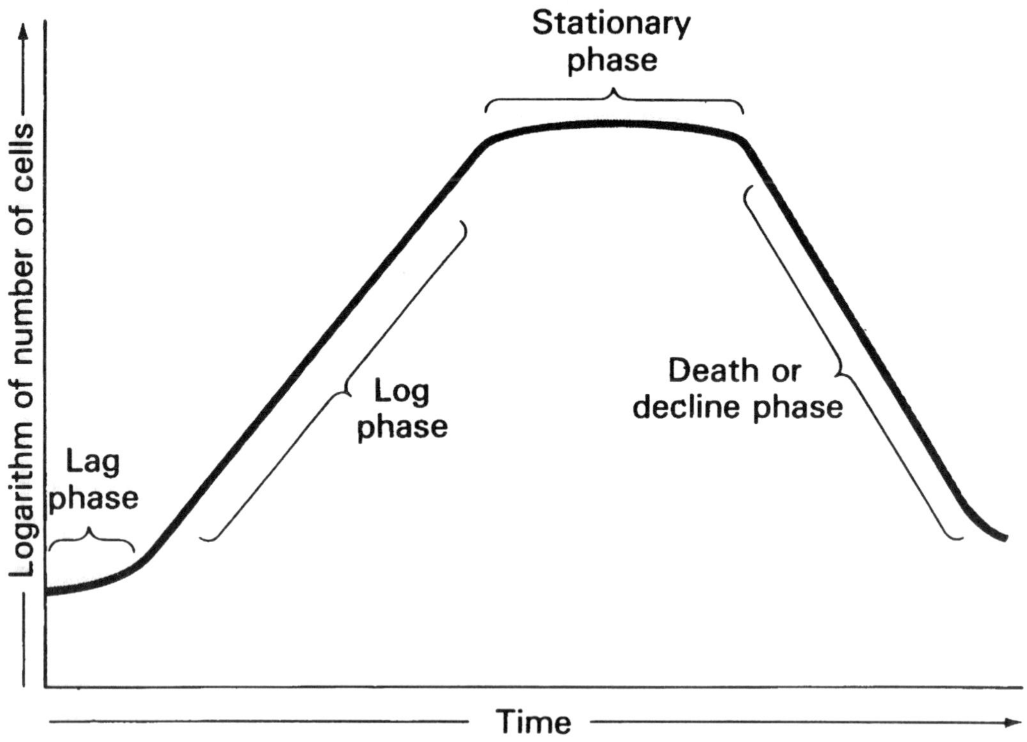

FIGURE 4.1 Bacterial growth curve.

more resistant cells or the population dies out entirely. Some species pass through the entire series of phases in only a few days, others retain some surviving cells almost indefinitely (see Figure 4.1).

4.2 PHYSICAL REQUIREMENTS

4.2.1 TEMPERATURE

Microorganisms are divided into three groups on the basis of their preferred range of temperature.

4.2.1.1 Cold Loving Microbes: Psychrophiles

This category is reserved for organisms with an optimum growth temperature of about 15°C and a maximum growing temperature of 20°C. Such organisms are found mostly in the ocean's depth or in certain arctic regions, and they seldom present concerns to humans in the preservation of food. However, according to some references, microbes belonging to this group have growth temperature ranging from 0 to 30°C. Only two fairly distinct groups of organisms are capable of growth at 0°C. Refrigeration is the most common method of preserving household food supplies. The temperature inside a properly set refrigerator will prevent the growth of most spoilage organisms and all of the pathogenic bacteria.

4.2.1.2 Moderate Temperature Loving Microbes: Mesophiles

Optimum growth temperature is between 25°C and 40°C. This is the most common type of microbe. Common spoilage and disease organisms belong to this group. The optimum temperature of many pathogenic bacteria is 37°C.

4.2.1.3 Heat Loving Microbes: Thermophiles

These organisms have optimum growth temperature between 50°C and 60°C. Many thermophiles are incapable of growth at temperature below about 45°C, and some of them grow well at temperature above 90°C (near to the boiling point of water) (see Figure 4.2).

4.2.2 pH and Buffers

Most bacteria grow best in a narrow range of pH, near neutrality, pH 6.5 to 7.5. Very few bacteria grow at an acid pH and are called *acidophiles*. That is the reason why a number of foods, such as sauerkraut, pickles, and many cheeses, are preserved by the acids of bacterial fermentation. One autotrophic bacterium, which is found in the drainage water from coal mines and which oxidizes sulfur to form sulfuric acid, can survive at a pH of 1. Alkalinity also inhibits bacterial growth, but is rarely used to preserve food. When bacteria are cultured in the laboratory, they produce acids that can interfere with desired bacterial growth. To avoid this unwanted side effect, chemicals called buffers are included in the growth medium.

A *buffer* is a solution that resists changes in pH when an acid or an alkali is added or when the solution is diluted. In the laboratory, buffers are used to prepare solutions of known stable pH. Natural buffers occur in living organisms where the biochemical reactions are very sensitive to change in pH. Buffer solutions are also used in medicine, in agriculture, and in many industrial processes.

Acidic buffers consist of a weak acid with a salt of the acid. The salt provides the negative ion A, which is the conjugate base of the acid, AH. An example is carbonic acid (H_2CO_3) and sodium hydrogen carbonate ($NaHCO_3$). In an acidic buffer, molecules HA and ions A^- are present. When an acid is added, most of the extra protons are removed by the base:

$$A^- + H^+ \longrightarrow HA \tag{4.1}$$

When a base is added, most of the extra hydroxide ions (OH^-) are removed by reaction with the undissociated acid:

$$OH^- + HA \longrightarrow A^- + H_2O \tag{4.2}$$

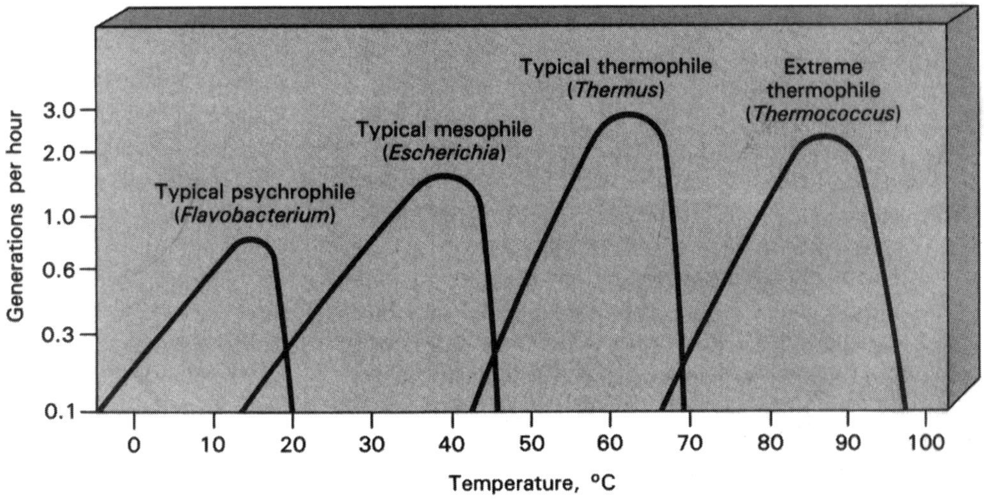

FIGURE 4.2 Comparison of the growth rates of typical psychrophile, mesophilic, and thermophilic organisms and the overlap of ranges at which those organisms can survive.

The hydrogen ion concentration (pH) in a buffer is given by the expression:

$$K_a = [H^+]/[HA] \quad (4.3)$$

Basic buffers have a weak base and a salt of the weak base (to provide the conjugate acid). An example is ammonia solution and ammonium chloride (NH_4OH/NH_4Cl).

4.2.3 Osmotic Pressure

Microbes obtain most of their nutrients in solution from the surrounding water. They, therefore, require water for growth and are actually about 80 to 90 percent water. When a microbial cell is in a solution that has a higher concentration of solute than in the cell (*hypertonic*), the cellular water passes out through the plasma membrane to the high salt concentration. This osmotic loss of water causes plasmolysis or shrinkage of the cell, and inhibits the growth of the cell. Thus, the addition of salt or other solutes to a solution and the resulting increase in osmotic pressure can be used to preserve food. For example, salted fish, honey, and sweetened condensed milk are preserved by this mechanism; the high salt or sugar concentrations draw water out from microbial cells that are present and, thus, prevent their growth.

When a microbial cell is in a solution that has lower concentration of solute than in the cell (*hypotonic*), the cell will gain water from outside of the cell through the cell membrane. This increases hydrostatic pressure inside of the cell causing cell rupture called *osmotic lysis*.

Some bacteria have adapted so well to an environment of high salt concentrations that they actually require salt concentrations for growth. These bacteria are called extreme *halophile*. To grow, bacteria from such saline waters as the Dead Sea require nearly 30 percent salt. The inoculating loop in the laboratory must first be dipped into a saturated salt solution before transferring the bacteria. Facultative halophile are grown at salt concentrations up to 2 percent, a concentration that inhibits the growth of many other bacteria.

The concentration of agar (a complex polysaccharide) used to solidify microbial growth media is usually 1.5 percent. If markedly higher concentrations are used, the growth of some bacteria can be inhibited by the higher osmotic pressure.

4.3 CHEMICAL REQUIREMENTS

4.3.1 Carbon

Besides water, one of the most important requirements for microbial growth is carbon. Carbon is the structural backbone of living matter. Carbon is needed for all the organic compounds that make up a living cell.

4.3.2 Nitrogen, Sulfur, and Phosphorus

Protein synthesis requires a considerable amount of nitrogen as well as some sulfur. The synthesis of DNA, RNA, and ATP also require nitrogen and some phosphorus.

Potassium, calcium, and magnesium are also examples of elements that microorganisms require, as co-factors for enzymes.

4.3.3 Trace Elements

Microbes require very small amounts of other elements, such as iron, copper, molybdenum, and zinc. These are referred to as trace elements. Most of these elements are essential for the activity of certain enzymes, usually as a co-factor.

4.3.4 Oxygen

Microbes that use molecular oxygen (O_2), called aerobes, produce more energy from nutrients than do microbes that do not use oxygen. Obligate aerobes are organisms that require oxygen to live.

Many aerobic bacteria have developed or retained the ability to continue growing in the absence of oxygen. Such organisms include some that exist in the intestinal tract called facultative anaerobes. In other words, facultative anaerobes can use oxygen when it is present, but are able to continue growth by using fermentation or anaerobic respiration when oxygen is not available. Obligate anaerobes are bacteria that are unable to use molecular oxygen (O_2) for energy yielding reactions. See Figure 4.3.

According to the oxygen requirement a microbe can be:

Aerobe—grows in the presence of free oxygen.
Obligate aerobe—cannot grow in absence of free oxygen.
Facultative aerobe—fundamentally an anaerobe, but can grow in presence of free oxygen.
Microaerophile—requires reduced amount of oxygen.
Anaerobe—grows in the absence of free oxygen.
Obligate anaerobe—cannot grow in the presence of free oxygen.
Facultative anaerobe—fundamentally an aerobe, but can grow in the absence of free oxygen.
Aerotolerant—does not grow well, but survives in the presence of free oxygen.
Capnophile—growth enhanced by increased carbon dioxide (CO_2).

4.3.5 Organic Growth Factors

Organic growth factors are essential compounds that the organism is unable to synthesize; they must be directly obtained from the environment. Organic growth factors for humans are vitamins. Many bacteria can synthesize all their vitamins and are not dependent on outside sources. However, some

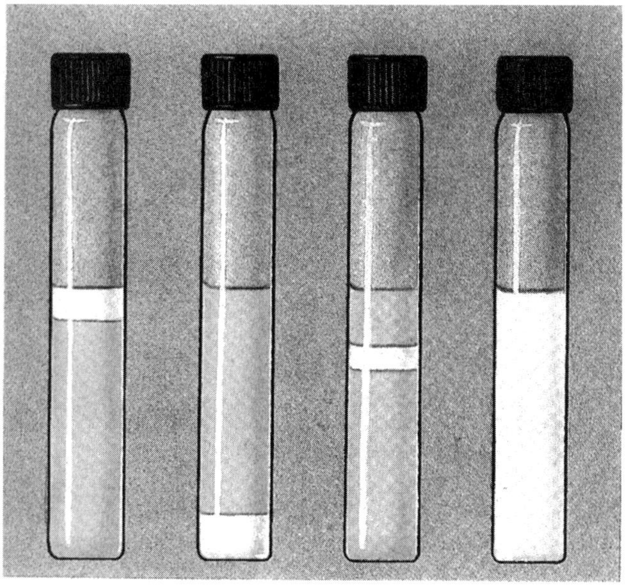

FIGURE 4.3 Different organisms incubated for 24 h in nutrient broth tubes accumulate in different regions depending on their need for or sensitivity to oxygen.

bacteria lack the enzymes needed for the synthesis of certain vitamins, and for them, those vitamins are organic growth factors. Most vitamins function as a co-enzyme, the accessories required by certain enzymes in order to function. Other organic growth factors required by some bacteria are amino acids, purines, and pyriminides.

4.4 CONTROL OF MICROBIAL GROWTH

The scientific control of microbial growth began only about 100 years ago. Prior to that time, epidemics commonly killed thousands of people. In some hospitals, 25 percent of delivering mothers died of infections carried on the hands and instruments of attending nurses and physicians. Ignorance of microbes was such that during the American Civil War, a surgeon might clean his scalpel on his bootsole between incisions.

Ignatz Semmelweis (1816–1865), a Hungarian physician working in Vienna, and *Joseph Lister* (1827–1912), an English physician, first introduced the concept of microbial control. At the obstetrics ward in the Vienna General Hospital, Semmelweis required that all personnel wash their hands in chlorinated lime: that procedure significantly lowered the infection rate. Lister meanwhile concluded that the number of infected surgical wounds (sepsis) could be decreased through procedures that prevented the access of microbes to the wound. This system, known as aseptic surgery, included the heat sterilization of surgical instruments, and, following surgery, the application of phenol (carbolic acid) to wounds. The pictures of Semmelweis and Lister are shown in Figure 4.4.

We have come a long way in controlling microbial growth since the time of Semmelweis and Lister. Today's procedures are far more sophisticated and effective.

Microbial growth can be controlled by physical methods including the use of heat, low temperatures, desiccation, osmotic pressure, filtration, and radiation. Chemical agents include several groups of substances that destroy or limit microbial growth.

(a) (b)

FIGURE 4.4 Two nineteenth-century pioneers in the control of infections: a. Ignaz Philipp Semmelweis, who died in an asylum before his innovations were widely accepted, is depicted on a 1965 Austrian postage stamp. b. Joseph Lister successfully carried on Semmelweis' work.

4.4.1 Terms Related to Destruction of Organisms

4.4.1.1 Sterilization

Sterilization is the process that eliminates living organisms from treated substances or objects. Sterilization is absolute; there are no degrees of sterilization.

Typical temperatures required are: moist heat, 121°C for 15 min; dry heat, 170°C for 120 min. Other methods are ionizing radiation and gases such as ethylene oxide.

4.4.1.2 Disinfection

Disinfection is the destruction or removal of the infectious agents by chemical or physical means. A disinfectant is an agent, usually a chemical, that kills the growing forms but not necessarily the resistant spore forms of disease-producing organisms. Disinfectants tend to reduce or inhibit growth; they usually do not sterilize. The term usually applied to use of liquid chemical solution on surfaces or to elimination of pathogens from water is chlorination.

4.4.1.3 Pasteurization

Pasteurization is a form of disinfection used for material which may be altered or damaged by excessive heat. Low heat is applied once or repeatedly to sensitive liquids to destroy vegetative cells. Pasteurization of milk is the process of exposing it to a temperature of about 63°C for 30 min or 72°C for only about 15 min.

4.4.1.4 Germicide

Germicide (cide means kill) is a chemical agent that rapidly kills microbes but not necessarily their endospores. Bactericide kills bacteria, a fungicide kills fungi, a virucide kills viruses, and an amebocyte kills amoebas.

4.4.1.5 Antiseptics

A chemical for disinfection of the skin, mucous membranes, or other living tissues is called an antiseptic; the term especially applied to treatment of wounds. Antisepsis is a special kind of disinfection.

4.4.1.6 Asepsis

Asepsis means without infection. It is the absence of pathogens from an object or area. Air filtration, UV light, personnel mask, gloves, and gowns, and instrument sterilization are all factors in achieving asepsis.

4.4.1.7 Degerming

The removal of transient microbes from the skin by mechanical cleansing or by the use of an antiseptic. For routine injections, alcohol swabs are often used. Before surgery, iodine-containing products are often used.

4.4.1.8 Sanitization

The reduction of pathogens to safe public health levels on eating utensils by mechanical cleansing or by chemicals is called sanitization. All chemicals must be compatible with safety and palatability of foods.

4.5 PHYSICAL METHODS OF MICROBIAL CONTROL

4.5.1 High Heat

High heat is the most common method by which microbes are killed. Heat appears to kill microbes by denaturing their enzymes.

4.5.1.1 Moist Heat Sterilization

One type of moist heat sterilization is boiling to 100°C at sea level which kills vegetative forms of bacterial pathogens, many viruses, and fungi, and their spores within about 10 min. Boiling is not always a reliable sterilization procedure. The hepatitis virus, for example, can survive up to 30 min of boiling, and some bacterial endospores have resisted boiling temperatures for more than 20 h.

Reliable sterilization with moist heat requires temperatures above that of boiling water. These high temperatures are most commonly achieved by steam under pressure in an autoclave. The higher the pressure in the autoclave, the higher the temperature. The autoclave is normally operated at 15 psi (pounds per square inch), that is about 1 atm (atmosphere) steam pressure for 15 min producing a temperature inside the autoclave of 121°C at sea level. The relationship between temperature and pressure is shown in Table 13.1.

4.5.1.2 Dry Heat Sterilization

One of the simplest methods of dry heat sterilization is direct flaming. For example, to sterilize inoculating loops in the laboratory, heat the wire to a red glow, which is 100°C.

Another effective way to sterilize and dispose of contaminated wastes (that are combustible) is incineration or burning.

Another form of dry heat sterilization is using a dry heat oven. Temperatures of 165° to 170°C (329° to 338°F) are required for 2 h. Do not use this method for the sterilization of liquids or other materials that will evaporate or deteriorate.

4.5.2 Filtration

Filtration is used to sterilize liquids that are heat resistant or heat sensitive such as some culture media, enzymes, vaccines, and so on. Membrane filters are composed of cellulose esters or plastic polymers. These filters are only 0.1 mm thick with uniform pore sizes. In some brands, plastic film is irradiated so that very uniform holes are etched in the plastic. The pores of membrane filters include the 0.22 μm and 0.45 μm sizes intended for bacteria, and range down to 0.01 μm which will retain viruses and even some large protein molecules.

4.5.3 Low Temperature

The effect of low temperatures on microbes depends on the particular microbe and the intensity of the application. Refrigeration can be used for short-term storage of bacterial culture. To preserve microbial cultures for a long period of time, two common methods are deep-freezing and lyophilization.

Deep freezing is a process in which a pure culture of microbes is placed in a suspending liquid and quick frozen at temperatures ranging from −50° to −95°C. The culture can be thawed and used several years later.

During lyophilization, a suspension of microbes is quickly frozen at temperatures ranging from −54° to −72°C, and the water is removed by a high vacuum. While under vacuum, the container is sealed by a high temperature torch. The remaining powder-like residue contains the surviving microbes and can be stored for years. The microbes can be revived at any time by hydration with a suitable liquid nutrient medium.

4.5.4 Desiccation

Microbes require water for growth and multiplication. In the absence of water—a condition known as desiccation—microbes are not capable of growth or reproduction, but they can remain viable for years. Then, when water is made available to them, they can resume their growth and division. This ability is used in the laboratory when microbes are preserved by lyophilization (freeze-drying). The process was discussed previously.

People use desiccation to preserve foods. If meat is sliced thin and sun dried (made into jerky), it can be preserved for long periods of time. Other examples include raisins and similarly dried fruits.

4.5.5 Radiation

Radiation has various effects on cells, depending on its wavelength, intensity, and duration. There are two types of sterilizing radiation: ionizing and nonionizing.

4.5.5.1 Ionizing Radiation

Ionizing radiation has a wavelength shorter than that of nonionizing radiation, less than about 1 nm. It is increasingly used for sterilizing pharmaceutical and disposable dental and medical supplies, such as plastic syringes, surgical gloves, suturing materials, and catheters.

4.5.5.2 Nonionizing Radiation

Nonionizing radiation has a wavelength longer than that of ionizing radiation, usually greater than about 1 nm. A good example for nonionizing radiation is ultraviolet (UV) light. UV light damages the DNA of exposed cells. The UV wavelength most effective for killing microorganisms is about 260 nm; these wavelengths are specifically absorbed by cellular DNA. UV radiation is also used to control microbes in the air. A UV or germicidal lamp is commonly found in hospital rooms, nurseries, operating rooms, and cafeterias. A potential problem is that UV light can damage the eyes, and prolonged exposure to UV light can cause burns and skin cancer. Sunlight contains some UV light, but the most effective wavelengths are screened out by the atmosphere.

4.6 CHEMICAL METHODS OF MICROBIAL CONTROL

4.6.1 Surfactants or Surface Acting Agents

These agents can decrease surface tension among molecules of a liquid. Such agents include soaps and detergents. Soap has little value as an antiseptic, but it has a very important function in the mechanical removal of microbes through scrubbing. The skin normally contains dead cells, dust, dried sweat, microbes, and oily secretions.

4.6.1.1 Soap

Soap breaks the oily film into tiny droplets (emulsification) and the water and soap lift up the emulsified oil and debris and float them away. In this sense, soaps are good degerming and emulsifying agents. Many so-called deodorant soaps contain compounds that strongly inhibit Gram-positive bacteria.

4.6.1.2 Acid–Anionic Surface Acting Sterilizers

Acid–anionic surface acting sterilizers are very important in the dairy industries for cleaning of utensils and equipment. They are nontoxic, noncorrosive, and fast acting.

4.6.1.3 Quaternary Ammonium Compounds (Quats)

Quats are cationic detergents. Their cleansing ability is related to the positively charged portion—cation—of the molecule. Quats are modifications of the ammonium ion (NH_4^+), and they are strongly bactericidal against Gram-positive bacteria and less bactericidal against Gram-negative bacteria. Two popular quats are Zephiran and Cepacol. They are strongly antimicrobial, colorless, odorless, tasteless, stable, easily diluted, and nontoxic. For example, mouth-wash solutions have quats. Certain bacteria, such as some of the Pseudomonas species, not only survive in quats, but actively grow in them.

4.6.2 Organic Acids

A number of organic acids are used as preservatives to control mold growth. *Sorbic acid*, or its salt *potassium sorbate*, prevents mold growth in cheese. *Benzoic acid*, or its salt *sodium benzoate*, is an antifungal agent. It is effective only in acidic pH levels. It has wide use in soft drinks. *Calcium propionate* prevents mold growth in bread. Their use is considered quite safe. The activity of these organic acids is not related to their acidity, but to their ability to inhibit enzymatic and metabolic activity.

4.6.3 Aldehydes

Aldehydes are among the most effective antimicrobials. *Formaldehyde* gas is an excellent disinfectant. However, it is more commonly used as formalin, a 37 percent aqueous solution of formaldehyde gas. *Formalin* was once used extensively to preserve biological specimens and inactivate bacteria and viruses in vaccines. Its sterilizing and disinfecting applications are limited by its tissue-irritating qualities, poor penetration, slow action, unpleasant odor, and its property of leaving a white residue on treated materials.

Glutaraldehyde is a chemical relative to formaldehyde that is less irritating and easier to handle than formaldehyde. When used in a 2 percent solution, it is a bactericidal, tuberculocidal, and virucidal in 10 min and a sporicidal in 3 to 10 h. Because such a long exposure time is required for sporicidal activity, chemical agents are not suitable for sterilization.

4.6.4 Ethylene Oxide

Ethylene oxide sterilizes in a closed chamber, similar to an autoclave. Its activity depends on the denaturation of proteins. It is toxic and explosive in its pure form, so it is usually mixed with a nonflammable gas, such as carbon dioxide or nitrogen. Its remarkable penetrating power is one reason why ethylene oxide was chosen to sterilize spacecraft sent to land on the Moon and other planets. Because of their ability to sterilize without heat, gases like ethylene oxide are also widely used on medical supplies and equipment.

Examples include disposable sterile plastic ware, for example, syringes, petri dishes, pipets, sutures, artificial heart valves, heart-lung machines, and even mattresses.

Propylene oxide and beta-propiolactone are also used for gaseous sterilization. All these gases are mutagens and carcinogens. Therefore, it is wise to minimize exposure.

4.6.5 Oxidizing Agents

Oxidizing agents exert antimicrobial activity by oxidizing cellular components of treated microbes.

Ozone, O_3, is a highly reactive form of oxygen. There is considerable interest in using ozone to replace chlorine in the disinfection of water. Although ozone is a more effective killing agent, its residual activity is difficult to maintain in water and it is more expensive than chlorine.

Hydrogen peroxide, H_2O_2, is not a good antiseptic for an open wound, but is effectively used to disinfect inanimate objects, an application where it is even sporicidal. The food industry uses hydrogen peroxide increasingly for aseptic packaging.

4.6.6 Phenol (Carbolic Acid)

Phenol is rarely used because it irritates the skin and has a disagreeable odor. Phenol is suitable for disinfecting pus, saliva, and feces. One commercial throat spray with a 1.4 percent phenol concentration has a significant effect on a sore throat.

Derivatives of phenol called phenolics with combination of a soap or detergent increase the antibacterial activity. The phenolics are stable and persist for long periods of time after application. The addition of halogens, especially chlorine, increase the antimicrobial activity.

4.6.7 Halogens

The halogen elements, particularly chlorine (Cl_2) and iodine (I_2), are effective antimicrobial agents.

Iodine is one of the oldest and most effective antiseptics. Iodine is effective against all kinds of bacteria, many endospores, various fungi, and some viruses. Iodine is available as an aqueous alcoholic solution and as an iodophor. *Iodophor* is a combination of iodine with one organic molecule. Iodophor has the antimicrobial effect of iodine but does not stain, is less irritating than iodine, and is commercially available as Betadine and Isodine. The main use of iodine is for skin disinfecting and wound treatment.

Chlorine as a gas or in combination with other chemicals is widely used. Its germicidal action is caused by the hypochlorous acid (HOCl) that forms when chlorine is added to water:

$$Cl_2 + H_2O \longrightarrow HCl + \underset{\text{hypochlorous acid}}{HOCl} \tag{4.4}$$

$$HOCl \longrightarrow H^+ + \underset{\text{hyochlorite ion}}{OCl^-} \tag{4.5}$$

Exactly how hypochlorous acid exerts its killing power is not completely known. It is a strong oxidizing agent that prevents the functioning of much of the cellular enzyme system. Hypochlorous acid has no electrical charge, therefore travels through the cell wall very rapidly. A liquid form of compressed chlorine gas is used extensively for disinfecting municipal drinking waters, sewage, and swimming pools. Several compounds of chlorine are also used as disinfectants in dairies, barns, restaurants, on eating utensils, and in slaughter houses. Examples are sodium hypochlorite, NaOCl, which is used as a household disinfectant. Bleach is a nine percent NaOCl solution.

4.6.8 Alcohols

These disinfectants have the advantage of acting and evaporating rapidly and leaving no residues. Alcohols effectively kill bacteria and fungi but not endospores and nonenveloped viruses. Two of the most commonly used alcohols are ethanol, or *ethyl alcohol*, and *isopropyl alcohol* (IPA), or propanol-2, called rubbing alcohol. The recommended optimum concentration is 70 percent. Pure alcohol is less effective than aqueous solution alcohols because denatured proteins need water.

4.6.9 Heavy Metals

Several heavy metals, such as silver, mercury, copper, and zinc, can be germicidal or antiseptic agents. The ability of very small amounts of heavy metals, especially silver and copper, to exert antimicrobial activity is referred to as oligodynamic action (oligo means few).

Silver is used as an antiseptic in 1 percent silver nitrate solution. Many states require that the eyes of newborn babies be treated with a few drops of silver nitrate against gonococcal infections in the eyes. In recent years, antibiotics have been replacing silver nitrate for this purpose.

Inorganic mercury compounds probably have the longest history of use as disinfectants. However, their use is now limited because of the high toxicity of mercury. Organic mercury compounds, such as *Mercurochrome* and *Merthiolate* are less toxic than inorganic mercury compounds. Mercurochrome and Merthiolate are antiseptics used on the skin and mucous membranes.

Copper, especially in the form of copper sulfate is used chiefly to control algae. To prevent mildew, copper compounds are added to paint.

Zinc chloride is the most common ingredient in mouthwash. Zinc oxide is the most widely used antifungal in paint.

5 Microorganisms in the Environment*

Microorganisms are found in every environment—in the air we breathe, the food we eat, the soil where food is grown, and the water we drink. Many microbes benefit humans, while only a few cause human diseases. Humans, being one of many organisms in the environment, affect and are affected by both living and nonliving components of the environment including its microorganisms. To control diseases, health scientists need to know how to control microorganisms in air, food, soil, and water. To do that they need to understand the roles of microorganisms in the environment.

5.1 BIOCHEMICAL CYCLES

As they carry on essential life processes, living organisms incorporate water, carbon, nitrogen, and other elements from their environment into their bodies. These materials must be recycled continuously to make them available to living organisms. The mechanisms by which such recycling occurs are referred to collectively as biochemical cycles. Without the activities of microorganisms in the various biochemical cycles, essential elements would become depleted and all life would cease.

5.1.1 THE WATER CYCLE OR HYDROLOGIC CYCLE

The water cycle recycles water. Water reaches the earth's surface as precipitation from the atmosphere. It enters living organisms during photosynthesis and by ingestion. It leaves living organisms through respiration, in wastes, and by evaporation from surfaces such as transpiration (water loss from pores in plant leaves). Like all other living things, microorganisms use water in metabolism, but they also live in water or very moist environments. Many form spores or cysts that help them survive periods of drought, but vegetative cells must have water. The hydrologic cycle is shown in Figure 5.1.

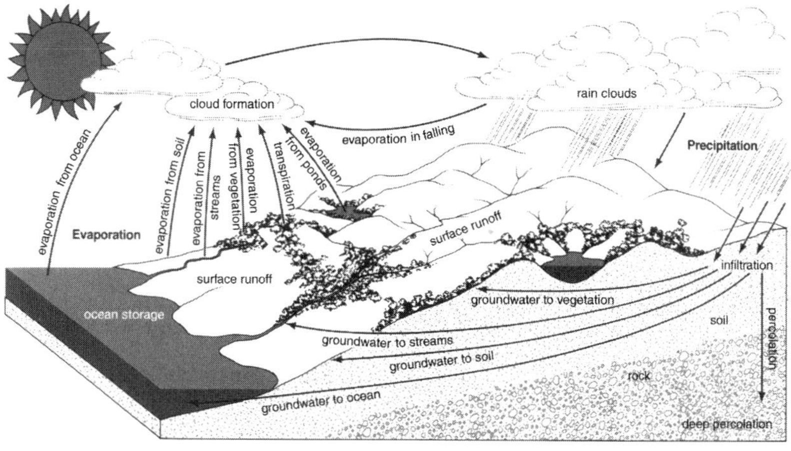

FIGURE 5.1 The hydrologic cycle.

* This chapter was contributed by Klara Ver, the Department of Microbiology, Janus Pannonius University, Pecs, Hungary.

5.1.2 The Carbon Cycle

All organic compounds contain carbon. Most of the inorganic carbon used in the synthesis of organic compounds comes from carbon dioxide (CO_2) in the atmosphere. Some CO_2 is also dissolved in water. In photosynthesis in the first step in the carbon cycle (see Figure 5.2), CO_2 is incorporated into organic compounds by photoautotrophs such as cyanobacteria, algae, and green and purple sulfur bacteria. In the next step in the cycle, chemoheterotrophs consume the organic compounds—animals eat photoautotrophs, especially green plants, as well as other animals. Thus, the organic compounds of the photoautotrophs are digested and resynthesized. In this way, the carbon atoms of CO_2 are transferred from organism to organism.

CO_2 is returned to the atmosphere by respiration and by the action of decomposers on the dead bodies and wastes of other organisms.

Carbon compounds can be deposited in peat, coal, and oil, and released from them during burning. The world gets most of its energy by burning fossil fuels—not only petroleum but also coal and natural gas. If the combustion is complete, all the carbon of these materials is converted into CO_2. The carbon in the fossil fuels has been stored in the earth for hundreds of millions of years. In the last century or two, large amounts of it have been converted to CO_2 in this way.

A small but significant quantity of CO_2 in the atmosphere comes from volcanic activity and from the weathering of rocks, many of which contain the carbonate ion, CO_3^{-2}. The oceans and carbonate rocks are the largest reservoirs of carbon, but recycling of carbon through these reservoirs is very slow.

All microorganisms require some carbon source to maintain life. Most carbon entering living things comes from CO_2 dissolved in bodies of water or in the atmosphere. Even the carbon in sugars and starches ingested by consumers is derived from CO_2. Because the atmosphere contains only a limited quantity of CO_2 (0.03 percent), recycling is essential for maintaining a continuous supply of atmospheric CO_2.

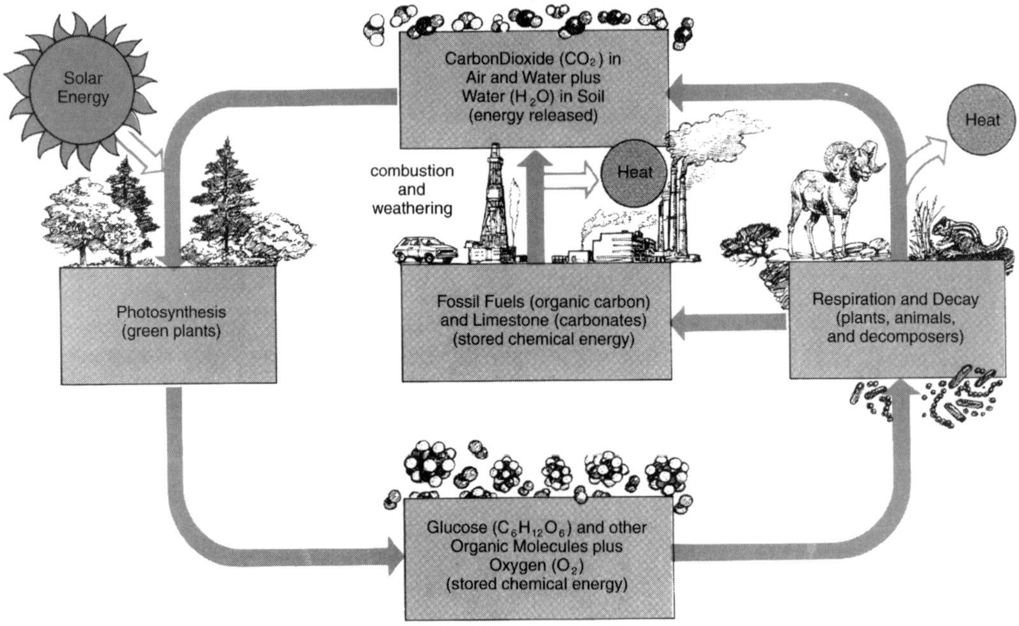

FIGURE 5.2 The carbon and oxygen cycles, showing chemical cycles (solid arrows) and one-way energy flow (open arrows).

5.1.2.1 The Greenhouse Effect

The CO_2 content of the atmosphere is slowly increasing every year and has been doing so for decades. Atmospheric carbon dioxide and water vapor form a blanket over the earth's surface, creating a greenhouse effect. These gases allow the sun's radiation to penetrate the atmosphere, thereby reaching the earth's surface, and warming it and the atmosphere. But they trap much of the infrared (heat) radiation produced by the warm surface, reflecting it back to the earth. Solar energy is thus captured within the greenhouse. We might, therefore, expect a gradual increase in the average temperature of the earth's surface, as the amount of CO_2 in the atmosphere increases every year.

It would not take a very large increase in the average temperature to cause substantial effects on the earth's climate. A rise of about 4°C would probably cause enough Antarctic ice to melt to flood most of the world's coastal cities, and possibly change the balance of organisms in ecosystems. This trend would make now-temperate regions too warm to grow wheat and other food crops, create droughts in other areas, and make new deserts.

Global warming may lead to an increase of incidence of infectious diseases as the range of the vectors (mosquitoes, snails, flies, and such) carrying the diseases expand. According to one computer based calculation, a global temperature rise of just 3°C over the next century could result in 50 to 80 million new malaria cases per year. Other diseases, including schistosomiasis, African trypanosomiasis, and dengue and yellow fevers are also likely to spread with global warming.

5.1.3 The Nitrogen Cycle

In the nitrogen cycle (see Figure 5.3), nitrogen moves from the atmosphere through various organisms and back into the atmosphere. This cyclic flow depends not only on decomposers but also on various nitrogen bacteria.

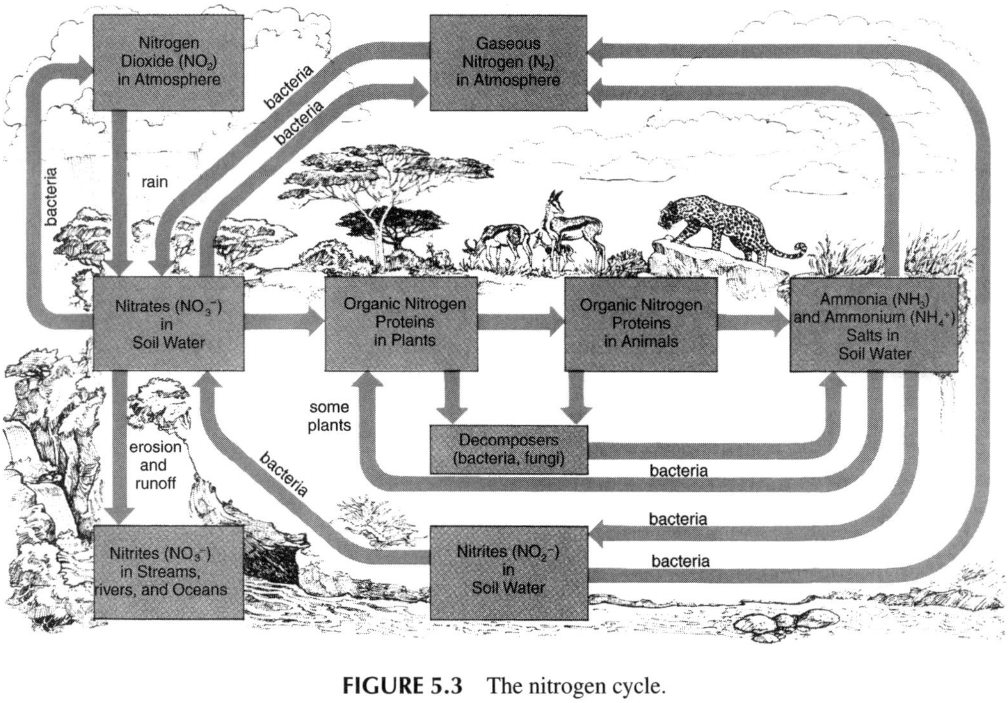

FIGURE 5.3 The nitrogen cycle.

Decomposers use several enzymes to break down proteins in dead organisms and their wastes, releasing nitrogen in much the same way they release carbon. Proteinases break large protein molecules into smaller molecules. Peptidases break peptide bonds to release amino acids. Deaminases remove amino groups from amino acids and release ammonia. Eventually, free nitrogen gas finds its way back into the atmosphere. Many soil microorganisms produce one or more of these enzymes. Clostridia, actinomycetes, and many fungi produce extracellular proteinases that initiate protein decomposition.

Nitrogen bacteria fall into one of three categories according to the roles they play in the nitrogen cycle:

1. Nitrogen-fixing bacteria.
2. Nitrifying bacteria.
3. Denitrifying bacteria.

5.1.3.1 Nitrogen-Fixing Bacteria

Nitrogen fixation is the reduction of atmospheric nitrogen gas (N_2) to ammonia (NH_3). Organisms that can fix nitrogen are essential for maintaining a supply of physiologically usable nitrogen on earth. About 255 million metric tons of nitrogen is fixed annually–70 percent of it by nitrogen-fixing bacteria. Bacteria and cyanobacteria fix nitrogen in many different environments–from Antarctica to hot springs, acid bogs to salt flats, flooded lands to deserts, in marine and fresh water, and even in the gut of some organisms.

The energy for nitrogen fixation can come from fermentation, aerobic respiration, or photosynthesis. The various organisms that fix nitrogen live independently, in loose associations, or intimate symbiosis. Regardless of the environment or the associations of the organisms, nitrogen-fixing bacteria must have a functional nitrogen-fixing enzyme called nitrogenase, a reducing agent that supplies hydrogen as well as energy from ATP. In aerobic environments, nitrogen fixers must also have a mechanism to protect the oxygen-sensitive nitrogenase from inactivation.

Free-living aerobic nitrogen fixers include several species of the genus azotobacter, some methylotrophic bacteria and cyanobacteria. *Azotobacter* is found in soils and its growth is limited by the amount of organic carbon available. *Methylotrophic bacteria* can fix nitrogen when provided with methane, methanol, or hydrogen from various substrates. *Cyanobacteria* can fix nitrogen by using hydrogen from hydrogen sulfide, so they increase the availability of nitrogen in sulfurous environments.

Rhizobium is the primary symbiotic nitrogen fixer. It lives in root nodules of certain plants, usually legumes (see Figure 5.4). In this symbiotic relationship, the plant benefits by receiving nitrogen in a usable form, and the bacteria benefit by receiving nutrients needed for growth. Farmers often fix nitrogen-fixing bacteria with seed peas and beans before planting to ensure that nitrogen fixation will be adequate for their crops to thrive. The nodules contain the enzyme nitrogenase which catalyzes the following reaction:

$$\underset{\substack{\text{nitrogen}\\\text{gas}}}{N_2} + \underset{\substack{\text{hydrogen}\\\text{gas}}}{3H_2} \xrightarrow{\text{Rhizobium}} \underset{\text{ammonia}}{2\,NH_3} \tag{5.1}$$

The nitrogenase enzyme is inactivated by oxygen, so nitrogen fixation can occur only if oxygen is not available. Nitrogenase is protected from oxygen by a kind of haemoglobin, a red pigment that binds oxygen. This particular haemoglobin is synthesized only in root nodules.

Although symbiotic nitrogen fixation occurs mainly in association between rhizobia and legumes, other such associations are known. The alder tree, which grows in soils low in nitrogen, has root nodules similar to those formed by rhizobia. These nodules contain nitrogen-fixing actinomycetes of the genus *Frankia*.

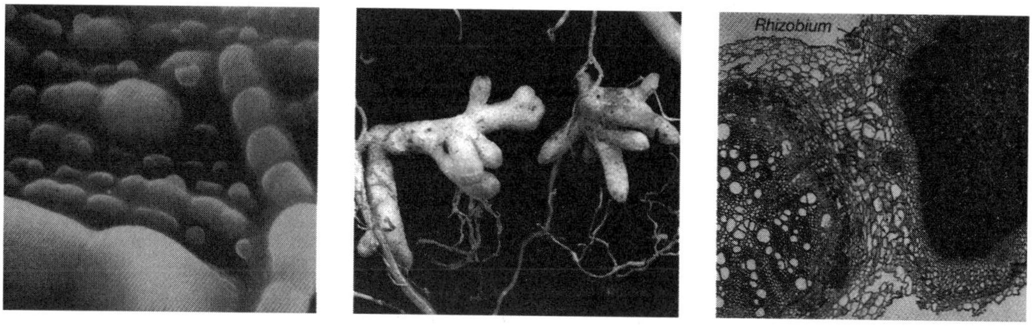

FIGURE 5.4 Nodules on the roots of a bean plant.

5.1.3.2 Nitrifying Bacteria

Nitrification is a process by which ammonia (NH_3) or ammonium ions (NH_4^+) are oxidized to nitrites (NO_2^-) and nitrates (NO_3^-). Carried out by autotrophic bacteria, nitrification is an important part of the nitrogen cycle because it supplies plants with nitrate (NO_3^-), the form of nitrogen most usable for metabolism. Nitrification occurs in two steps. Various species of nitrosomonas (see Figure 5.5) and related genera, gram-negative rod-shaped bacteria produce nitrite (NO_2^-).

$$\underset{\text{ammonium ion}}{NH_4^+} + \underset{\text{oxygen gas}}{\tfrac{1}{2}O_2} \xrightarrow{\text{nitrosomonas}} \underset{\text{nitrite}}{NO_2^-} + \underset{\text{hydrogen ions}}{2H^+} + \text{energy} \quad (5.2)$$

Species of nitrobacter and related genera, also gram-negative rods, produce nitrates.

$$\underset{\text{nitrite}}{NO_2^-} + \underset{\text{oxygen}}{\tfrac{1}{2}O_2} \xrightarrow{\text{nitrobacter}} \underset{\text{nitrate}}{NO_3^-} + \text{energy} \quad (5.3)$$

Furthermore, because nitrite is toxic to plants, it is essential that these reactions be carried out in sequence to provide nitrates and to prevent excessive accumulation of nitrites in the soil.

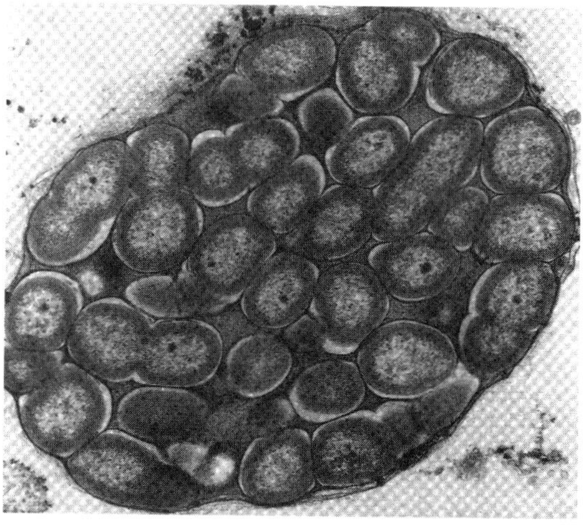

FIGURE 5.5 Microcolony of the nitrifying bacterium, *Nitrosomonas*.

5.1.3.3 Denitrifying Bacteria

Denitrification is a process by which nitrates are reduced to nitrous oxide (N_2O) and nitrogen gas (N_2).

$$\underset{\text{nitrate}}{NO_3^-} \longrightarrow \underset{\text{nitrous oxide}}{N_2O} + \underset{\text{nitrogen gas}}{N_2} \tag{5.4}$$

Although this process does not occur to any significant degree in well-oxygenated soils, it does occur in oxygen depleted, waterlogged soils. Most denitrification is performed by the pseudomonas species, but it can also be performed by thiobacillus denitrificans, micrococcus denitrificans, and several species of serratia and achromobacter. These bacteria, although usually aerobic, used nitrate instead of oxygen as a hydrogen acceptor under anaerobic conditions.

Another process that reduces soil nitrate is the reduction of nitrate to ammonia. Several anaerobes carry out this process (called dissimilative nitrate reduction) in a complex reaction that can be summarized as follows (unbalanced equation):

$$\underset{\text{nitrate}}{NO_3} \longrightarrow \underset{\text{hydrogen gas}}{H_2} \longrightarrow \underset{\text{ammonia}}{NH_3} \longrightarrow \underset{\text{nitrous oxide}}{N_2O} \tag{5.5}$$

Denitrification is a wasteful process because it removes nitrate from the soil and interferes with plant growth. It is responsible for significant losses of nitrogen from fertilizers applied to soils. Another unfortunate effect of denitrification is the production of nitrous oxide, which is converted to nitric oxide (NO) in the atmosphere. Nitrous oxide, in turn, reacts with ozone (O_3) in the upper atmosphere.

5.1.4 The Sulfur Cycle

During the sulfur cycle, the sulfhydryl (–SH) groups in proteins of dead organisms are converted to hydrogen sulfide (H_2S) by a variety of microorganisms. This process is analogous to the release of ammonia from proteins during the nitrogen cycle. Hydrogen sulfide is toxic to living things and, thus, must be oxidized rapidly. Oxidation to elemental sulfur is followed by oxidation to sulfate (SO_4^{2-}), which is the form of sulfur most usable by both microorganisms and plants. This process is analogous to nitrification. The sulfur cycle is of special importance in aquatic environments, where sulfate is a common ion, especially in ocean water.

5.1.4.1 Sulfate Reducing Bacteria

Sulfate reduction is the reduction of sulfate (SO_4^{2-}) to hydrogen sulfide (H_2S). The sulfate reducing bacteria are among the oldest life forms, probably more than 3 billion years old. They include the closely related genera *Desulfovibrio, Desulfomonas*, and *Desulfotomaculum.* By reducing sulfate, these bacteria produce large amounts of hydrogen sulfide. Sulfate reducing bacteria are strict anaerobes.

5.1.4.2 Sulfur Reducing Bacteria

Sulfur (S) reduction is the reduction of sulfur to hydrogen sulfide (H_2S). The sulfur can be in elemental form or in disulfide bonds of organic molecules. This process provides energy for the organisms when sunlight is not available for photosynthesis. Sulfur reducing bacteria are anaerobes.

5.1.4.3 Sulfur Oxidizing Bacteria

Sulfur oxidation is the oxidation of sulfur to sulfate. *Thiobacillus* and similar bacteria oxidize hydrogen sulfide (H_2S), ferrous sulfide (FeS), or elemental sulfur (S) to sulfuric acid (H_2SO_4). When this acid ionizes, it greatly decreases the pH of the environment, sometimes lowering the pH to 1 or 2. Sulfur oxidizing compounds are responsible for oxidizing ferrous sulfide in coal-mining wastes, and the acid they produce is extremely toxic to fish and other organisms in streams fed by such wastes.

5.1.5 The Phosphorus Cycle

The phosphorus cycle involves the movement of phosphorus among inorganic and organic forms.
Soil microorganisms are active in the phosphorus cycle in at least two important ways :

1. They break down organic phosphates from decomposing organisms to inorganic phosphates.
2. They convert inorganic phosphates to orthophosphate (PO_4^{-3}), a water-soluble nutrient used by both plants and microorganisms. These functions are particularly important because phosphorus is often the limiting nutrient in many environments.

Phosphorus is essential to the growth of organisms and can be the nutrient that limits the primary productivity of a body of water. Phosphorus has been identified as the most critical growth factor in most waters. The discharge of wastewater, agricultural drainage, or certain industrial wastes to a water system may stimulate growth of photosynthetic aquatic micro- and macro-organisms. Sediments play a major role in the availability of phosphorus in many aquatic areas.

Phosphate is slowly dissolved (leached) from rocks by rain and is carried to waterways. Dissolved phosphates are incorporated by plants and passed to animals by the food web. Phosphorus reenters the environment directly by animal excretum and by detritus decay. Each year large quantities of phosphates are washed into the oceans, where much of it settles to the bottom and is incorporated into the marine sediments. Sediments may release some of the phosphate needed by aquatic organisms, and the rest may become buried. Phosphate is a major component of fertilizer. By applying excess fertilizer, farmers may alter the phosphorus cycle.

5.2 MICROORGANISMS FOUND IN AIR

Microorganisms do not grow in air, because it lacks the nutrients needed for metabolism and growth. However, spores are carried in air, and vegetative cells are carried on dust particles and on water droplets in the air. The kinds and numbers of airborne microorganisms vary greatly in different environments. Large numbers of many different kinds of microorganisms are present in indoor air where humans are crowded together and building ventilation is poor. Small numbers have been detected at altitudes of 3,000 m.

Indoor biological pollution should receive the attention of environmental and health agencies. The airborne bioflora is inherently complex and variable to the point that defies quantification. The air in a single room in a clean house may contain hundreds of different kinds of biological particles. Health effects of biological aerosols are basically different from their chemical counterparts. The majority of bioaerosols are nonpathogenic and cause disease only in sensitized people. Even pathogenic microorganisms are usually able to infect only susceptible hosts.

Microorganisms may cause allergic respiratory reactions as well as respiratory infections. Respiratory infections account for about 50 to 60 percent of all community acquired illness. Although most of these infections are of a viral etiology, bacterial diseases do occur and have caused substantial problems in hospitals, schools, and day-care centers. Outbreaks of hypersensitivity pneumonitis caused by microorganisms in both large and small office buildings and residences are well-documented.

The increased use of recirculated air in buildings where susceptible individuals are present 24 h per day (hospitals, retirement homes) can become a serious problem if microorganisms are not removed by cleaning or preventive maintenance.

Asthma may be caused by pollens, mites, animal dander, fungi, and insect emanations.

One of the most prominent examples of infections caused by microorganisms in buildings would be the occurrence of Legionnaires' disease. For the most part Legionnaires' disease has been associated with the infiltration into the building environment of aerosols containing *Legionella pneumophila* from external sources such as cooling towers. Other airborne bacteria such as *Neisseria meningitididis* and *Hemophilus influenza* are responsible for illness outbreaks in day-care centers and schools, but they do not appear to be transmitted by recirculated, conditioned air.

Pathogenic microorganisms may also be disseminated into the intramural environment by use of hot tubs (dermatitis caused by pseudomonas), whirlpools, cold mist vaporizers, and nebulizers.

Among the organisms found in air, mold spores are the most numerous, and the predominant genus is usually *Cladosporium*. Bacteria commonly found in air are both aerobic sporeformers, such as *Bacillus subtilis* and nonspore forming such as *Micrococcus* and *Sarcina*. Algae, protozoa, yeasts, and viruses also have been isolated from air.

While coughing, sneezing, or talking, infected humans can expel pathogens along with water droplets.

5.2.1 Sources of Indoor Pollution

5.2.1.1 Outdoor Environment

The majority of biological particles found in interior situations come from the outdoor environment.

Pollen is almost entirely from outdoor sources. The majority of fungus spores encountered indoors are derived either directly or indirectly from outdoor air. Many bacteria, including some that cause human disease, thrive in outdoor reservoirs is *Legionella*, which is basically soil borne. A wide variety of other bacteria, most of which are nonpathogenic to people, are abundant in outdoor substrates and may penetrate and possibly grow on interior surfaces, for example, *Actinomycetes*.

Other biological particles that are occasionally common in outdoor air and that can contribute to indoor aerosols include algae, insects, and arachnids. Algae can be abundant outdoors near ponds or lakes and become airborne and enter interiors through wind. Outdoor insect infestations can result in airborne fragments small enough to penetrate even closed buildings. Other insects (e.g., cockroaches) and spiders, while primarily outdoor creatures, can abundantly colonize interiors and produce body fragments, excrements, and web material that can accumulate indoors.

5.2.1.2 Indoor Contamination

Although most indoor contamination results from outdoor sources, interior situations can become heavily contaminated with biological entities to the point of presenting severe health hazards.

Materials that are shed and accumulate indoors include particulates from human and animal skin scales, bacteria and dermatophytes shed from animal surfaces, and insect and arachnoid fragments. Bacterial endotoxins and fungi mycotoxins can accumulate indoors from microorganisms growing on interior surfaces.

Most severe indoor biological pollution problems result from the growth of the offending organism on surfaces. Virtually any substrate that includes both a carbon source and water will support the growth of some microorganism. Human or animal skin scale support not only the mite population, but mesophilic bacteria and fungi such as *Aspergillus amstelodami* and *Wallemia sebi*, both of which can withstand relative dryness. Cellulose or lignin based materials, when damp, provide ideal enrichment media for a wide variety of bacteria and fungi. Leather can also support fungal growth. Plastics and nylon are long-chain compounds, with similar structure to cellulose, utilized by some microorganisms. Soap, grease, and other hydrocarbon films on surfaces can support a variety of microorganisms.

The absolute requirement for all of these kinds of interior contamination is a more or less consistent source of moisture. Relative humidity directly affects fungal spore levels, probably by increasing growth on surfaces by absorbing water. Above 70 percent is apparently optimal. Unfortunately, a variety of modern appliances provide standing water reservoirs which, if not absolutely clean, provide excellent enrichment situations for microorganisms. Humidifiers, evaporative coolers, self-defrosting refrigerators, and flush toilets all have water reservoirs with the potential of contamination. Air conditioners have cold surfaces often bearing an abundant and constant supply of water condensed from the air. When contaminated, microorganisms are readily blown from these surfaces into room air.

5.2.2 Factors Affecting Indoor Microbial Levels

Indoor surface contamination by bacteria, fungi, insects, arachnoids, or other biological particles is dangerous for the most part only when the particles become airborne and are inhaled.

Air currents produced by convection from radiant heat, of course, by air mechanically circulated by forced air systems are more than adequate to spread dust as well as mobilizing surface growth. Many of the appliances mentioned previously can diffuse microbial aerosols into the air. Clothes dryers present a different problem. Warm humid air passes through exhaust pipes often up on pipe surfaces. Thermotolerant and thermophilic organisms can thrive in this situation, and produce abundant spores which enter the air stream during dryer operation. Dishwashing equipment, especially in hospitals, poses particular risks for kitchen workers, as disease causing organisms can reach potentially infective levels when such units are improperly sealed.

Any human or pet activity can increase airborne microbial loads, especially vacuuming, sweeping, dusting, scrubbing contaminated surfaces, and bed-making, etc. None of these activities should be undertaken by or in the presence of sensitive or susceptible individuals.

5.2.3 Controlling Microorganisms in Air

Chemical agents, radiation, filtration, and laminar air-flow all can be used to control microorganisms in the air.

5.2.3.1 Chemical Agents

Triethylene glycol, resorcinol, and lactic acid dispersed as aerosols kill many microorganisms in air. These agents are highly bactericidal, remain suspended long enough to act at normal temperature and humidity, are nontoxic to humans, and do not damage or discolor objects in the room.

5.2.3.2 Ultraviolet Radiation

Ultraviolet radiation has little penetrating power and is bactericidal only when rays make direct contact with microorganisms. They are most useful in maintaining sterile conditions in rooms only sporadically occupied by humans. They can be turned off when humans are working in the area and turned on again when the room is not in use. Humans entering a room while ultraviolet (UV) lights are on must wear protective clothing and special glasses to protect the eyes.

5.2.3.3 Air Filtration

Air filtration involves passing air through fibrous materials, such as cotton or fiberglass. It is useful in industrial processes when sterile air is needed. Cellulose acetate filters can be installed in a laminar air-flow system to remove microorganisms that may have escaped into the air underneath the laminar flow hood. The air is suctioned away from the opening, filtered, and then returned to the room.

To briefly summarize, prevention of outdoor influx, maintenance of low humidity, and elimination of dampness or standing water of any kind will go far to prevent serious indoor air biopollution problems.

5.3 MICROORGANISMS IN SOIL

Soil, in fact, is teeming with microscopic and small macroscopic organisms, and it receives animal wastes and organic matter from dead organisms. Microorganisms act as decomposers to break down this organic matter into simple nutrients that can be used by plants and by the microbes themselves. Soil microorganisms are, thus, extremely important in recycling substances in the ecosystems.

Soil is a unique medium that contains a diverse community of organisms, representing many morphological and physiological types. Organisms in soil are never static in numbers of activity. Variation of microbial numbers and activities can occur with depth and soil type. Soil structure, texture, and moisture levels can drastically affect aeration within the soil environment from one characterized as aerobic to one with anaerobic properties, thus influencing the distribution of physiological types.

At least as important as soil physical properties on microbial diversity is the season of the year. Soil fertility also plays a role in governing microbial development.

Finally, the diversity of organisms is dependent on the type of plant and the proximity of an organism to the plant root itself.

5.3.1 COMPONENTS OF SOIL

Soil is divided into a number of layers called soil horizons. These horizons include the topsoil, subsoil, and parent material, which overlay bedrock. Most plant roots grow in topsoil, the loose surface layer of the soil. Topsoil usually lays on top of subsoil, which consists of tightly packed rock particles that only the toughest roots can penetrate. Subsoil contains much less organic matter and air than topsoil. Under the subsoil lays bedrock, solid rock that may break down to form soil over a long period of time.

Soil is a somewhat complex mixture of solid inorganic matter (rocks and minerals), water, air, living organisms, and the products of their decay. Numerous physical and chemical changes take place in this mixture.

Soil contents consist of inorganic and organic components. The inorganic components are rocks, minerals, water, and gases. The organic components are humus (nonliving organic matter) and living organisms. Soils differ greatly in the relative proportions of these components.

5.3.1.1 Inorganic Components

The most abundant inorganic components of soils are pulverized rocks and minerals. Weathering of rocks–mechanical or chemical breakdown–releases minerals, and the elements that make up those minerals, into the soil. The most abundant elements in many soils are silicon, aluminum, and iron. Small quantities of other elements, such as calcium, potassium, magnesium, sodium, phosphorus, nitrogen, and sulfur also are present in soil. In addition to rock and minerals, soils contain water and gases such as carbon dioxide, oxygen, and nitrogen.

Water molecules adhere to soil particles or are interspersed among them. The amount of water in soil is highly variable and depends on the climate, the quantity of recent rainfall, and the drainage of the soil. Through the solvating property of the water, various inorganic and organic constituents of soil are dissolved in the soil water and are made available to the living inhabitants of the soil.

Gases are dispersed among soil particles or dissolved in water. The concentration of gases in the soil also varies with the metabolic activity of soil organisms. Compared with atmospheric air, soil is lower in oxygen and higher in carbon dioxide.

5.3.1.2 Organic Matter

The organic matter in soil consists of carbohydrates, proteins, lipids, and other materials. Organic matter comprises 2 to 10 percent of most agriculturally important soils. Swamps and bogs, by contrast, contain a higher content of organic matter, up to 95 percent in some peat bog soils. The waterlogged soil of swamps and bogs is an anaerobic environment, and the microbial decomposition of organic matter there only occurs slowly.

All organic matter in soil is derived from the remains of microorganisms, plants and animals, their waste products, and the biochemical activities of various microorganisms. A great portion of the organic matter is of plant origin, mostly dead roots, wood, bark, and fallen leaves. A second source is the vast number of bacteria, fungi, algae, protozoans, small animals, and viruses that can total in the billions per gram of fertile soil. Through the actions of these organisms, the breakdown of organic substances produces and maintains a continuous supply of inorganic substances that plants and other organisms require to grow. Much of organic matter is ultimately decomposed into inorganic substances, such as ammonia, water, carbon dioxide, and various compounds of nitrate, phosphate, and calcium.

A considerable part of the organic matter in soil occurs as humus, a dark material composed chiefly of organic materials that are relatively resistant to decay. In this regard, humus is partially decomposed organic matter. Humus is constantly changing as organisms die and as decomposers degrade complex molecules to simpler ones. Soils differ greatly in the amount of humus they contain. Most soils contain 2 to 10 percent humus, but a peat bog can be 95 percent humus.

The addition of organic matter, either completely or partially decomposed, is essential to the continuation of soil fertility. Fertile soils are those in which most plants are healthy and grow rapidly. Fertile soil is composed of rock particles, decaying organic matter, living organisms, water, and air.

5.3.1.3 Root Systems

Soil also contains the root systems of higher plants and enormous number of microorganisms.

5.3.1.4 Living Organisms

Soil contains a great many animals, ranging from microscopic forms, including numerous nematodes, to larger forms, including insects, millipedes, centipedes, spiders, slugs, snails, earthworms, mice, moles, gophers, and reptiles. Most of these animals are beneficial in that they promote some mechanical movement of the soil, thereby helping to keep the soil loose and open. All soil organisms also contribute to the organic matter of soils in the form of their waste products and eventual remains.

Living organisms are so important to the formation of soil that it may take 10 million years for 1 centimeter of soil to form in dry, cold areas where soil organisms grow and reproduce slowly. In contrast, 1 centimeter of soil may form in a year or less in warm, wet tropical regions where organisms grow rapidly.

Despite the great diversity of soil organisms, microorganisms are the most numerous both in total numbers and in the number of species present. Microorganisms convert humus to nutrients that are usable for other organisms. They also make great demands on soil nutrients. Topsoil, the surface layer of the soil contains the greatest number of microorganisms because it is well supplied with oxygen and nutrients. Lower layers of soils (subsoil and parent material) and soils depleted of oxygen and nutrients contain fewer organisms.

5.3.2 Microorganisms in Soil

All the major groups of microorganisms–bacteria, fungi, algae, and protista, as well as viruses–are present in soil, but bacteria are the most numerous, as shown in Figure 5.6.

5.3.2.1 Bacteria

Without the microorganisms, especially bacteria, the soil would soon become unfit to support life. Among the bacteria in soil are autotrophs, heterotrophs, aerobes, anaerobes, and depending on the soil temperature, mesophiles and thermophiles. There are nitrogen-fixing, nitrifying, and denitrifying

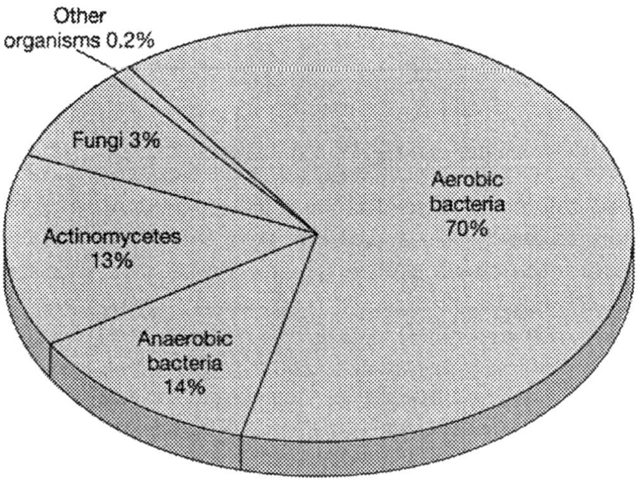

FIGURE 5.6 Relative proportions of various kinds of organisms found in soil. Other organisms include algae, protists, and viruses.

bacteria and bacteria that digest special substances such as cellulose, protein, pectin, butyric acid, and urea. When usable nutrients are added to soil, the microbial population and their activity rapidly increase until the nutrients are depleted, and then the microbial activity returns to the lower levels. The most numerous organisms in the soil are bacteria. Table 5.1 shows the distribution of microorganisms in the number per gram of typical garden soil at various depths.

5.3.2.2 Actinomycetes

Actinomycetes are bacteria, but they are usually considered separately in enumerations of soil populations. Found in large numbers, the actinomycetes produce a gaseous substance called *geosmin*, which gives fresh soil its characteristic musty odor (see Chapter 22). Interest in these bacteria was greatly stimulated by the discovery that some genera, particularly streptomycetes produce valuable antibiotics.

5.3.2.3 Fungi

Fungi are found in the soil in much smaller numbers than bacteria and actinomycetes. Soil fungi are mostly molds. Both mycelia and spores are present mainly in topsoil, the aerobic surface layer of the soil. Fungi decompose plant tissues such as cellulose and lignin, and their mycelia form networks around soil particles, giving the soil a crumbly texture. In addition to molds, yeast are abundant in soil in which grapes and other fruits are growing.

TABLE 5.1
Distribution of Microorganisms in Numbers per Gram of Typical Garden Soil at Various Depths

Depth (cm)	Bacteria	Actinomycetes	Fungi	Algae
3–8	9,750,000	2,080,000	119,000	25,000
20–25	2,179,000	245,000	50,000	5,000
35–40	570,000	49,000	14,000	500
65–75	11,000	5,000	6,000	100
135–145	1,400	—	3,000	—

5.3.2.4 Cyanobacteria, Algae, Protista and Viruses

Small numbers of cyanobacteria, algae, protista, and viruses are found in most soils.

Algae are found only on the surface, where they can carry on photosynthesis. In the desert and other barren soils algae contribute significantly to the accumulation of organic matter in the soil. Algae and cyanobacteria sometimes form visible blooms (abundant growths) on the surface of moist soil; these are also found in dry desert soils.

Protistas are mostly amoebas and flagellated protozoa in the soil, they feed on bacteria and help to control the bacterial population.

Soil viruses infect mostly bacteria, but attack plants and animals, too.

5.3.3 Abiotic Factors Influence Microorganisms in the Soil

The growth of microorganisms is influenced by abiotic factors and by other organisms. Abiotic factors include moisture, pH, temperature, and oxygen content.

5.3.3.1 Water and Oxygen

Spaces among soil particles ordinarily contain water and oxygen. Therefore, aerobic organisms live there. In waterlogged soils, all the spaces are filled up with water and without oxygen. Therefore, anaerobic organisms grow there.

5.3.3.2 Soil pH

Soil pH is an important property of soil, and it can vary from pH 2.00 to pH 9.00. Most soil bacteria prefer the pH range of 6.00 to 8.00, but some mold can live at any pH. Molds thrive in acidic soil, because there is no competition from bacteria for available nutrients. Lime is used to neutralize the high acidity in soils and to increase the bacterial population.

5.3.3.3 Temperature

Soil temperature varies seasonally from below freezing to as high as 60°C at soil surfaces exposed to intense summer sunlight. *Mesophilic* and *thermophilic* bacteria are quite numerous in warm to hot soils, whereas *cold-tolerant mesophiles* are present in cold soils. Most soil molds are mesophilic and are found in soils of moderate temperature.

Exceedingly wide variations exist in the physical characteristics of the soil and in the number and kinds of organisms that live in the soil. The interaction among organisms and between the organisms and the environment can be quite different in different microenvironments, no matter how close together they are.

5.3.4 Importance of Decomposers in the Soil

Soil microorganisms, that are decomposers, are important in the carbon cycle, because of their ability to decompose organic matter. Organic substances include cellulose, lignins, and pectins in the cell walls of plants, glycogen from animal tissues, and proteins and fats from plants and animals.

Cellulose is mostly decomposed by bacteria, special genus *Cytophaga*, and by various fungi.

Lignins and pectins are mostly digested by fungi, but protozoa and nematodes also can participate in the degradation of these materials.

Proteins are degraded to individual amino acids by fungi, actinomycetes, and clostridia.

Under the anaerobic conditions of waterlogged soils in marshes and swamps, methane is the main product of the degradation. It is produced by three genera of strictly anaerobic bacteria—*Methanococcus, Methanobacterium,* and *Methanosarcina.* In addition to degrading carbon compounds to methane, anaerobic bacteria also obtain energy by oxidizing hydrogen gas.

In one way or another, organic substances are metabolized to carbon dioxide, water, and other small molecules and one or more organisms that can decompose it. Thus, carbon is continuously recycling. However, certain organic compounds manufactured by humans resist the action of microorganisms. Accumulation of these synthetic substances create environmental hazards.

Nitrogen enters the soil through the decomposition of protein from dead organisms and through the action of nitrogen-fixing bacteria. In addition to protein decomposition which introduces nitrogen to the soil, gaseous nitrogen is fixed both by free-living microorganisms and by symbiotic microorganisms associated with the roots of legumes, as previously mentioned.

5.3.5 Soil Pathogens

Soil pathogens are usually plant pathogens. A few soil pathogens can affect humans and animals. Most human pathogens that can survive in soil are endospore-forming bacteria.

Endospores of *Bacillus anthracis*, which causes anthrax in animals, can survive in certain soils for decades. Grazing animals can contract anthrax from spores of *Bacillus anthracis* in the soil.

The main human pathogens found in soil belong to the genus *Clostridium*. All are anaerobic spore formers. *Clostridium tetani* causes tetany and can be introduced easily into a puncture wound. *Clostridium botulinum* causes botulism. Its spores, found on many edible plants, can survive in incompletely processed foods to produce a deadly toxin. *Clostridium perfringers* causes gas gangrene in poorly cleaned wounds.

In fact, most soil organisms that infect warm-blooded animals exist as spores because soil temperatures are usually too low to maintain vegetative cells of these pathogens.

5.4 AQUATIC MICROBIOLOGY

All aquatic ecosystems—fresh water, ocean water, even rainwater—contain microorganisms.

5.4.1 Freshwater Environment

Freshwater environments include surface waters such as lakes, ponds, rivers, streams, and groundwater.

5.4.1.1 Groundwater

Groundwater contains only a few microorganisms.

5.4.1.2 Surface Water

Surface water contains large numbers of many different kinds of microorganisms. Ponds and lakes are divided vertically into zones.

The shoreline, or *littoral zone*, is an area of shallow water near to the shore where light penetrates to the bottom.

The *limnetic zone* comprises the sunlit water away from the shore. Resident microorganisms include algae and cyanobacteria.

Between the limnetic zone and the lake sediment is the *profundal zone*. When organisms in the limnetic zone die, they sink into the profundal zone, where they provide nutrients for other organisms.

The sediment, or *benthic zone*, is composed of organic debris and mud.

5.4.2 Factors Affecting Microorganisms in Aquatic Environments

5.4.2.1 Temperature

In aquatic environments, water temperatures vary from 0°C to 100°C. Most microorganisms thrive in water at moderate temperature. However, some thermophilic bacteria have been found in geysers at a water temperature of 90°C and psychrophilic fungi and bacteria have been found in water at 0°C.

5.4.2.2 pH

The pH of fresh water varies from 2 to 9. Although most microorganisms grow best in waters with a neutral pH, a few have been found in extremely acidic or extremely alkaline waters.

5.4.2.3 Nutrients

Most natural waters are rich in nutrients, but the quantities of various nutrients can vary considerably. Sometimes nutrients become so abundant that a bloom, or sudden proliferation of organisms in a body of water, occurs.

5.4.2.4 Oxygen

In aquatic environments, oxygen can be the limiting factor in the growth of microorganisms. Because of oxygen's slow solubility in water, its concentration never exceeds 0.007 g per 100 g of water.

When water contains large quantities of organic matter, decomposers rapidly deplete the oxygen supply as they oxidize the organic matter. Oxygen depletion is much more likely in standing water in lakes and ponds than in running water of rivers and streams because the movement of running water causes it to be continuously oxygenated.

5.4.2.5 Depth

Another factor affecting microorganisms in aquatic environments is the depth to which the sunlight penetrates the water. It is not important in fresh waters (except deep lakes), but it is very important in the ocean.

Photosynthetic organisms are limited to locations with adequate sunlight.

Aerobic bacteria are found where oxygen supplies are adequate, whereas anaerobic bacteria are found in oxygen-depleted waters.

Eucaryotic algae, cyanobacteria, and sulfur bacteria only live in water that receives adequate sunlight.

5.4.3 Marine Environments

The ocean environment covers about 70 percent of the earth's surface and is, therefore, larger than other environments combined. Compared with freshwater, the ocean is much less variable in both temperature and pH.

5.4.3.1 Temperature

Except in the vicinity of volcanic vents in the sea floor, where water reaches 250°C, ocean water ranges from 30°C to 40°C at the surface near the equator to 0°C in polar regions and in the lowest depths. At any one location and depth, the temperature is nearly constant.

5.4.3.2 pH

The pH of ocean water ranges from nearly neutral to slightly alkaline (pH 6.5 to 8.3). This pH range is suitable for the growth of many microorganisms, and sufficient carbon dioxide dissolves in the water to support photosynthetic organisms.

5.4.3.3 Salinity

Ocean water, which is much more salty than fresh water, displays a remarkably constant concentration of dissolved salts, 3.3 to 3.7 percent w/v or 33,000 to 37,000 ppm. Thus, organisms that live in marine environments must be able to tolerate high salinity but need not tolerate variations in salinity.

5.4.3.4 Hydrostatic Pressure

Marine environments display a far greater range of hydrostatic pressure than do fresh waters. Hydrostatic pressure increases with depth at a rate of approximately 1 atm per 10 m, so the pressure at a depth of 1000 m is 100 times that at the surface. A few microorganisms, including archeobacteria, have been isolated from Pacific Ocean trenches at depths greater than 1000 m.

5.4.3.5 Sunlight and Oxygen Concentration

Other factors that vary with the depth of ocean water are penetration of sunlight and oxygen concentration. Sunlight of sufficient intensity to support photosynthesis penetrates ocean water to depths of only 50 to 125 m, depending on the season, latitude, and transparency of the water.

5.4.3.6 Nutrient Concentrations

Nutrient concentrations vary in ocean water, depending on depth and proximity to the shore. Nutrients are more plentiful near the mouths of rivers and near the shore, where runoff from land enriches the water.

However, ocean water is lower in phosphates and nitrates than fresh waters.

Photosynthetic organisms near the surface serve as food for heterotrophic organisms at the same or deeper level.

Decomposers are usually found in bottom sediments, where they release nutrients from dead organisms.

Large numbers of many different kinds of microorganisms live in the ocean. The primary producers of the ocean are photosynthetic microorganisms called phytoplankton. They are motile and contain oil droplets or other devices for buoyancy, allowing them to remain in sunlit waters. Phytoplankton include cyanobacteria, diatoms, dinoflagellates, chlamydomonas, and a variety of other protists and eukariotic algae.

Many of the consumers of the ocean are heterotrophic bacteria. Members of the genera pseudomonas, vibrio, achromobacter, and flavobacterium are common in ocean water.

Protozoa, especially radiolarians and foraminiferans, and a variety of fungi also feed on producers in the open ocean. However, most of these organisms inhabit an aquatic zone beneath the region of intense sunlight.

In the sediment at the bottom of the ocean, the number of microorganisms increases. They are usually strict or facultative anaerobic decomposers. Many of them contribute significantly to the maintenance of biochemical cycles and produce substances such as ammonium ion, hydrogen sulfide, and nitrogen gas.

5.4.4 Water Pollution

Water is considered polluted if a substance or condition is present that renders the water useless for a particular purpose. For example, human drinking water is considered polluted if it contains pathogens or toxic substances. Water that is fit for human consumption is termed potable water.

5.4.4.1 Organic Wastes

The major types of water pollutants are organic wastes, such as sewage and animal manures, industrial wastes, oil, radioactive substances, sediments from soil erosion, and heat. Organic wastes suspended in water are decomposed by microorganisms provided that the water contains sufficient oxygen for oxidation of the substances. The oxygen required for such decomposition is called biological oxygen demand (BOD). The test of BOD is outlined in Appendix D. If BOD is high, the water can be depleted of oxygen rapidly. Anaerobes increase in number while populations of aerobic decomposers decrease, leaving behind large quantities of organic wastes. Organic wastes also can contain pathogenic organisms from among the bacteria, viruses, and protozoa.

5.4.4.2 Industrial Wastes

Industrial wastes contain minerals, inorganic and organic compounds, and synthetic chemicals. The metals, minerals, and other inorganic substances can alter pH, and osmotic pressure of the water, and some are toxic to humans and other organisms.

5.4.4.3 Synthetic Chemicals

Synthetic chemicals can persist in water because most decomposers lack the enzymes to degrade them. Oil is another important water pollutant.

5.4.4.4 Radioactive Substances

These substances are released into water and persist as a hazard to living organisms until they have undergone natural radioactive decay.

5.4.4.5 Agricultural, Mining, and Construction Activities

Soil particles, sand, and minerals from erosion enter water from agricultural, mining, and construction activities. Nitrates, phosphates, and other nutrients enter the water from detergents, fertilizers, and animal manures.

Abundant nutrient enrichment of water, called *eutrophication*, leads to excessive growth of algae and other plants. Eventually, the plants become so dense that sunlight cannot penetrate the water. Many algae and other plants die, leaving large quantities of dead organic matter in the water. The high BOD of this organic matter leads to oxygen depletion and the persistence of undecomposed matter to the water.

5.4.4.6 Heat

Even heat can act as a water pollutant when large quantities of heated water are released into rivers, lakes, or oceans. Increasing the temperature of water decreases the solubility of the oxygen limit. The altered temperature and the decreased oxygen supply significantly change the ecological balance of the aquatic environment.

5.4.5 Pathogens in Water

Human pathogens in water supplies usually come from contamination of water with fecal material. Many pathogens that leave the body through the feces—many bacteria, viruses, and some protozoa—can be present. The most common pathogens transmitted in water are listed in Table 5.2.

Water is usually tested for fecal contamination by isolating *Escherichia coli* from the water sample. *Escherichia coli* or *E. coli* is called an indicator organism because it is a natural inhabitant of the human digestive tract. Its presence indicates that the water is contaminated with fecal material.

Chief sources of water pollution include municipal and industrial wastes, residential septic tank systems, active and abandoned oil and gas wells, active and inactive coal and mineral mines, active and inactive underground storage tanks at gas stations, and a vast quantity of pesticides, fertilizers,

TABLE 5.2
Human Pathogens Transmitted in Water

Organisms	Diseases Caused
Salmonella typhy	Typhoid fever
Other Salmonella species	Salmonellosis (gastrointerities)
Shigella species	Shigellosis (bacillary dysentery)
Vibrio cholerae	Asiatic cholera
Vibrio parahaemolyticus	Gastroenteritis
Escherichia coli	Gastroenteritis
Yersinia enterocolitica	Gastroenteritis
Campylobacter fetus	Gastroenteritis
Hapatitis A virus	Hepatitis
Poliovirus	Poliomyelitis
Giardia intestinalis	Giardiasis
Balantidium coli	Balantidiasis
Entamoeba hystolitica	Amoebic dysentery

and improperly disposed of motor oil. Table 5.3 summarizes the effects of water pollution. A detailed discussion of pathogens in water supplies is presented in Section 7.3 and drinking water related regulations and standards are outlined in Section 7.7.

5.5 WATER PURIFICATION

Purification procedures for human drinking water are determined by the degree of purity of the water at its source. Water from deep wells or from reservoirs fed by clean mountain streams require very little treatment to make it safe to drink. In contrast, water from rivers that contain industrial and animal wastes and even sewage from upstream towns requires extensive treatment before it is safe to drink.

5.5.1 FLOCCULATION

Such water is first allowed to stand in a holding reservoir until some of the particulate matter settles out. Then alum (aluminum potassium sulfate) is added to cause flocculation or precipitation of suspended colloids. Many organisms are also removed from the water by flocculation.

5.5.2 FILTRATION

Following flocculation treatment, water is filtered by passing it through beds of sand. Filtration removes nearly all the remaining microorganisms. Charcoal can be used instead of sand; it has the advantage of removing organic chemicals that are not removed by sand.

5.5.3 CHLORINATION

The addition of chlorine to water readily kills bacteria, but chlorine is less effective in destroying viruses and cysts of pathogenic protozoa. The amount of chlorine required to destroy microorganisms is increased, and chlorine can combine with methane in the water to form trihalomethanes (THMs), a carcinogenic substance.

TABLE 5.3
Effects of Water Pollution

Pollutant	Effects	Comment
Organic wastes	Increased BOD in water	If adequate oxygen is available, these substances can be degraded microorganisms. If oxygen is depleted, decomposition is limited to anaerobic decomposer.
Pathogenic organisms	Cause diseases	Most bacterial agents are well-controlled in public drinking water, but viruses and specially those that cause hepatitis cause human disease.
Inorganic chemicals and minerals	Increase salinity and acidity of water and render it toxic	Such chemicals should be removed during waste treatment.
Synthetic organics (herbicides, pesticides, detergents, plastics, and other wastes)	Birth defects, cancer, nerve damage, and other illnesses	These substances are not biodegradable and can be removed only by physical and chemical means. Their effects are magnified through the food chain.
Plant nutrients	Excessive growth of aquatic plants; undesirable odor and taste to drinking water	Removal of excess phosphates and nitrates from water during waste treatment is costly and difficult.
Sediments from land erosion	Siting of waterways and destruction of hydroelectric equipment near dams; reduced light reaching plants and oxygen in water	
Radioactive wastes	Cancer, birth defects	Effects can be magnified through the food chain.
Heated water	Reduces oxygen solubility in water; alters habitats and kinds of organisms	

5.6 SEWAGE TREATMENT

Sewage is about 99.9 percent water and about 0.1 percent solid or dissolved wastes. These wastes include household wastes, industrial wastes, and wastes carried by rainwater that enters sewers.

Complete sewage treatment consists of three steps: primary, secondary, and tertiary treatment.

5.6.1 Primary Treatment

There are several physical processes to remove solid wastes. Screens remove large pieces of floating debris, and skimmers remove oily substances. Water is then directed through a series of sedimentation tanks, where small particles settle out. Flocculating substances can be used to increase the amount of solids that settle out and, thus, the proportion of solids removed by primary treatment. Sludge is removed from the sedimentation tanks intermittently or continuously depending on the design of the treatment plant.

5.6.2 Secondary Treatment

The effluent from primary treatment flows into secondary treatment systems. These systems are of two types: trickling filter systems and activated sludge systems. Both systems use the decomposing activity of aerobic microorganisms. The BOD (see Appendix D) is high in secondary treatment systems, so the systems provide for continuous oxygenation of the wastewater.

5.6.2.1 Trickling Filter Systems

Sewage is spread over a bed of rocks about 2 m deep. The individual rocks are 5 to 10 cm in diameter and are coated with a slimy film of aerobic organisms such as *Spherotylus* and *Beggiatoa*. Spraying oxygenates the sewage so that the aerobes can decompose the organic matter in the sewage.

Such a system is less efficient but is subject to fewer operational problems than an activated sludge system. It also removes about 80 percent of the organic matter in the water. A trickling filter system is shown in Figure 5.7 and the bacteria used in trickling filters is shown in Figure 5.8.

5.6.2.2 Activated Sludge Systems

The effluent from the primary treatment is constantly agitated, aerated, and from a settling tank the solid material is added to the solid material remaining from the previous treatment. This sludge contains large numbers of aerobic organisms that digest organic material. However, filamentous bacteria multiply rapidly in such systems and cause some of the sludge to float on the surface of the water instead of settling down. This phenomenon, called bulking, allows the floating matter to contaminate the effluent. The sheathed bacterium *Spherotylus*, which sometimes proliferates rapidly on decaying leaves in small streams and causes a bloom, can interfere with the operation of sewage systems in this way. Its filament clogs filters and create floating clumps of undigested organic matter.

Sludge from both primary and secondary treatments can be pumped into sludge digesters. Here, oxygen is virtually excluded, and anaerobic bacteria partially digest the sludge to simple organic molecules and the gases carbon-dioxide and methane.

Methane can be used for heating the digester and providing for other power needs of the treatment plant.

Undigested matter can be dried and used as a soil conditioner or as landfill.

5.6.3 Tertiary Treatment

The effluent from secondary treatment contains only 5 to 20 percent of the original quantity of organic matter and can be discharged into flowing rivers without causing serious problems. However, this effluent can contain large quantities of phosphates and nitrates, which can increase the growth rate of plants in the river. Tertiary treatment is an extremely costly process that involves phys-

FIGURE 5.7 Trickling filters used in secondary sewage treatment.

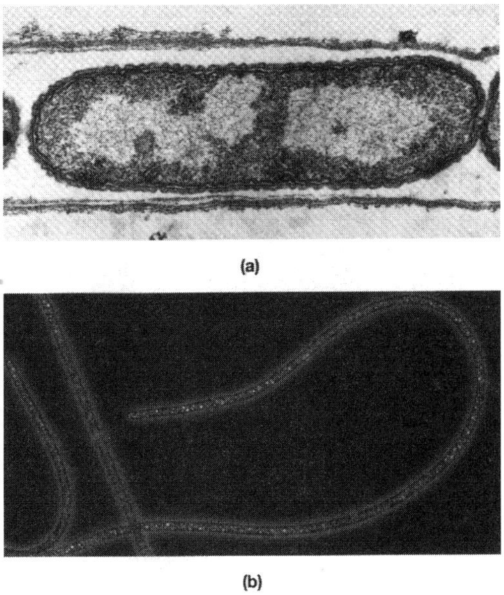

FIGURE 5.8 Two sheathed bacteria used in trickling filters: a. Spherotylus (29,300×). b. Beggiatoa (400×).

ical and chemical methods. Fine charcoal and sand is used for filtration. Various flocculating chemicals precipitate phosphates and particular matter. Denitrifying bacteria convert nitrates to nitrogen gas. Finally, chlorine is used to destroy any remaining organisms. Water that has received tertiary treatment can be released into any body of water without danger to cause eutrophication. Such water is pure enough to be recycled into the domestic water supply. However, the chlorine-containing effluent, when released into streams and lakes, can react and produce cancer causing compounds that may enter the food chain or be ingested directly by humans in their drinking water. It would be safer to remove the chlorine before releasing the effluents, but this is rarely done today, although the cost is not great. Ultraviolet lights are now replacing chlorination as the final treatment of effluent. They destroy microbes without adding carcinogens to our streams and drinking waters.

The schematic picture of a sewage treatment plant is shown in Figure 5.9.

5.6.4 Septic Tanks

A number of rural homes do not have access to city sewer connections or their treatment facilities. These homes rely on backyard septic tank systems. Homeowners must be careful not to flush or put materials such as poisons and grease down the drain, as these might kill the beneficial microbes in the septic tank that decompose sludge solids that accumulate there. This would necessitate the immediate pumping of the tank by a vehicle known as the honey wagon to prevent sewage from backing up into the house. Even with normal operation, it, occasionally, is necessary to pump the sludge from the tank and to haul it to the sewage treatment plant.

Soluble components of the sewage continue from the septic tank into the drainage (leaching) field. There they seep through perforated pipe, past a gravel bed, and into the soil which filters bacteria and some viruses and binds phosphate. Soil bacteria decompose organic materials. Drainage fields must be placed where they will not allow seepage into wells, a difficult problem on hills or in densely populated areas. Drainage fields cannot be used where the water table is too high or the soil is insufficiently permeable, such as in rocky areas. After 10 years or more, the average drainage field clogs up and can no longer be used.

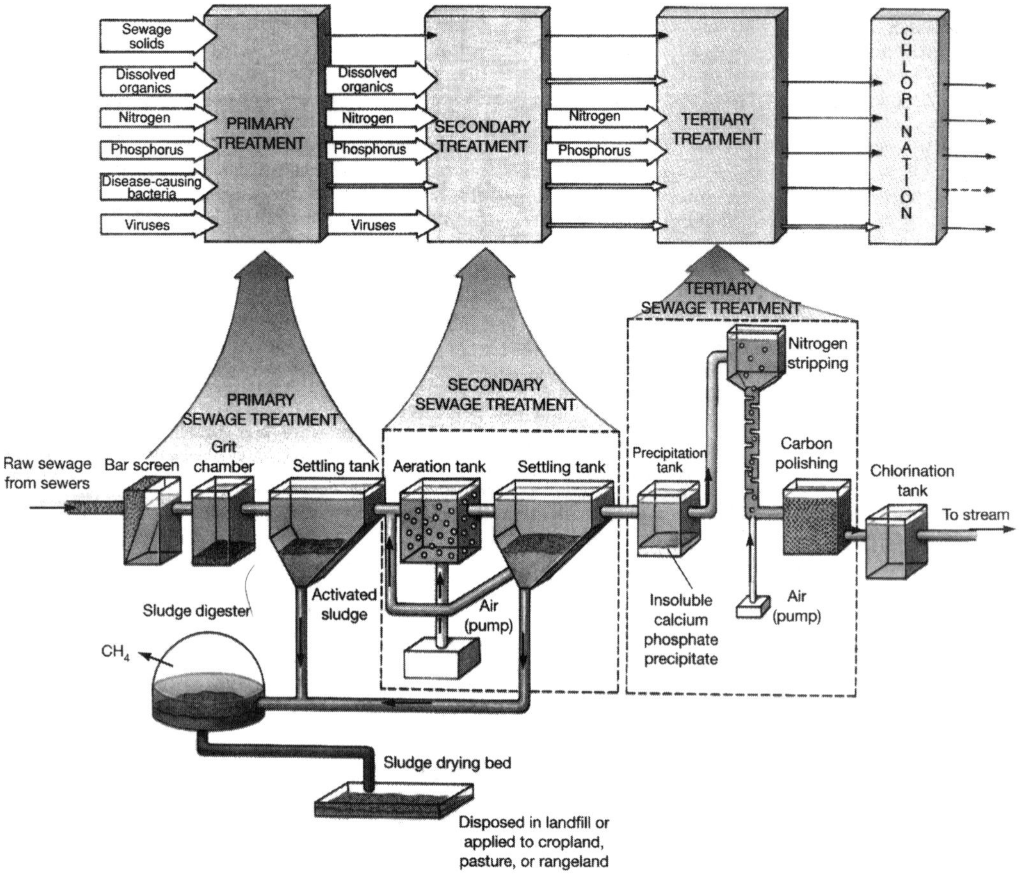

FIGURE 5.9 An overview of a sewage treatment plant showing primary, secondary, and tertiary treatment facilities.

6 Bioremediation of Organic Contaminants

Environmental biotechnology for pollution control uses a biological process to remove pollutants from contaminated water, wastewater, or other media. A biological process is an engineered system that accumulates desired microorganisms to a sufficiently large mass that the pollutant concentration is lowered to meet treatment standards during the time that the microorganisms are in contact with the material being treated.

Over the past few decades, the extensive production and use of organic compounds as solvents, fuel additives, pesticides, plasticizers, plastics, dyes, and chemical feedstocks has led to a wide spread of these substances into the environment. Well waters and groundwaters are polluted with highly dangerous, cancer causing chemicals including pesticides found in groundwaters at concentrations in the 0.1 to 5 ppb level. The most commonly found pesticides in groundwater are soil fumigants and nematocides such as ethylene dibromide (EDB), *1,2-dichloropropane* and herbicides such as *alachlor, atrazine*, and *2,4-dichlorophenoxyacid* (2,4-D).

The effect of organic contaminants dictates the necessity of developing suitable mitigation techniques. Processes used in aquatic environments against these organic pollutants are sorption, volatilization, abiotic transformation (chemical or photochemical), and biotransformation.

Sorption and volatilization concentrate contaminants and transfer them to another medium, but do not destroy them. Abiotic transformation of organic pollutants are slow and the photochemical reaction in the subsurface is insignificant. Biotransformation of many organic contaminants by bacteria in the subsurface is a useful process.

6.1 BIOREMEDIATION

Bioremediation is a process that uses naturally occurring or genetically engineered microorganisms such as yeast, fungi, and bacteria to transform harmful substances into less toxic or nontoxic substances.

Microorganisms break down a variety of organic compounds in nature to obtain nutrients, carbon, and energy for growth and survival.

6.1.1 The Objective of Bioremediation

The objective of bioremediation is to stimulate the growth of indigenous or introduced microorganisms in regions of subsurface contamination and, thus, provide direct contact between microorganisms and the dissolved and sorbed contaminants for biotransformation. Many organic contaminants are subject to attack by microorganisms.

Biotransformation of organic chemicals is facilitated by enzymes during normal metabolic functions of microorganisms.

6.1.2 Advantage of Bioremediation

The advantage of bioremediation is that it is a natural process, it destroys target chemicals at the contamination site, and the process is usually less expensive than other methods used for cleanup of hazardous wastes.

6.1.3 Disadvantage of Bioremediation

The disadvantage of bioremediation is that it takes longer than other remediation methods. More research is needed to perfect this technology.

Nevertheless, bioremediation holds enormous promise for the future to cleanup the environment.

6.1.4 Development of Bioremediation

Bioremediation has been used since the late 1970s to degrade petroleum products and hydrocarbons. A few species of bacteria, such as some members of the genus pseudomonas, can use crude oil for energy. These microbes produce oil-eating enzymes effective for cleaning oil spills. They can grow in seawater with only oil, potassium phosphate, and urea (a nitrogen source) as nutrients. These organisms act as bioremediators by cleaning up oil spills in the ocean. They have also proved useful in degrading oil that remains in the water carried by tankers as ballast after the tankers have unloaded their cargo of oil. Then the water pumped from the tankers into the sea in preparation for a new cargo of oil does not pollute the seawater. A detergent-like substance recently has been isolated from these organisms. When the detergent is added to a quantity of oil sludge, it converts 90 percent of the sludge into usable petroleum in about four days, thereby reducing waste and providing a convenient means of cleaning oil-fouled tanks.

In March 1989, the supertanker Exxon Valdez ran aground in Alaska. The leaking tanker flooded the shoreline with crude oil. The gravel and sand beaches were saturated with oil more than $\frac{1}{2}$ m thick. The effort to cleanup this pollution with conventional treatments, such as high-pressure hot water spray, skimmers, and manual scrubbers, was unable to remove the oil. Bioremediation action, such as spraying the area with nutrients to enhance the growth of native microorganisms, has worked beautifully. The sprayed area quickly became free from oil. Measurement of the success of the cleanup shows that about 60 percent of the *total hydrocarbons* and 45 percent of the *polycyclic aromatic hydrocarbons* (PAHs) were degraded by bacteria within 3 months.

During the 1991 Gulf War, oil wells and pipelines were damaged and the spilled oil soaked the desert. In spite of this fact, the roots of desert flowers were healthy and oil-free. Cultures made of bacteria and fungi from the sand revealed several types of known oil-eating organisms, such as arthrobacter. This proved that a new, natural way had been found to cleanup oil spills on land also.

6.1.5 Biotransformation by Subsurface Microorganisms

Bacteria are the predominant microorganisms found in subsurface ecosystems although fungi and protozoa have also been observed. Diverse, metabolically active microorganisms exist in subsurface sediments at depths of more than 500 m. Morphologically subsurface bacteria differ very little from terrestrial bacteria except with respect to size (less than 1 μm). Gram-positive forms are more prevalent in many uncontaminated soils. Because of their smaller size, subsurface bacteria have a higher surface-to-volume ratio which allows them to take up and utilize nutrients effectively from dilute solutions. Classes of organic compounds known to be biotransformed by subsurface microorganisms are listed on Table 6.1. The favorable and unfavorable chemical and hydrogeological site condition for applying biotransformation in a cleanup operation are listed on Table 6.2.

6.2 APPLICATION OF BIOREMEDIATION

The applications for bioremediation are rapidly expanding. It is used successfully for decomposing wastes in the landfill. It is suitable to clean soil and groundwater contamination from leaking underground petroleum, heating oil, and other material storage tanks.

TABLE 6.1
Organic Compounds Known to Be Biotransformed by Subsurface Microorganisms

Alkylbenzenes	Phenols
Benzoates	Phenoxy acid
Halogenated aliphatics and aromatics	Herbicides
	Phthalate esters
Nitro-substituted aromatics	Polycyclic aromatics
PCBs	Some pesticides

TABLE 6.2
Favorable and Unfavorable Chemical and Hydrological Site Conditions for in Situ Bioremediation

Favorable Factors	
Chemical Characteristics	*Hydrogeological Characteristics*
Small number of compounds	Granular porous media
Nontoxic concentration	High permeability
Diverse microbial population	Uniform mineralogy
pH 6–8	Homogeneous media
	Saturated media
Unfavorable Factors	
Chemical Characteristics	*Hydrogeological Characteristics*
Numerous contaminants	Fractured rock
Complex mixture of inorganic and organic compounds	Low permeability
	Complex mineralogy
	Heterogenous media
Toxic concentration	Unsaturated–saturated conditions
Sparse microbial activity	
pH extremities	

6.2.1 Wood Preservative Industries

Wood preservative industries contaminate soils and groundwaters with creosote, an oily liquid distilled from coal-tar. Creosote is used for wood preservation and the pollution originates from leaking holding tanks. The main contaminant from wood preservation is pentachlorophenol which can be degraded with a fungus, *Phanerochaete chrysosporium*. The same fungus is used successfully in degradation of other toxic compounds such as *dioxin*, *polychlorinated biphenyls* (PCBs) and *polycyclic aromatic hydrocarbons* (PAHs).

6.2.2 Solid and Hazardous Waste Bioremediation

At present, naturally occurring microbes are used for bioremediation. Scientists are experimenting with genetic engineering techniques to develop microorganisms for use at hazardous waste sites. Before they use these new organisms, it is necessary to undergo a safety review to evaluate any possible risks to human health or to the environment.

The technologies to manage hazardous and solid wastes fall into four categories:

1. Thermal treatment.
2. Biological treatment.
3. Physical and or chemical treatment.
4. Methods for containment and or disposal.

The effectiveness of the application of each of these technology groups to a specific waste varies depending on the type of waste, the physical phase (solid or liquid) of the material, the media (if any) in which the waste is contained, the desired level of treatment, and the final method of disposal of any remaining residue. Another consideration in selecting a treatment technology is where the wastes are to be treated. Wastes may be treated:

1. In place *(in situ)*, within the confines of the site.
2. At a facility off-site *(ex situ)*.

Much of the current concern over the treatment of hazardous wastes centers on the remediation of waste disposal sites required under the Comprehensive Environmental Response Compensation and Liability Act (CERCLA), commonly known as the Superfund Law in 1980, and the Resource Conservation and Recovery Act (RCRA) in 1976. The technologies currently being applied to hazardous wastes focus on the destruction of specific contaminants within the waste while solid waste management technologies focus primarily on volume reduction. The viability of biological treatment for VOCs (volatile organic compounds) and SVOCs (semivolatile organic compounds) has dramatically increased.

6.2.2.1 Bioreactors

Bioreactors have been the cornerstone of wastewater treatment processes. The wastes are introduced to a biomass of microorganisms within the reactor which metabolize the soluble organic components. In most instances, additional nutrients and oxygen (aeration) must also be added.

6.2.2.2 Solid Phase Bioremediation

The wastes are placed directly on the ground or in shallow tanks. Nutrients and microorganisms are normally added to the wastes which are routinely tilled during the treatment process. This tilling improves aeration and the contact of the organisms with the wastes.

6.2.2.3 Soil Heaping

This treatment is piling wastes in heaps several feet high on an asphalt or concrete pad. Nutrients, microorganisms, and air are provided through perforated piping placed throughout the pile.

6.2.2.4 Composting

Composting is another application of bioremediation. In this process, the wastes are normally mixed with a structurally firm bulking material such as chopped hay and wood chips. As with the other bioremediation technologies, nutrients, air, and microorganisms must be added.

6.2.3 Degradation of Synthetic Chemicals in Soil

Many synthetic chemicals, such as pesticides, are highly resistant to degradation by microbial attack. A well-known example is the pesticide *dichloro-diphenyl-trichloroetane* (DDT). When DDT was first introduced, its property of resistance to degradation was considered quite beneficial because one application remained effective in the soil for an extended time. However, it was soon found that the chemical tended to accumulate and concentrate in parts of the foodchain because of its solubility in fat.

Not all synthetic chemicals are as resistant to degradation as DDT and are subject to attack by bacterial enzymes. Small differences in chemical structure can make large differences in biodegradability. For example, in *2,4 dichloro-phenoxyacetic acid (2,4-D)*, the common chemical used to kill lawn weeds, and *2,4,5 trichloro-phenoxyacetic acid (2,4,5-T or Silvex)*, which is used to kill shrubs, the addition of a single chlorine atom to the structure of 2,4-D extends its life in soil from a few days to an indefinite period.

The growing problem is the leaching into groundwaters of toxic chemicals that are not biodegradable or ones that degrade very slowly. The sources of these materials are landfills, illegal industrial dumps, or pesticides applied to agricultural crops.

6.2.4 Biological Control of Groundwater Pollution

Biological treatment of groundwater uses naturally occurring bacteria to break down complicated organic compounds into carbon dioxide and water resulting in the growth of new bacteria.

Bacteria can grow in either aerobic (uses oxygen) or anaerobic (oxygen not used) environments. Aerobic environments are most commonly used when treating for organic compound removal. The necessary inputs to the system are oxygen and nutrients (nitrogen and phosphorus) for the bacteria. Additionally, dissolved oxygen levels should be kept above 1 part per million (ppm), pH between 6 and 9, and temperature above 25°C (77°F).

Certain organic compounds such as most petroleum related compounds degrade very easily in a biological system, others are harder to degrade (refractory), and others will not biodegrade. A list of compounds and their biodegradabilities are presented in Table 6.3.

TABLE 6.3
Biodegradability of Volatile Organic Pollutants

Compound	Biodegradability
Acetone	Degradable
Benzene	Degradable
Carbon tetrachloride	Nondegradable
Chloroform	Nondegradable
Methylene chloride	—
Chlorobenzene	Degradable
Ethylbenzene	Degradable
Hexachloro benzene	Nondegradable
Ethylene chloride	Refractory
1,1,1-Trichloroethane	Refractory
1,1,2-Trichloroethane	Refractory
Trichlorethylene	Refractory
Tetrachloroethylene	Refractory
Phenol	Degradable
2-Chloropohenol	Degradable
Pentachlorophenol	Refractory
Toluene	Degradable
Methyl-ethyl-keton	Degradable
Naphtalene	Degradable
Vinyl chloride	—

In situ biological treatment of groundwater is growing in popularity. Oxygen is provided by pumping in hydrogen peroxide, ozone, or by blowing in air (aeration). Nutrients are also pumped into the aquifer. However, if the addition of oxygen or nutrients is stopped, the bacteria will die and stop degrading the contaminants.

6.2.5 Modern Microbiological Concepts and Pollution Control

In the past decade, the field of environmental microbiology has made significant progress in the struggle against environmental contamination. During this period, molecular genetics has become a valuable source of new techniques for the detection of microorganisms, the degradation of hazardous chemicals, and as a means of safely controlling agricultural pests.

Environmental microbiology offers an in-depth examination of microbiological processes related to environmental deterioration, involving contamination of water, soil, and the atmosphere.

Although biological processes have been used for many decades, innovations nowadays are occurring at an increasing rate. Three forces are driving the rapid pace of innovation: regulations, economic pressures, and scientific and technological advances.

7 Microbiological Quality of Environmental Samples

7.1 MONITORING MICROBIOLOGICAL QUALITY

Microbiological parameters supply unique information on water and wastewater quality and public health risks from waterborne diseases. Microbiological monitoring of drinking water has been practiced in the United States and other countries since the beginning of the century. Potable water systems can become polluted with coliform and pathogenic bacteria from normal, diseased, or carrier human and animal excrements. This can occur by cross connections between a water main and a sewer, especially when the pressure in the water main is lower than the atmospheric pressure, or from the entry of sewage water through leaks in damaged pipes. Also, deficiencies in water treatment may allow unharmed or injured organisms to escape. Some microorganisms may remain unaffected by the chlorine treatment. For example, the Legionella species not only survive but multiply in storage tanks and other water systems. Recreational water quality is also affected by sewage contamination and nonpoint sources of human and animal wastes.

7.1.1 Waterborne Disease Outbreaks

Waterborne disease outbreaks in Public Community and Noncommunity water systems in the United States during 1986 and 1988 are listed in Table 7.1, Table 7.2 and Table 7.3 lists the bacterial, viral, and parasitic diseases transmitted by contaminated drinking water. Table 7.4 shows the list of the recreational water-associated disease outbreaks. The inactivation of these agents requires high doses of chlorine and extended periods of contact with the chemical. Worldwide, the most common bacterial diseases transmitted through water are caused by shigella, salmonella, enterotoxigenic *E. coli, Campylobacter jejuni,* and *Vibrio cholera.* Viral infections include hepatitis A, rotavirus and Norwalk-like agent. Common parasites include *Giardia lamblia, Cryptosporidium,* and *Entamoeba histolytica.* The first water-borne outbreak caused by cryptosporidium occurred in Texas in 1985, and in 1988 this organism was added to the Drinking Water Priority list.

TABLE 7.1
Water-borne Disease Outbreaks Reported During 1971–1988

Year	Number of Outbreaks	Cases of Illnesses	Cases per Outbreak
1971–1975	124	27,838	224
1976–1980	202	50,590	251
1981–1985	176	32,807	186
1986–1988	46	25,852	562

Source: From Mitchell R., *Environmental Microbiology,* Wiley-Liss, 1992.

TABLE 7.2
Bacterial Diseases Generally Transmitted by Contaminated Drinking Water

	Community		Noncommunity		Totals	
Agent	**No. Of outbreaks**	**No. of cases**	**No. of outbreaks**	**No. of cases**	**No. of outbreaks**	**No. of cases**
AGI	4	1,467	18	1,423	22	2,890
Giardia	8	1,146	1	123	9	1,269
Chemical	4	102	0	0	4	102
Shigella	2	1,825	1	900	3	2,725
Norwalk	0	0	3	5,474	3	5,474
Salmonella	2	70	0	0	2	70
Campylobacter	1	250	0	0	1	250
Cryptosporidium	1	13,000	0	0	1	13,000
CGI	0	0	1	72	1	72
Combined total					46	25,852

Source: From Mitchell, R., *Environmental Microbiology,* Wiley-Liss, 1992, 128.

TABLE 7.3
Viral and Parasitic Diseases Generally Transmitted by Drinking Water

Agent	Disease	Incubation Period	Symptoms
Hepatitis A	Hepatitis	15–45 days	Fever, malaise, anorexia, jaundice
Norwalk-like agent	Gastroenteritis	1–7 days	Diarrhea, abdominal cramps, headache, fever, vomiting
Virus-like 27 nm particles	Gastroenteritis	1–7 days	Vomiting, diarrhea, fever
Rotavirus	Gastroenteritis	1–2 days	Vomiting followed by diarrhea for 3–8 days
G. Lamblia	Giardiasis	7–10 days or longer	Chronic diarrhea, abdominal cramps, flatulence, malodorous stools, fatigue, weight loss
Cryptosporidium	Cryptospordiosis	5–10 days	Abdominal pain, anorexia, watery diarrhea, weight loss; immunocompromised individuals may develop chronic diarrhea
Entamoeba histolytica	Amebiasis	2–4 weeks	Vary from mild diarrhea with blood and mucus to acute or fulminating dysentery with fever and chills

Source: From Mitchell, R., *Environmental Microbiology,* Wiley-Liss, 1992, 130.

7.1.1.1 Swimming Associated Outbreaks

Swimming associated outbreaks caused by Shigella, Giardia, and Norwalk-like viruses are well known. In addition, outbreaks of *Pseudomonas dermatitis* associated with use of hot tubes, whirlpool baths, and swimming pools have been reported.

TABLE 7.4
Recreational Water-Associated Disease Outbreaks in the United States During 1986 and 1988

Agent	Illness	Source	No. of outbreaks	No. of cases
Pseudomonas	Dermatitis	Whirlpool, tub	10	153
UE	Dermatitis	Whirlpool, hot tub	2	15
Shigella sonnei	Gastroenteritis	Lake, pond, pool	3	280
Giardia	Gastroenteritis	Pool	2	300
Norwalk-like	Gastroenteritis	Lake	1	41
UE	Gastroenteritis	Lake, pond, pool	4	512
Legionella	Pontiac fever	Whirlpool	1	14
Leptospira	Leptospirosis	Stream	1	8
UE	Aseptic meningitis	Creek	1	4
UE	Enterovirus, like infection	Pool	1	26
Total			26	1363

Source: From Mitchell, R. *Environmental Microbiology,* Wiley-Liss, 1992, 131.

7.1.2 Recovery of Pathogens from Environmental Samples

Recovery of pathogens from environmental samples is difficult and involved and the methods are quite often inadequate. The low microbial densities usually found in water samples necessitate the analysis of large volumes of water. For example, in potable waters the virus levels are so low that hundreds or thousands of liters must be sampled to increase the probability of the virus detection. Also, the detection of *Giardia* cysts and *Cryptosporidium* requires filtering of a minimum of 380 to 1000 liters of water, respectively.

To ensure safe recreational water and a continued supply of potable water, frequent monitoring of both raw water sources and finished products for the presence of pathogens is very important. However, the detection and enumeration of all pathogenic bacteria that may potentially be present in water is not only expensive and time consuming but impracticable to perform on a regular basis.

7.1.3 Indicator Bacteria

The most effective microbiological monitoring of water sources is the simple, rapid, and inexpensive determination of the presence of indicator bacteria.

7.1.3.1 Coliform Group

The coliform group of bacteria is the principal indicator of suitability of a water for domestic, industrial, or other uses. The coliform group of the Enterobacteriaceae that ferments lactose constitutes a part of the normal mammalian intestinal flora.

The density of the coliform group is a significant criterion of the degree of pollution and, thus, sanitary quality. The detection and enumeration of the coliform group have been used as a basis for standard monitoring of bacteriological quality of water supplies.

7.1.3.2 Fecal Coliform Bacteria

The fecal coliform bacteria, predominantly *E. coli*, which originate primarily in the mammalian intestine, are detected by their ability to ferment lactose at 44.5°C.

These bacteria are not usually long-term occupants of aquatic systems. Thus, their presence in water serves as a useful indicator of recent fecal contamination, which is the major source of many enteropathogenic diseases transmitted through water.

The density of the indicator bacteria is a criterion of the degree of pollution and, thus, of sanitary quality.

7.1.4 Standards on Microbiological Quality

Standards on microbiological quality were based previously on the detection and enumeration of the coliform group and were related to enteropathogens such as cholera and typhoid bacilli. Most recently, the National Primary Drinking Water Regulation (NPDWR) increased the drinking water standard by identifying 83 new pollutants with maximum contaminant levels (MCLs) including new microbiological parameters such as Giardia lamblia, Legionella, viruses, heterotrophic plate count bacteria and *Cryptosporidium*.

7.2 INTRODUCTION TO MICROBIOLOGICAL PARAMETERS OF THE SANITARY QUALITY OF ENVIRONMENTAL SAMPLES

7.2.1 Heterotrophic Plate Count (HPC)

Heterotrophic plate count (HPC) was formerly known as standard plate count determination. The test provides an approximate enumeration of live heterotrophic bacteria. It may yields useful information about bacterial quality and may provide supporting data for further investigations. Colonies may arise from pairs, chains, clusters, or single cells, all of which are included in the term colony forming units (CFU) which is used to report the result.

7.2.2 Coliform Bacteria Group

7.2.2.1 Total Coliform Group

This group includes all of the aerobic, anaerobic, gram-negative, nonspore forming and rod-shaped bacteria that ferment lactose in 24 to 48 h at 37°C. *Escherichia coli* with hairlike pili shown in Figure 2.8. The density of the coliform group is the principal indicator of the bacteriological quality of water supplies.

There are several criteria for an indicator organism. The most important criterion is that the organism be consistently present in human feces in substantial numbers so that its detection will be a good indication that human wastes are entering water supplies. The indicator organisms should also survive in the water at least as well as the pathogenic organisms would. Therefore, the presence of any significant number of coliform is evidence that the water is contaminated with fecal material and any pathogens that leave the body through the feces can be present.

The coliform group includes the genera: *Escherichia, Citrobacter, Enterobacter,* and *Klebsiella.*

7.2.2.2 Fecal Coliform, *Escherichia coli (E. coli)*

The fecal coliform bacteria are part of the total coliform group. The major species is *Escherichia coli* or *E. coli*, a species indicative to fecal pollution and the possible present of enteric bacteria. The total coliform group includes organisms of fecal and nonfecal origins.

Test for fecal coliform bacteria makes it possible to separate the members of the coliform group found in the feces of warm-blooded animals from those found from other environmental sources.

Organisms of fecal origin are detected at 44.5 ± 0.2°C incubation by using an enriched lactose medium. The fecal coliform test is applicable to investigations of stream pollution, raw water sources, wastewater treatment systems, bathing water, seawater, and monitoring general water quality. For potable water, the recommended test is the detection of the total coliform group.

7.2.2.3 Fecal Streptococcus

The terms fecal streptococcus and Lancefield's Group D streptococcus have been used synonymously. The members of the groups are *Streptococcus fecalis, S. faecium, S. avium, S. bovis, S. equinus, and S. gallinarum.* They give a positive test with Lancefield's Group D antisera.

The normal habitat of fecal streptococci is the gastrointestinal tract of warm-blooded animals.

Fecal coliform/fecal streptococcus (FC/FS) ratios may provide information on possible sources of pollution. A ratio greater than 4.4 is considered indicative of human fecal pollution, and a ratio less than 0.7 is a sign of nonhuman pollution (see Chapter 19.1.1).

The fecal streptococci are valuable pollution indicators in the study of rivers, streams, lakes, and marine systems, especially when used with fecal coliform bacteria.

7.2.3.4 Klebsiella

Klebsiella is included in the coliform group and may be associated with regrowth in water distribution systems. It is also the major component of the coliform population of industrial wastewater containing high bacterial nutrients; for example, in paper mills, textile and sugar cane factories, farm production units, and other industrial food processing systems. These wastewaters contain great quantities of carbohydrates and can support these organisms in the effluent and receiving waters.

7.3 PATHOGENIC MICROORGANISMS

A wide variety of enteric pathogenic organisms may occur in environmental samples. Human pathogens that are transmitted in water and the effects they cause are as follows:

Salmonella typhi	typhoid fever.
Other salmonella species	salmonellosis (gastroenteritis).
Shigella species	shigellosis (bacillary dysentery).
Vibrio cholera	cholera.
Escherichia coli	gastroenteritis.
Yersinia enterocolitica	gastroenteritis.
Campylobacter fetus	gastroenteritis.
Hepatitis A virus	hepatitis.
Poliovirus	poliomyelitis.
Giardia intestinalis	giardiasis.
Balantidium coli	balantidiasis.
Entamoeba histolytica	amoebic dysentery.

A more serious problem is that several pathogens are more resistant to disinfection than coliforms. Chemically disinfected water samples that are free from coliform bacteria are often contaminated with enteric viruses. The cysts of *Giardia lamblia* are so resistant to chlorination that eliminating them with this method is impractical. Mechanical methods, such as filtration and flocculation, are necessary to remove colloidal particles because the microorganisms are trapped mostly by surface adsorption in the sand beds.

7.3.1 PATHOGENIC BACTERIA

The most common pathogenic microorganisms that can be demonstrated in wastewater and under certain conditions in surface and groundwaters are: *Salmonella*, *shigella*, enteropathogenic *E. coli*, *Leptospira*, and *Vibrio cholera.*

Routine examination of water and wastewater for pathogenic microorganisms is not recommended because very well-equipped laboratories with well-trained personnel are needed.

7.3.1.1 Salmonella

The salmonella bacteria (named for *Daniel Salmon*) are Gram-negative, facultative anaerobic, motile rods that ferment glucose to produce gas. The normal habitat of this bacteria is the intestinal tract of humans and many animals. The salmonella species causes *salmonellosis*, a gastroenteric infection characterized by moderate fever accompanied with nausea, abdominal pain, cramps, and diarrhea. The severity of the infection depends on the number of bacteria digested. Meat products are particularly susceptible to contamination by salmonella. If these products are mishandled, the bacteria can grow to infective numbers very quickly. Meat can be contaminated during processing. Poultry, eggs, and egg products are often contaminated by salmonella. The organisms are generally killed by normal cooking. Consumption of raw, unpasteurized milk can also be a good source of the infection.

7.3.1.2 *Salmonella typhi*

The organism is a species of the salmonella family, and it produces *typhoid fever*. Before the days of proper disposal, drinking water treatment, and food sanitation, typhoid was a very common disease and was responsible for many deaths. After recovery, many people were carriers of the bacteria and continued indefinitely to shed the organism.

7.3.1.3 Shigella

Shigellosis, or *bacillary dysentery*, is a severe form of diarrhea caused by the bacteria, *Shigella*, named after the Japanese microbiologist *Kiyoshi Shiga*. Shigella produce an exotoxin that inhibits protein synthesis. In the intestine, shigella produce tissue destruction and cause severe diarrhea with blood and mucus in the stool. While most shigellosis epidemics are food-borne or spread by person to person contact, it may be caused by contaminated drinking water. Water-borne shigellosis may result from accidental interruption of water treatment, an untreated water supply, contamination of well water, and cross connection between contaminated water pipe and potable water supply lines.

7.3.1.4 *Vibrio cholera*

During the 1800s, a bacterial infection called Asiatic cholera crossed Europe and North America in repeated epidemics. Today, cholera is epidemic in Asia, particularly in India, and very rare in western countries. The spread of cholera in 1970 may reflect the international quarantine enforcement by some countries having primitive public water supplies and inadequate sanitary regulations. The organism that causes the sickness is *Vibrio cholera*, a slightly curved, Gram-negative rod with a single polar flagellum, as shown in Figure 7.1. It grows in the small intestine and produces an enterotoxin that causes the intensive secretion of water and minerals. Because of the loss of water and electrolytes (12 to 20 l daily), the results are shock, collapse, and often death. In the United States, the water of the Gulf of Mexico and the Pacific coasts support the population of *Vibrio cholera* that has been the cause of a number of outbreaks of gastroenteritis, usually associated with the ingestion of contaminated seafood.

7.3.1.5 Enteropathogenic *Escherichia coli*

Escherichia coli is the most common and well-known microorganism in the human intestinal tract. It is usually harmless, but some species produce toxins that cause diarrhea. One group of the pathogenic *E. coli* group, the enterotoxigenic *E. coli*, causes travelers diarrhea, and the common infant diarrhea in developing countries.

7.3.1.6 Yersinia

Yersinia enterocolitica and *Yersinia pseudotuberculosis* are gram-negative organisms and intestinal inhabitants of many domestic animals and are often transmitted in meat and milk. Both organisms have the ability to grow at refrigerator temperature (4°C). The sickness caused by the bacteria is *yersiniosis*. Its symptoms are diarrhea, fever, headache, and abdominal pain.

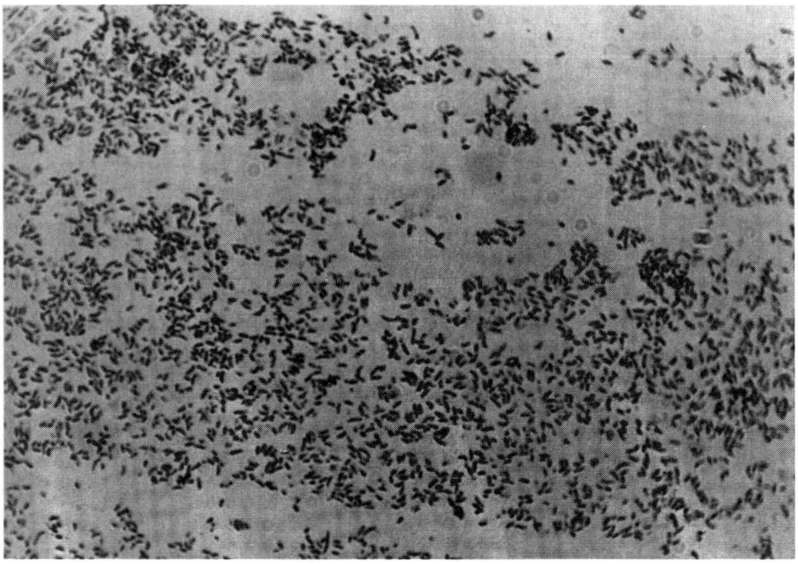

FIGURE 7.1 Gram stain of smear made from *Vibrio Cholera* colony on TCBS agar. (TCBS = thiosulphate citrate-bile salts agar).

7.3.1.7 Campylobacter

Campylobacter are Gram-negative, spirally curved rods that cause gastroenteritis with mild symptoms. Like salmonella, the bacteria are common in the intestinal tract of certain animals, especially sheep and cattle. *Campylobacter* are the second most common casual agent of diarrhea.

7.3.1.8 *Legionella pneumophila*

A type of pneumonia was identified as *legionellosis* or *Legionnaire's disease*, which received attention in 1976 when a series of deaths occurred among the members of the American Legion who had attended a meeting in Philadelphia. A total of 182 persons become sick with pneumonia at this meeting and 29 died. Close investigation identified an aerobic, Gram-negative, rod shaped, unknown bacterium, with polar, subpolar, and lateral flagella as the causative agent of the tragedy (see Figure 7.2). Now the bacterium is called *Legionella pneumophila*, and the disease is legionellosis which is commonly characterized by very high fever, cough, and general symptoms of pneumonia. Recent studies show that the organism can grow well in the water of air conditioning cooling towers, which might explain some epidemics in hotels, business districts, and hospitals. The organism is transmitted via the airborne route and it is ubiquitous in the moist environment. It was also isolated from contaminated water distribution systems and from lakes, streams, reservoirs, and wastewaters. *Legionella* is resistant to chlorine and can survive in waters with a low chlorine level.

7.3.1.9 *Clostridium*

Members of the genus *Clostridium* are obligate anaerobes. They vary in length from 3 to 8 μm and in most cases the cells containing endospores appear swollen. Diseases associated with *Clostridia* are tetanus (caused by *C. tetani*), botulism (caused by *C. botulinum*), and gas gangrene (caused by *C. perfingers*) and other clostridia.

The causative agent of tetanus, *Clostridium tetani,* is an obligately anaerobic, endospore-forming, Gram-positive rod. It is especially common in soil contaminated with animal fecal wastes. The symptoms of tetanus are caused by an extremely potent neurotoxin. It is enters the central nervous

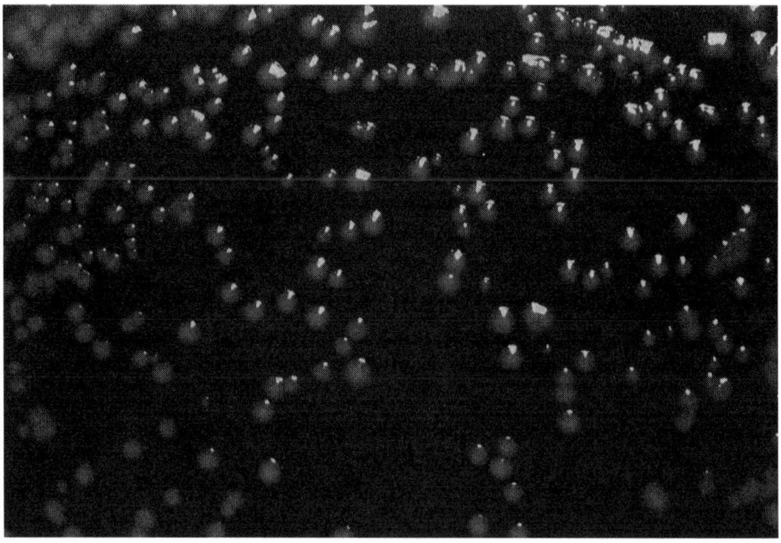

FIGURE 7.2 Colonies of *Legionella pneumophila* on BCYE agar (BCYE = buffered charcoal-yeast agar).

system via peripheral nerves or the blood. The result is uncontrollable muscle contraction, producing the convulsive symptoms of tetanus or lockjaw.

The bacteria themselves do not spread from the infection site, and there is no inflammation. Because the organism is an obligate anaerobe, the wound through which it enters the body must provide an anaerobic growth condition. The first vaccination in our life is DPT (diphtheria, pertussis, tetanus) and is nearly 100 percent effective.

The causative agent of botulism is *Clostridium botulinum* and has been associated with a variety of improperly processed meats and vegetables. We now know that botulism is caused by ingesting food contaminated with the neurotoxin produced by *Clostridium botulinum*. This toxin is one of the most potent toxins known. However, the toxin can be inactivated by boiling food for one minute before it is eaten. Treatment for botulism usually requires the early administration of an antitoxin, intensive care, and the use of a respirator to assist breathing.

Clostridium perfringens is the species most commonly involved in gangrene, but other clostridia and several other bacteria can also grow in such wounds. Once ischemia and the subsequent necrosis have developed, gas gangrene can develop, especially in muscle tissue. As the *C. perfringens* microorganisms grow, they ferment carbohydrates in the tissue and produce gases (carbon dioxide and hydrogen) that swell the tissue. The bacteria produce toxins that move along muscle bundles, killing cells, and producing necrotic tissue that is favorable for further growth. The surgical removal of necrotic tissue or amputation are the most common medical treatments for gas gangrene.

7.3.2 Viruses

Virus is the latin name for poison. Viruses are found as parasites in all types of organisms. Viruses cause diseases in humans, animals, and plants. They also infect fungi, bacteria, and protozoa. Viruses differ in size from bacteria. Although most viruses are much smaller than bacteria, some of the larger viruses are about the same size as very small bacteria; the range of the diameter for viruses is 20 to 300 nm. Viruses are not normal flora of the intestinal tract, and they are excreted only by infected individuals. These viruses are transmitted most frequently from person to person by the fecal–oral route. However, they are also present in domestic sewage which, after various degrees of treatment, is discharged to either surface waters or land. Consequently, enteric viruses may be present in sewage contaminated surface and groundwaters that are used as sources of drinking water. The viruses known to be excreted in relatively large numbers with feces include poliovirus, coxsackievirus, echovirus, and other enterovirus, adenovirus, reovirus, rotavirus, the hepatitis A (infectious hepatitis viruses), and the Norwalk-like agents that can cause acute infectious nonbacterial gastroenteritis. More than 100 different human enteric viruses are recognized. In temperate climates enteroviruses occur at peak levels in sewage during the late summer and early fall. However, the incidence of the diseases caused by the hepatitis A virus (HAV), Norwalk-like viruses and rotaviruses occur in increased numbers in the colder months. Most recognized water-borne virus disease outbreaks in the United States have been caused by sewage contamination of untreated or inadequately treated private and semipublic water supplies.

7.3.2.1 Hepatitis A Virus (HAV)

Hepatitis A virus (HAV) is the causating agent of infectious hepatitis. After a typical entrance via the oral route, the hepatitis virus multiplies in the intestinal tract and then spreads to the liver, kidney, and spleen. The virus is shed in the feces and can also be detected in blood and in urine. Recent studies indicate that the virus might also be carried in oral secretions. The virus probably survives for several days on surfaces such as cutting boards. The hepatitis A virus is very resistant to disinfectants such as chlorine at the concentration used in drinking water systems. Oysters that live in contaminated waters are also a good source of the infection. The incubation time of two to six weeks makes it difficult to find the source of infection. The symptoms are loss of appetite, nausea, diarrhea, abdominal discomfort, fever, and, in some cases, jaundice, with the yellowing of the skin and the whites of the eyes and dark urine which is typical of liver infections. Persons at risk for exposure to the virus and travelers in high risk areas can be given immunoglobulin which gives protection for several months.

7.3.2.2 Norwalk Agent

A number of viruses, such as the polioviruses, echoviruses, and coxsackieviruses, are transmitted by the feces–oral route. About 90 percent of viral gastroenteritis are caused by the rotavirus (named after its wheel shape, rota means wheel) or by the Norwalk agent (named after the outbreak in Norwalk, OH, in 1968). The symptoms are nausea, diarrhea, and vomiting with a low fever for a few days.

7.3.3 Pathogenic Protozoa

Protozoans are one-celled, eucaryotic organisms that inhabit water and soil and feed upon bacteria and small particulate nutrients. From the large number of protozoans, only a few are pathogenic.

7.3.3.1 *Giardia lamblia*

One group of protozoans is called *flagellates*. These organisms are typically spindle-shaped with flagella projections from the front end. An example of a flagellate that is a human parasite is *Giardia lamblia*. The parasite is found in the small intestine of humans and other mammals. It is passed out of the intestine and survives as a cyst before being ingested by the next host. It is the cause of a prolonged diarrheal disease in humans called *giardiasis*. The disease often persists for weeks and is characterized by weakness, nausea, weight loss, and abdominal cramp. The parasite sometimes occupies a very large place in the intestinal wall and disturbs absorption. Most outbreaks in the United States are transmitted by contaminated water supplies. The problem is that the cyst is not sensitive to the regular disinfection by chlorine in the concentration recommended for drinking water. The best treatment is the filtration of the water supply to remove the cysts. *Giardia lamblia* trophozoite and cysts are shown in Figure 7.3.

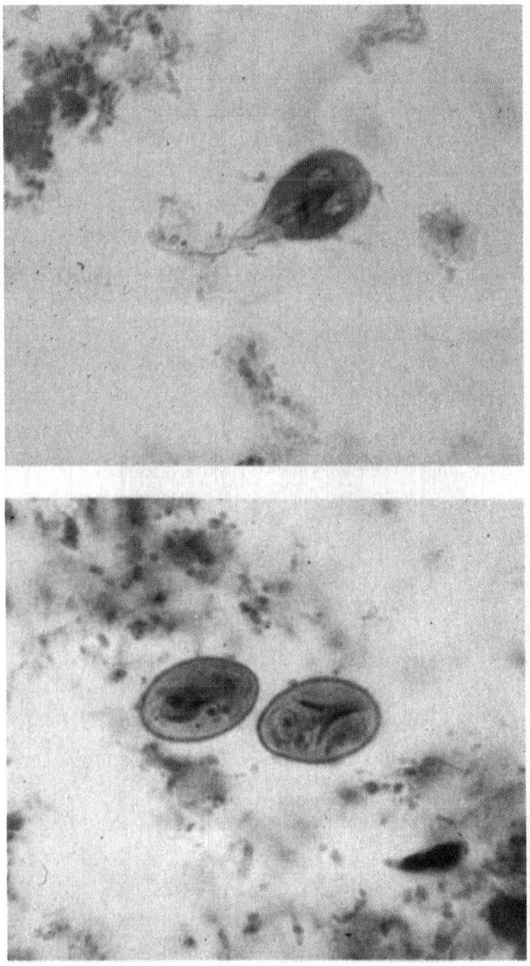

FIGURE 7.3 *Giardia lamblia* trophozoite (top) and *Giardia lamblia* cysts (bottom).

7.3.3.2 *Entamoeba histolytica*

Amoebas move by extending usually blunt, lobe-like projections of the cytoplasm called pseudopods. Any number of pseudopods can flow from one side of the amoeba cell, and the rest of the cell will flow toward the pseudopods. Amoebic dysentery or *amoebiasis* is found worldwide and is spread by food and water contaminated by cysts of the protozoan *Entamoeba histolytica*. Although stomach acid can destroy vegetative cells, it does not affect the cysts. It causes severe dysentery and the feces contain blood and mucus (the vegetative form feed on red blood cells). In severe cases, the intestinal wall is perforated and invasion of other organs (for example, the liver) is possible. The major source of amoebiasis is drinking water contamination by sewage, the fecal–oral route, and consumption of uncooked polluted vegetables. Trophozoites of *E. histolytica* are shown in Figures 7.4 and 7.5.

7.4 IRON AND SULFUR BACTERIA

The group of nuisance organisms collectively designated as iron and sulfur bacteria has the ability to transform or deposit significant amounts of iron or sulfur, usually in the form of objectionable slime. However, iron and sulfur bacteria are not the sole producers of bacterial slime.

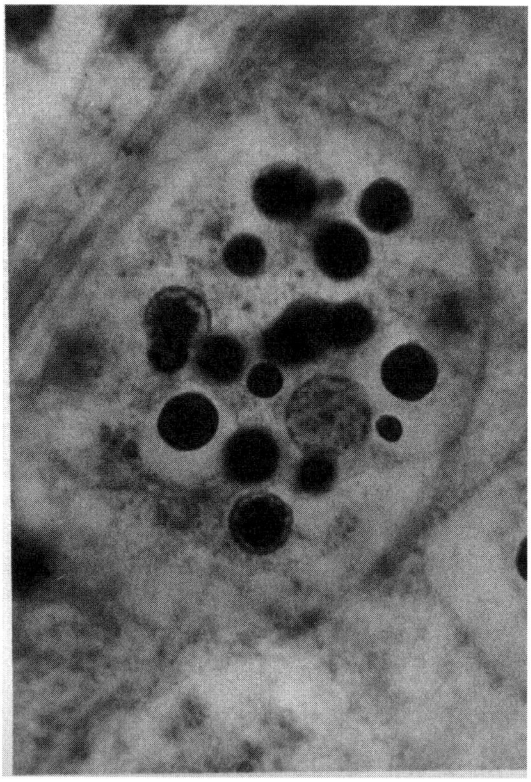

FIGURE 7.4 *Entamoeba histolytica* trophozoite containing ingested red blood cells.

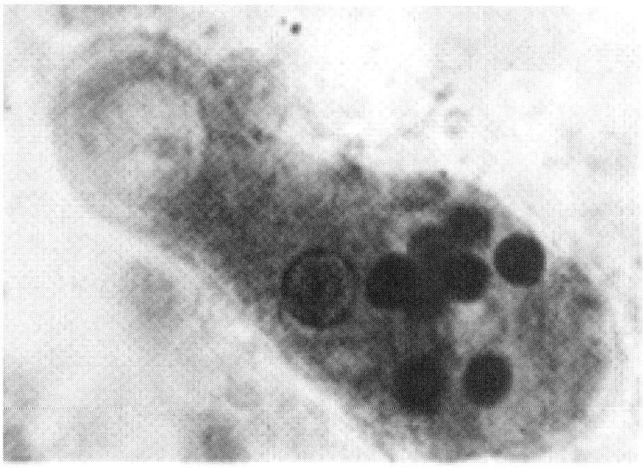

FIGURE 7.5 *Entamoeba histolytica* trophozoite.

The organisms in this bacterial group may be filamentous or single-celled, autotrophic or heterotrophic, or aerobic or anaerobic.

According to the conventional bacterial classification, these organisms are assigned to a variety of orders, families, and genera.

Iron bacteria may cause fouling and plugging of wells and distribution systems, and sulfate-reducing bacteria may cause rusty water and tuberculation of pipes. These organisms also may cause odor, taste, frothing, color, and increases in turbidity in waters.

Temperature, light, pH, and oxygen supply affect the growth of these organisms. Under different environmental conditions some bacteria may appear either as iron or as sulfur bacteria.

More information on iron and sulfur bacteria is provided in Chapter 21.

7.5 ACTINOMYCETES

Actinomycetes are filamentous bacteria, but their morphology resembles that of the filamentous fungi, or molds. Actinomycetes are very common inhabitants of soil, lake, and river mud. Actinomycetes may cause problems in wastewater treatment by developing a thick foam during the activated sludge process. An actinomycetes colony shows a fuzzy appearance. This is caused by the mass of branching filaments.

The best known genus of actinomycetes is *Streptomycetes*, which produces a gaseous compound called *geosmin* that gives fresh soil its typical musty odor. Traces of this compound can create a disagreeable odor to water and give fish a muddy flavor.

Discussion of actinomycetes and actinomycete detection is provided in Chapter 22.

7.6 FUNGI

Fungi include unicellular yeasts and multicellular filamentous molds. Fungi are heterotrophic organisms, and may be found wherever nonliving organic matter occurs. Fungi may be useful indicators of pollution, because their appearance is associated with organic matter (for example, pulp and paper mill wastes). Other species live in warmer waters, therefore, they are indicators of thermal pollution.

Some pathogenic fungi live in recreational waters, such as pools and beaches, and they are also found in beach sands and shower stalls.

7.6.1 *ASPERGILLUS*

Most common pathogenic fungi are *Aspergillus*. They cause pulmonary aspergillosis by breathing contaminated air.

7.6.2 *CANDIDA ALBICANS*

Candida albicans possibly live in wastewater effluents and may pollute recreational waters and streams. A large number of women have vaginal candidiasis from contaminated baths and beaches. Vaginal candidiasis causes irritation, severe itching, and a thick, yellow discharge with a yeasty odor.

7.7 REGULATIONS FOR DRINKING WATER QUALITY

Potable or drinking water can be defined as the water delivered to the consumer that can be safely used for drinking, cooking, and washing. Public health control helps to guarantee a continuous supply of water maintained within safe limits according to updated drinking water standards. In other words, potable water must meet the physical, chemical, bacteriological, and radiological parameters when supplied by an approved source, delivered to a treatment and disinfection facility of proper design, construction and operation, and, in turn, be delivered to the consumer through a protected distribution system in sufficient quantity and pressure.

Throughout human history, the search for clean, fresh, and palatable water was man's priority. When the relationship between water-borne diseases and drinking water was established, then the technology for treatment and disinfection developed rapidly and standards were established by health authorities.

The first water treatment in the United States was a slow sand filtration system developed in New York in 1870. Chlorination for microbiological disinfection was introduced at the beginning of this century. Since 1925, most public water supplies were equipped with some form of treatment.

The prevention of water-borne diseases left too many precautions to be taken from the source of raw water to the ultimate consumer. With water-borne diseases in mind, the emphasis should be given to the following:

1. Disinfection, continuous, effective and adjustable.
2. Protection of the source.
3. Protection of the distribution system, cross connection survey, prevention of contamination.
4. Water treatment, adequate to the quality of the raw water and suitable to changes required by the variations in the raw water quality.

7.7.1 Monitoring Agencies

7.7.1.1 World Health Organization (WHO)

The World Health Organization (WHO) is a specialized agency of the United Nations dedicated to public health and international matters related to health. The WHO stated: "The Guidelines for Drinking Water Quality are intended for use by countries as a basis for the development of standards, which if properly implemented, will ensure the safety of drinking water supplies." These guidelines are intended to supersede both the European and International Standards for Drinking Water. The WHO also states that these standards are a basis for the development of standards. "Although the main purpose of the guidelines is to provide a basis for the development of standards, the information given may also be of assistance to countries in developing alternative control procedures where the implementation of drinking water standards is not feasible."

7.7.1.2 European Economic Community (EEC)

The European Economic Community (EEC) is the Health Authority that does issue standards to 12 countries in Europe indirectly. Member states must promulgate the directive issued in 1980 and enforce it to protect the residents and travelers of this large community.

7.7.1.3 Environmental Protection Agency (EPA)

In the United States, the Environmental Protection Agency (EPA) is mandated by Congress to protect the environment. The Agency acts in water programs according to the Safe Drinking Water Act (SDWA). The EPA issues proposed regulations containing recommended maximum contaminant levels, with the intention to publish after a period of comments. The EPA reviews the final Maximum Contaminant Levels (MCLs) or the drinking water standards that are official and enforceable. The EPA can not enforce these standards without the help and contribution of the state. The state also can issue standards that are at least as restrictive as the federal standards. Primary Drinking Water Regulations provide for a minimum number of samples to be analyzed per month and to establish the maximum number of coliform bacteria (Maximum Contaminant Level, MCL) allowable per 100 ml of finished water. The regulations also require that analyses be made in certified laboratories. SDWA Drinking Water Standards for microbiological quality is provided in Table 7.5.

TABLE 7.5
SDWA Drinking Water Standard for Microbiological Quality

Name	MCL	Monitoring Frequency	Trigger for Increased Monitoring
Total Coliform	Less 40 samples per month with no more than 1 positive greater 40 samples per month with no more than 5 percent positive.	Compliance is based on the presence or absence of total coliforms. All coliform positives must be tested for presence of fecal coliform *E. coli.* The total number of samples is based on the population.	Each positive coliform sample triggers a repeat sampling requirement. If any one or more of the repeat samples in a set are also coliform-positive, an additional set of repeat samples is required. For the system of collecting more than one routine sample per month a set of free repeat samples for each positive sample must be collected within 24 h of the time that the lab notifies the system of the positive result.
Giardia lamblia	3 Log (99.9%) removal	Based on calculated residual disinfectant contact time (CT) values. Disinfectant CT must be determined for each disinfection sequence during peak hourly flow each day.	If residual is less than 0.2 mg/l, then sampling must be every 4 h until residual is restored.
Enteric viruses			
Legionella	4 Log (99.9%) removal		
Heterotrophic Bacteria			

7.7.2 Monitoring Requirements

Examination of routine bacteriological samples cannot be regarded as providing complete or final information concerning sanitary conditions surrounding the source of any particular sample. The results of the examination of a single sample from a given source must be considered inadequate. The final evaluation must be based on examination of a series of samples collected over a known and protracted period of time.

7.7.3 Minimum Coliform Monitoring Requirements

Drinking water source is groundwater:

No disinfection:

- 25–500 persons: five samples per month and a sanitary survey every five years.
- 501–3300 persons: five samples per month and a sanitary survey every three years.
- Over 3300 persons: monitoring frequency is provided in Table 10.1 and a sanitary survey every three years.

With disinfection:

25–500 persons:	five samples per month or a sanitary survey every five years and one sample per month.
501–3300 persons:	five samples per month or a sanitary survey every five years and three samples per month.
Over 3300 persons:	monitoring frequency is provided in Table 10.1.

Drinking water source is surface water:

With disinfection only, no filtration:

25–500 persons:	five samples per month and an annual sanitary survey.
501–3300 persons:	five samples per month and an annual survey.
Over 3300 persons:	monitoring frequency is provided in Table 10.1, with an annual sanitary survey. One additional sample each day if the turbidity exceeds 1 NTU (Nephelometric Turbidity Unit).

With filtration and disinfection:

25–500 persons:	five samples per month or a sanitary survey every five years and one sample per month.
501–3300 persons:	five samples per month or a sanitary survey every three years and three samples per month.
Over 3300 persons:	monitoring frequency is provided in Table 10.1.

Table 10.1 provides sampling requirements based upon population as recommended by the EPA Drinking Water Regulations Final Coliform Rule (dated June 29, 1989). The minimum number of monthly samples could be reduced, substituting the bacteriological samples with chlorine residual samples at representative points of the distribution system, with increased frequency or with larger number of chlorine determinations. The goal for MCL (Maximum Contaminant Level) for total coliform should be zero. A realistic MCL could be adopted as 95 to 98 percent of all samples examined monthly if the samples are negative for coliforms.

7.7.4 Standards

The 1962 EPA standards officially introduced the limit of one coliform per 100 ml as the mean coliform density of all samples examined per month, and for a single sample of 4 coliform per 100 ml.

In 1985, the proposed rule of the National Primary Drinking Water Regulations, the EPA proposed a recommended MCL of zero for total coliform. Turbidity regulation was coupled with the microbiological parameters setting a recommended limit of 0.1 NTU. The European Community Standards use the same zero limit for total coliform bacteria and the World Health Organization (WHO) has similar standards.

According to the microbiological quality standard for drinking waters accepted by European communities the total coliform number should be 0 per 100 ml sample and the value of the heterotrophic bacteria count is a maximum 20 in a 1 ml sample. The results of the examination of *E. coli* per 100 ml, *Fecal streptococci* per 100 ml, *Pseudomonas aeruginosa* per 100 ml, *Clostridium* per 50 ml, enterics and pathogenic bacteria per 5000 ml and bacteriophages per 100 ml should be zero. Enterics and pathogenic microorganisms are possible *Campylobacter*, *Salmonella*, *Shigella*, *Staphylococcus aureus*, pathogenic fungi, protozoa, and viruses (Hungarian standard, MSZ 450-3, 1992).

7.7.5 Disinfection Requirements for All Public Water Supplies

As required by the Safe Drinking Water Act (SDWA) Amendment of 1986, the EPA must promulgate disinfection requirements for all public water supplies. On July 31, 1992, the EPA published a notice announcing the availability of the Draft Ground Water Disinfection Rule (Fed. Reg. 33,960,

TABLE 7.6
Drinking Water Standards for Disinfection By-Products

Name	MCL	Monitoring Frequency
Total Trihalomethanes (TTHMs)	0.1 mg/l Applies to all systems serving more than 10,000 people and also to all surface water systems that meet the criteria to avoid filtration.	Quarterly for chlorinated surface and groundwaters.

July 31, 1992). This rule includes disinfection, performance monitoring, reporting, and variance requirements for public water systems that use groundwater not under the influence of surface water as a source. This rule requires disinfection as a treatment technique and controls certain microbial contaminants, respectively, to the 1986 Amendments of the Safe Drinking Water Act. By Court order, the EPA issued a proposed rule that all public water systems using groundwater use disinfection treatment processes unless they meet specified criteria. The deadline for the final action on this rule was August 30, 1997.

7.7.6 Disinfection By-Product Regulations

Water systems that require disinfection are also subject to a number of Disinfection By-Product Regulations. The rule of Drinking Water Standards for Disinfection By-Products has a deadline of December 15, 1996. The rule proposed MCLGs (Maximum Contaminant Level Goals) for 25 disinfectants and by-products resulting from disinfection of public water supplies. The 25 contaminants come from two groups in the 1991 Drinking Water Priority List: disinfection by-products (Phase VIA) and organic and inorganic chemicals (Phase VIB). According to the EPA rule, public water systems that serve 10,000 people or more, have to generate specific monitoring data to ascertain microbial and disinfection by-product level in their water supplies. Drinking water standards for disinfection by-products shown in Table 7.6.

7.8 GROUNDWATER

Groundwater is the primary source of drinking water. All groundwater shall at all places and at all times be free from domestic, industrial, agricultural, or other man-induced nonthermal components of discharges in concentration alone or in combination with other substances or components of discharges (whether thermal or nonthermal).

The frightening truth is that groundwater is not safe from contamination. Groundwater contamination is a national problem and the sources of potential contaminants are everywhere. The present and future most beneficial uses of groundwater shall be protected to ensure the availability and utility of this invaluable resource. To achieve such protection, the groundwater of different states are classified and appropriate water quality criteria for those classes are set. The criteria are minimum levels which are necessary to protect the designated use of groundwater. These quality standards are designed to protect public health or welfare. The groundwater categories have been established by taking into consideration the use and value of water for public water supply, agricultural, industrial, and other purposes.

7.8.1 Classification of Groundwater

Groundwater classification refers to a comprehensive system to classify waters for different groundwater protection strategies. States can use a variety of protection mechanisms to implement classification systems. In the United States, examples are taken from the Environmental Protection Agency (EPA) "Selected State and Territory Ground Water Classification Systems," dated May 1985, Office of Groundwater Protection, Washington D.C.

7.8.2 Groundwater Standards

Protection and appropriate management of the quality of groundwater resources are significant environmental issues. Development of laws and regulations serve to evaluate the quality of groundwater, control pollution sources, and design remediation activities. Monitoring programs have been developed to collect and analyze data on groundwater quality, to find the actual source(s) and cause(s) of pollution, and to compile geologic and hydrologic information. The objectives of the monitoring wells are to determine which pollutants are present and at which concentration and to give information on the distribution of contaminants. States design their own monitoring program, and by summarizing the results on the water quality, build up their effective groundwater protection and quality management.

Groundwaters classified as a suitable source for public water systems should meet the primary and secondary drinking water quality standards (see Table 7.7), except that the total coliform bacteria count shall be 4 per 100 ml instead of 0 per 100 ml.

7.9 SURFACE WATERS

Fresh water ecosystems fall into two categories: standing systems, such as lakes and ponds, and flowing systems, such as rivers and streams. Lakes and ponds are more susceptible to pollution because the water is replaced at a slow rate. A complete replacement for a lake's water may take 10 to 100 years or more, and during these years pollutants may build up to toxic levels. In rivers and streams the flow of the water is fast and easily purges out pollutants. If the pollution is continuous and distributed uniformly along the banks, the cleaning effect by purging does not work well. Rivers, streams, and lakes contain many organic and inorganic nutrients needed by the plants and animals that live in them. These nutrients in higher concentrations may become pollutants. Nutrients include nitrogen as ammonia, nitrite and nitrate and phosphorus as phosphate. The increased organic matter will stimulate the growth of the bacterial population, and the increased bacteria will consume the organic matter and help to clean out pollution. Unfortunately, bacteria also use up oxygen and decrease the dissolved oxygen content of the water. The lack of dissolved oxygen will cause the death of fishes and other aquatic organisms, and the aerobic (oxygen requiring) bacteria population will change to anaerobic (nonoxygen requiring) bacteria. Anaerobic bacteria will produce foul smelling and toxic gases such as methane (CH_4) and hydrogen sulfide (H_2S). This process in rivers and streams occurs more readily in the hot summer months.

Groundwater contamination by flow from surface water is well known. Surface waters from open bodies (rivers and lakes) can enter into aquifers when groundwater levels are lower than surface water levels. The reverse situation, when groundwater contaminating surface water is also possible, occurs when the water table is higher or the surface water is lowered by pumping wells.

7.9.1 Surface Water Classification

Classification of surface water is based on the water quality and on the use. The five main groups of surface waters are:

Class I	Potable water supplies.
Class II	Shellfish propagation or harvesting.
Class III	Recreation-propagation and maintenance of a healthy and well-balanced population of fish and wildlife.
Class IV	Agricultural water supplies.
Class V	Navigation, utility, and industrial use.

TABLE 7.7
Primary and Secondary Drinking Water Standards

Primary Standards		Secondary Standard	
	MCLs	**Parameter**	**Recommended Level**
Inorganics		Chloride	250 mg/l
Arsenic	0.05 mg/l	Color	15 C.U.
Barium	1.00 mg/l	Copper	1.0 mg/l
Cadmium	0.01 mg/ l	Corrosivity	noncorrosive
Chromium	0.05 mg/l	Foaming agent	0.5 mg/l
Lead	0.05 mg/l	Hardness, as $CaCO_3$	50 mg/l
Mercury	0.002 mg/l	Iron	0.3 mg/l
Nitrate-nitrogen	10.0 mg/l	Manganese	0.05 mg/l
Selenium	0.01 mg/l	Odor	3 T.O.N.
Fluoride	4.00 mg/l	PH	6.5–8.5 pH unit
Silver	0.05 mg/l	Sulfate	250 mg/l
Turbidity	1-5 NTU	Total dissolved solids (TDS)	500 mg/l
Organics		Zinc	5.0 mg/l
Endrine	0.0002 mg/l		
Lindane	0.004 mg/l		
Methoxychlor	0.10 mg/l		
Toxaphene	0.005 mg/l		
2,4-D	0.10 mg/l		
2,4,5-T (Silvex)	0.01 mg/l		
Volatile Organics (VOCs)			
Trichloroethylene	0.005 mg/l		
Carbontetrachloride	0.005 mg/l		
Vinyl chloride	0.002 mg/l		
1,2-Dichloroethane	0.005 mg/l		
Benzene	0.005 mg/l		
para-Dichlorobenzene	0.075 mg/l		
1,1-Dichloroethylene	0.007 mg/l		
1,1,1-Trichloroethane	0.20 mg/l		
Total trihalomethanes	0.10 mg/l		
Microbiology			
Total Coliform bacteria	zero counts/100 ml		
Radionuclides			
Radium 226 and 228 (total)	5 pCi/l		
Gross alpha particle activity	15 pCi/l		
Gross beta particle activity	50 pCi/l		

Note: Recommended levels for these substituents are mainly to provide aesthetic and taste characteristics. Secondary drinking water regulations are not health related. They are intended to protect public welfare by offering unenforceable guidelines on the taste, odor, or color of drinking water.

Abbreviations: MCLs means maximum contaminant levels; N.T.U. means nephelometric turbidity unit; PCi/L means picoCurie per liter (definition is in Table 18-3.2); C.U. means color unit; T.O.N. means threshold odor number; 2,4-D means 2,4-dichlorophenoxyacetic acid; 2,4, 5-T means 2,4,5-trichlorophenoxyacetic acid (Silvex).

7.9.2 Surface Water Treatment Rule (SWTR)

In June 1989, the EPA promulgated the Surface Water Treatment Rule (SWTR). This rule contains disinfection requirements for surface water supplies and groundwater under the direct influence of surface water. The treatment techniques prescribed in this rule were designed to protect against the

TABLE 7.8
Bacteriological Criteria for Surface Water Quality Classification

	Fecal Coliform, Counts/100 ml
Class I water	MPN (Most Probable Number) and MF (Membrane Filter) counts shall not exceed a monthly average of 200, not exceed 400 in 10 percent of the samples, nor exceed 800 on any one day. Monthly averages shall be expressed as geometric means based on a minimum of 5 samples taken over a 30 day period.
Class II water	MPN shall not exceed a median value of 14 with no more than 10 percent of the samples exceeding 43, nor exceed 800 on any one day.
Class III water	MPN and MF counts shall not exceed a monthly average of 200, nor exceed 400 in 10 percent of the samples, nor exceed 800 on any one day. Monthly averages shall be expressed as a geometric means based on a minimum of 10 samples taken over a 30 day period.
Class IV water	—
Class V water	—

potential adverse health effects of exposure to *Giardia lamblia*, viruses, *Legionella*, and heterotrophic bacteria as well as other pathogenic organisms.

In the wake of the 1993 outbreak of *Cryptosporidium* in Milwaukee, WI, the EPA proposed to amend the SWTR to provide additional protection against disease causing organisms (pathogens) in drinking water such as *Giardia, Cryptosporidium,* and viruses (59 Fed. Reg. 38,832, July 29, 1994). The proposed rule includes stricter standards for system using surface water or groundwater under the influence of surface water, the Maximum Contaminant Level (MCL) of zero for *Cryptosporidium*, and several alternative requirements for enhancing treatment control.

7.9.3 Surface Water Quality Standards

Monitoring, maintaining, and regulating the quality of surface waters is the responsibility of state programs.

Requirements for microbiological quality of surface waters are provided in Table 7.8.

7.10 WASTEWATER

7.10.1 Industrial Wastewater

Quality requirements for industrial uses vary widely according to potential use. Industrial process waters must be of much higher quality than cooling waters. Municipal supplies are generally good enough to satisfy the quality requirements of most process waters, except for those waters used for boilers.

Boiler waters are specially checked and treated for quality. The silica content of water is very sensitive. Silica is an important constituent of the incrusting material or scale formed by many waters. As deposited, the scale is commonly calcium or magnesium silicate. Silicate scale cannot be dissolved by acids or other chemicals. Therefore, silica-rich water that is used for boilers must be treated.

Sanitary requirements for water used in processing milk, canned goods, meats, and beverages exceed even those for drinking water.

7.10.2 Permit for Industrial and Domestic Wastewater Effluents

To construct, modify, or operate a domestic or industrial wastewater facility or activity which discharges wastes into waters or which will be reasonably be expected to be a source of water pollution should have a permit. The permit contains requirements for monitoring and reporting, and the forms needed for the results of testing. Each wastewater should have a Federal National Pollution Discharge Elimination System (NPDES) permit and State issued requirements.

7.10.3 Minimum Treatment Standard of Domestic Wastewater Effluents

7.10.3.1 Effluents for Direct Surface Water Disposal or Surface Water Disposal via Ocean Outfall

All domestic wastewater facilities are required, at a minimum, to provide secondary treatment of the wastewater (secondary treatment facility was discussed previously in Chapter 5). The treatment plant should be designed to achieve an effluent after disinfection containing not more than 20 mg per l CBOD.MDSD/5 (carbonaceous biological oxygen demand, see Appendix D) and 20 mg per l TSS (total suspended solids, see Appendix E), or 90 percent removal of each of these pollutants from the wastewater influent, whichever is most stringent. Appropriate disinfection and pH control of effluents shall also be required.

7.10.3.2 Discharging to Open Ocean Water

All domestic wastewater plants discharging to open ocean water shall be designed to achieve an effluent prior to discharge containing not more than 30 mg per l CBOD (Carbonaceous Biological Oxygen Demand), and TSS (Total Suspended Solids) 30 mg per l, or 85 percent removal of these pollutants from the wastewater influent.

7.10.3.3 Effluent Disinfection

All wastewater treatment facilities shall be designed and operated such that disinfection to the extent necessary to protect public health is provided and the microbiological pollutants shall not violate this criteria. The regulators should be aware of the possible harmful effect of chlorine.

7.10.3.3.1 Basic disinfection

Basic disinfection shall be designed to result in not more than 200 fecal coliforms per 100 ml of effluent sample. Where chlorine is used as disinfection, the design shall include provisions for rapid and uniform mixing.

- A total chlorine residual of at least 0.5 mg per l shall be maintained after at least 15 min contact time at the peak hourly flow.
- The arithmetic mean (the value computed by dividing the sum of a set of terms by the number of terms) of the monthly fecal coliform values collected during an annual period shall not exceed 200 per 100 ml of water.
- The geometric mean (the *n*th root of the sum of *n* numbers) of the fecal coliform values for a minimum of 10 samples of effluent, each collected on a separate day during the period of 30 consecutive days (monthly), shall not exceed 200 per 100 ml of sample.
- No more than 10 percent of the samples collected during a period of 30 consecutive days shall exceed 400 fecal coliform values per 100 ml of sample.
- Any one sample shall not exceed 800 fecal coliform values per 100 ml of sample.

7.10.3.3.2 High level disinfection

Facilities to provide high-level disinfection shall include additional TSS control to maximize disinfection effectiveness to the fecal coliform value in 100 ml effluent are below the detectable limit (BDL).

- The total residual chlorine of at least 1.0 mg per l shall be maintained at all times with a minimum 15 min contact time at the peak hourly flow. Higher residual or longer contact time may be needed to meet operational criteria.
- It is designed for effluent containing 1000 fecal coliforms per 100 ml before disinfection, the product of the total residual chlorine used for design shall be at least 25 mg per l with a contact time at peak hourly flow of 40 mg per l and 40 min, respectively.
- For an effluent containing greater than 1,000 and up to and including 10,000 fecal coliform per 100 ml before disinfection, the total residual chlorine and the contact time at the peak hourly flow shall be 40 mg per l and 40 min, respectively.
- For an effluent containing greater than 10,000 fecal coliform per 100 ml before disinfection, the product of the total residual chlorine shall be 120 mg per l with a contact time of 120 min at the peak hourly flow.
- For samples collected over a 30 day period, 75 percent of the fecal coliform values shall be below the detection limit.
- Any one sample shall not exceed 25 per 100 ml fecal coliform values.
- Any one sample shall not exceed 5.0 mg per l TSS at a point before disinfection.

7.10.3.3.3 Intermediate disinfection

Facilities shell be designed to result in not more than 14 fecal coliform bacteria per 100 ml effluent sample.

- The total residual chlorine shall be at least 1.0 mg per l after 15 min contact time at the peak hourly flow.
- The arithmetic mean of the monthly total coliform values collected during an annual period shall not exceed 14 per 100 ml of effluent sample.
- The median value of the total coliform values for a minimum number of 10 samples, each collected on a separate day during a monthly period shall not exceed 14 per 100 ml sample.
- No more than 10 percent of the samples collected during a monthly period shall exceed 43 fecal coliform per 100 ml of sample.
- Any one sample shall not exceed 86 fecal coliform per 100 ml of sample.

7.10.3.3.4 Low level disinfection

Facilities shall be designed to result in an effluent containing not more than 2400 fecal coliform per 100 ml. These facilities shall be operated under highly controlled conditions.

7.11 SOIL AND SEDIMENT

The microbiological population is discussed in Chapter 5. The bacterial population of solid samples calculate and report as bacterial count per g sample as discussed in the determination of individual microorganisms.

7.12 AIR

Air density plates reports should contain the time at which the plate was exposed and the count of bacterial colonies or CFU (colony forming units) per plate per minute of exposure. Microbiological laboratories should be monitoring air quality by maintaining high standards of cleanliness in work areas. Air should be monitored at least monthly with air density plates and bench tops should be monitored with RODAC plates or by the swab method. The number of colonies on the air density plate test should not exceed 16O counts per m^2 per 15 min exposure (15 colonies per plate per 15 min). Discussion of the measurement and expression of the microbiological quality of air is discussed in Section 9.2.5.

8 Safety in Environmental Microbiology Laboratory

Individuals who work in microbiological laboratories face a number of special problems when they work with organisms that are infectious to humans. These persons are usually well aware of the risks to themselves posed by the infectious agents. Some of this awareness can be taught, some can be dictated by firm rules of laboratory practice, but some must be gained by experience. Each laboratory must have safety standards containing the basic requirements about which all employees must be made aware and provide documented training programs about the hazards associated with materials with which they work.

8.1 LABORATORY FACILITIES

A microbiological laboratory should be designed for the level of activity expected to be housed within. It is a major factor in the protection of employees and it must be designed properly to provide for safe working conditions.

- The ventilation should be a 100 percent fresh air system in most cases, with perhaps more stringent temperature and humidity controls than other types of laboratories.
- The temperature should be controlled over a relatively narrow range around 22°C (72°F) and the humidity should be maintained between 45 and 60 percent.
- The interior layout should be conducive to free movement of personnel. Aisles should be sufficiently wide so the stools, chairs, or equipment placed temporarily in the aisles will not block them.
- Floors, walls, and surfaces of the equipment should be of easily decontaminated material.
- Each laboratory must contain a sink for hand washing only.
- The laboratory is designed so that it is easily cleaned. Seamless or poured floor covering are recommended. Epoxy paint is recommended for the walls.
- Laboratory furniture should be sturdy and incorporate as few seams and cracks as possible.
- The junctures of the floors with the walls and the equipment should be designed to be as seamless as possible to avoid cracks in which organic materials could collect and microorganisms thrive.
- There should be self-contained areas, neither near the entrance to the laboratory or adjacent to it, where paperwork and records can be kept and processed, and where other social activities (conversation, studying, etc.) can be done.
- Adequate utilities should be provided.
- Proper experimental equipment, which will allow the laboratory operations to be done safely, should be provided and properly maintained. A preventive maintenance plan must be available.
- An eyewash fountain and deluge shower should be provided, and emergency equipment likely to be needed in the laboratory should be kept in a readily accessible place. A deluge shower should be located where water will not splash on electrical equipment or circuits.
- The laboratory should be at a negative pressure with respect to the corridor servicing it and the airflow within the laboratory should be away from the social area toward the work area.
- Hoods should be located in low traffic draft-free zones. Hoods should not be used for storage of surplus materials.

Equipment is considered the primary barrier for protection of the employees. Items such as biosafety cabinets, safety centrifuges, enclosed containers, impervious work surfaces, autoclaves, foot-operated sinks, and other equipment specifically designed to prevent direct contact with infectious organisms or with aerosols must be available. Personal protective equipment can also be considered as a primary barrier. These items include a lab coat, possibly gloves, masks, or respirators, goggles, and head and foot covers.

8.2 LABORATORY SAFETY CONSIDERATIONS

Microbiology laboratories also contain the safety risks associated with any laboratory and institutional environment; these include fire, electrical hazards, chemical hazards, hazardous environmental situations, and equipment malfunction.

8.2.1 General Laboratory Safety Rules

General laboratory safety rules include the following:

- Each laboratory has to have a formal documented safety program (written standard operating procedures, material safety data sheets, emergency response and evacuation plan).
- Each individual working in the laboratory has to have a copy of the safety program and each individual must know how to report accidents or any unsafe conditions.
- Records are maintained of accidents and the consequences.
- Name and phone number of the safety supervisor and an alternate person should be posted at the door of the laboratory so that he or she may be contacted in the case of an emergency.
- Emergency telephone numbers for fire, ambulance, health centers, and poison control center are placed in a conspicuous location near the telephone.
- Everybody working in the laboratory must know the location of first aid supplies. The contents of first aid kit is given in Table 8.1.
- Emergency first aid charts and hazardous agents charts should be posted in the laboratory.
- Minor spills should be cleaned up immediately by laboratory personnel. A basic emergency spill kit should always be maintained within the facility.
- Everyone should be familiar with the location and use of all safety equipment in the laboratory area. This includes the means to initiate an evacuation (fire alarm pull stations, etc.), fire extinguisher, fire blankets, eyewash stations, deluge showers, first aid kits, spill kit materials, respiratory protective devices, and any other materials normally kept for emergency response.

TABLE 8.1
Contents of a Laboratory First Aid Kit

Adhesive bandages, various sizes	Antiseptic wipes
Sterile pads, various sizes	Cold packs
Sterile sponges	Burn cream
Bulk gauze	Antiseptic cream
Eye pads	Absorbent cotton
Adhesive tape	Scissors
First aid booklet	Tweezers

Note: There are no tourniquets, aspirin, or other medication taken internally, and no iodine or merthiolate. A specific person should be designated to be responsible for maintaining the supplies of the first aid kit.

- Foam and carbon dioxide fire extinguisher are installed within easy access to the laboratory, near a door, or on a path.
- Fire evacuation plan is established for the laboratory and is posted in a conspicuous place. Everyone should be familiar with the primary and secondary evacuation routes from the area to the nearest exit or an alternative one if the primary exit is blocked. Everyone should be told what method is used to signal a building evacuation, where to go, to check in with a responsible person, and not to reenter the building until an official clearance to do so is given.

8.2.2 Standard Safety Practices in a Microbiological Laboratory

Standard safety practices in a microbiological laboratory include:

- Personal clothing and other belongings are stored outside the laboratory.
- It is recommended that laboratory coats, gowns, or uniforms be worn to prevent contamination or soiling of street clothes.
- Gloves should be worn if the skin on the hands is broken or if a rash exists.
- Germicidal soap or other disinfectants should be available. Persons must wash their hands with a disinfectant soap or detergent after they handle viable materials and before leaving the laboratory.
- Work surfaces are to be decontaminated before and after work each day in which operations are performed and after any spill of viable material.
- All contaminated liquid or solid wastes must be decontaminated before disposal by an approved decontamination method such as autoclaving. Materials to be decontaminated off-site from the laboratory must be packaged in accordance with applicable local, state, and federal regulations before removal from the facility.
- Eating, drinking, smoking, handling contact lenses, and applying cosmetics are not permitted in the laboratory.
- Chewing gum or licking labels is not permitted in the laboratory.
- Food or drink are not stored in the laboratory refrigerator.
- Laboratory coats are not worn outside the laboratory.
- Persons who have cuts, abrasions, etc. on the face, hands, arms, etc. do not work with infectious agents.
- A suitable disinfectant should be available for immediate use.
- Proper disinfectant is used routinely to disinfect table tops and carts before and after laboratory work.
- There must be effective insect and rodent control.
- Microscopes, colony counters, etc. are kept out of the work area.
- Autoclaves, ovens, water distilling equipment, and centrifuges are checked routinely for safe operation.
- Laboratory workers are instructed in the operation of the autoclave and operating instructions are posted near the autoclave.
- Performance checks of autoclave and hot air ovens are conducted with the use of spore strips, spore ampuls, indicators, etc.
- An important safety precaution for those operating the autoclave is to open the door slowly after completion of a cycle. Several workers have been severely burned by the rush of steam that escapes through the crack in the autoclave door as it is first being opened. The use of a protective face mask for this procedure will prevent such burn incidents.
- No broken, chipped or scratched glassware is to be used.

- Broken glass is discarded in designated containers. These containers are sturdy kraftboard boxes, taped closely, with the words "BROKEN GLASS" written prominently.
- Water taps are protected against back-siphoning.
- Receptacles of contaminated items should be marked.
- Safety glasses are provided to each person in the laboratory. Safety glasses are used with toxic or corrosive agents and during exposure to UV radiation. Eye protection devices are shown in Figures 8.1 and 8.2.
- Hazard warning signs should be posted at entrance(s). Some warning signs are shown in Figure 8.3.
- Safety cabinets of appropriate type and class are provided.
- Floors are wet-mopped weekly with a disinfectant solution.
- If open flame burners are used, care must be taken to prevent spattering of material from inoculating needles and loops.
- Mouth pipetting is strictly prohibited. Mechanical devices are used to drawing liquids into pipettes.
- Workbenches and horizontal surfaces should be decontaminated at least after work and immediately after every spill by washing with a liquid antiseptic agent such as a phenolic compound, 70 percent alcohol, or a 0.5 percent solution of sodium hypochlorite (10 percent solution of household bleach in water).

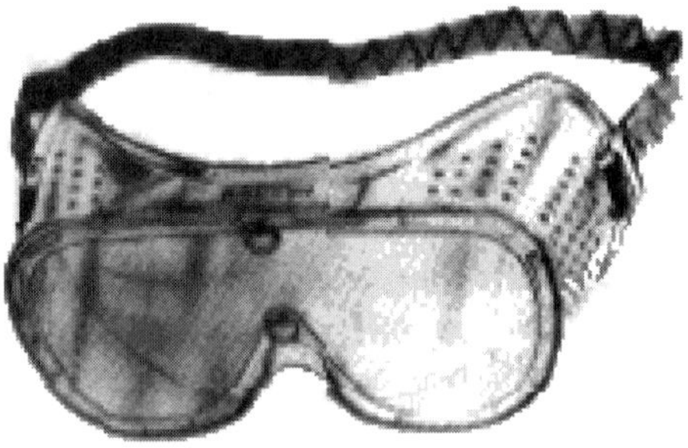

FIGURE 8.1 Eye protective device.

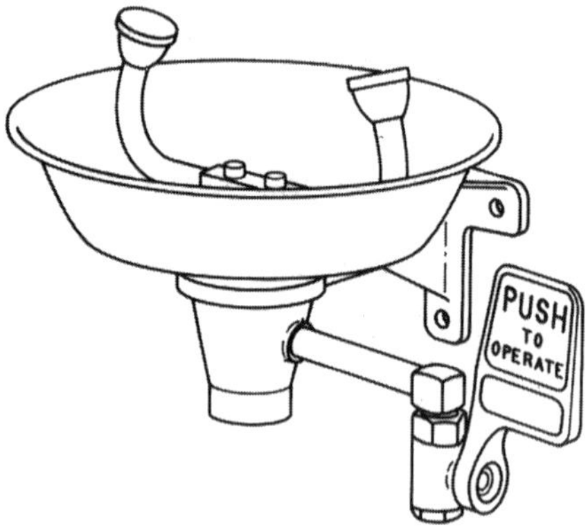

FIGURE 8.2 Eye wash fountain.

FIGURE 8.3 Hazard warning signs.

8.2.3 General Handling and Storage of Chemicals and Gases

General handling and storage of chemicals and gases include:

1. Containers of reagents and chemicals are labeled properly. Tables 8.3 and 8.4 show the LabGuard Safety Label Systems.
2. Chemicals should be stored at safe levels, in cabinets, or on stable shelving; no chemicals should be stored on the floor.
3. Chemicals should be stored according to compatibility.
4. Toxic chemicals are clearly marked poison or toxic.
5. Material safety data sheets (MSDS) for each chemical used have to be available for each employee. Typical MSDS are shown in Table 8.2.
6. Flammable solvents are stored in an approved storage cabinet or well-ventilated area away from burners, hot plates, etc.
7. Refrigerators used for flammables should be unflammable material storage units or explosion-free.
8. Flammables should not be stored along a path of egress.
9. Bottle carriers are provided for hazardous substances.
10. Gas cylinders should be strapped firmly in place; cylinders not in use should be capped; oxidizing and reducing gases should be properly segregated.

8.2.4 Electrical Precautions

The following electrical precautions should be taken:

1. All electrical circuits should be three wire.
2. No circuits should be overloaded with extension cords or multiple connections.
3. No extension cords should be used unsafely, cords should be protected or in raceways.
4. Apparatus should be equipped with three-prong plugs or should be double-insulated.
5. Motors are nonsparking.
6. Heating apparatus should be equipped with redundant temperature controls.
7. There should be adequate lighting, lights in hoods should be protected from vapors.
8. Circuits and equipment should be provided with ground-fault interrupters as needed.
9. Electrical equipment should be properly covered.
10. Breaker panel should be accessible.
11. Deluge shower should be located so water will not splash on electrical equipment or circuits.

8.3 SUMMARIZED SAFETY CHECK LIST FOR ENVIRONMENTAL MICROBIOLOGY LABORATORIES

8.3.1 Administrative Considerations

The following procedures should be established:

1. The laboratory should have a formal documented safety program.
2. Each worker should have a copy of the safety program.
3. Employees should be aware of procedures for reporting accidents and unsafe conditions.
4. New employees are instructed on laboratory safety.
5. Joint supervisor-employee safety committee has been established to identify potential laboratory hazards.

TABLE 8.2
Material Safety Data Sheet (MSDS)

Material Safety Data Sheet
LaMotte Chemical Products Company
PO Box 329 • Chestertown • Maryland • 21620
Telephone # for information 301-778-3100
In an emergency: Local Poison Control Center

I • Product Identification

Name: Nitric Acid *Code #:* 3933

II • Hazardous Ingredients

Name	*CAS #*	*%*	*PEL*	*TLV*
Nitric Acid	7697-37-2	0.6	1 mg/m^3	1 mg/m^3

III • Non-Hazardous Ingredients except water (7732-18-5)

Name	*CAS#*	*%*

IV • Physical Data

Appearance: Clear, colorless liquid *Odor:* None
Solubility in water: Soluble *pH:* 1
Boiling point: unknown *Melting point:* N/A
*Vapor pressure (mmHg):*Unknown *Vapor Density (Air=1):* Unknown

V • Fire & Explosion Data

Flash point (method used):
Flammable limit: LEL: *UEL:*
Extinguishing Media: Not a fire hazard
Special Fire Fighting Procedures:
Unusual Fire & Explosion Hazard:

VI • Reactivity Data

Stability: Stable [X] Unstable [ ]
Conditions to Avoid: N/A
Incompatability (Materials to Avoid): N/A
Hazardous Decomposition Products: N/A

6. Records are maintained of accidents and the consequences.
7. Name and phone number of the supervisor and an alternate are posted at door of the laboratories so he or she may be contacted in case of an emergency.
8. Laboratory supervisor and at least one other permanent employee should have attended an appropriate first aid course. Refresher first aid courses should be taken in a timely manner.
9. Emergency telephone numbers for fire, ambulance, health centers, and poison control center are placed in a conspicuous location near the phone.
10. Employees know the location of the first aid supplies.
11. Emergency First Aid Charts and Hazardous Agent Charts are posted in the laboratory.
12. Fire evacuation plan is established for the laboratory and is posted in a conspicuous location.

TABLE 8.2 Material Safety Data Sheet (Continued)

VII • Health Hazard Data

Toxicity: Unknown

Primary Route of Entry: Inhalation [ ] Skin [X] Ingestion [ ]

Target Organ: N/A

Signs & Symptoms of Exposure: Irritating to eyes, nose and skin.

Medical Condition Aggravated by Exposure: N/A

Carcinogenicity: [X] N/A [ ] NTP [ ] IARC

VIII • Emergency First Aid Procedures

Eye Contact: Flush thoroughly with water for 15 minutes. Consult a physician.

Skin Contact: Flush with water, remove affected clothing and wash skin with soap and water.

Ingestion: Rinse mouth, drink glass of water and consult physician.

Inhalation: N/A

IX • Spill & Disposal Procedures

Spill & Leak: Cover spill with sodium bicarbonate. Scoop up slurry and wash down drain with excess water.

Disposal: Add slowly to a solution of soda ash and slaked lime. Pour this neutralized solution down drain with excess water.

X • Precautionary Measures

In Handling: Gloves [X] Eye Protection [X] Other: Lab Coat

Ventilation: Normal [X] Mechanical [ ] Respiratory Protection [ ]

Work/Hygienic Practices: Avoid contact with skin and clothing

XI • Special Precautions: N/A

Date: 8/6/90

The above information is believed to be correct but does not claim to be all inclusive and should be used only as a guide.

† This is a toxic chemical subject to reporting requirements of section 313 of EPCRA and 40CFR372.

TABLE 8.3
Lab Guard Safety Label System

Color Code	Hazard	Storage
WHITE	Contact hazard	Corrosion proof area
YELLOW	Reactivity hazard	Store separately from flammables and combustible
BLUE	Health hazard	Store in a secure "poison" marked area
RED	Flammable hazard	Store in an area segregated for flammables
GRAY	Minimum or no hazard	Store in a general storage area

Note: The storage code color band runs the entire width of the label of the chemical container, so it is perfectly visible regardless of container's orientation on the shelf. Store chemicals with like colors together according to recommendation. Incompatible substances that should not be stored alongside chemicals with the same color labels would carry a blue color band at the bottom of the label. Recommended protective equipment, warning statements, and first aid instructions are carried within the label's safety information portion.

TABLE 8.4
Lab Guard Safety Label System: Acids and Bases

Color Code	Chemical
YELLOW	Sulfuric acid
RED	Nitric acid
BLUE	Hydrochloric acid
BROWN	Acetic acid
BLACK	Phosphoric acid
GREEN	Ammonium hydroxide

Note: Beside the colored label, they also have color-coded polystyrene screw caps that simplify identification and prevent contamination. Acid bottles have dripless pouring ring and finger grip for easy handling as well as impact resistant plastic coating. Although the bottle can break, the coating keeps glass fragments and reagent trapped within.

8.3.2 Personal Conduct

The following is a listing of acceptable personal conduct:

1. Personal clothing is stored outside of the microbiology laboratory.
2. Laboratory coats are worn at all times in the laboratory.
3. Germicidal soap is available for employees use.
4. Preparing, eating, or drinking food and beverages is not permitted in the laboratory.
5. Smoking or chewing gum is not permitted in the laboratory.
6. Food or drink are not stored in the laboratory refrigerator.
7. Reading materials are not kept in the laboratory.
8. Laboratory coats are not worn outside the laboratory.
9. Employees who have cuts, abrasions, etc. on the face, hands, arms, etc. do not work with infectious agents.

8.3.3 Laboratory Equipment

Standard laboratory procedures include the following:

1. Bulb or mechanical device is used to pipet.
2. Pipets are immersed in disinfectant after use.
3. Benches are maintained in a clear and uncluttered condition.
4. Centrifuge caps and rubber cushions are in good condition.
5. A suitable disinfectant is available for immediate use.
6. Blender is used with sealed container assembly.
7. Microscopes, colony counters, etc. are kept out of the work area.
8. Water baths are clean and free from growth or deposits.
9. Employees are instructed in the operation of autoclave and operating instructions are kept posted near the autoclave.
10. Autoclaves, hot air sterilizing ovens, water distilling equipment, and centrifuges are checked routinely for safe operation. Record frequency and date of the maintenance.
11. No broken, chipped, or scratched glassware are in use.
12. Broken glass is discarded in designated containers.
13. Electrical circuits are protected against overload with circuit breakers and ground-fault breakers.
14. Power cords, control switches, and thermostats are in good working order.
15. Water taps are protected against back-siphoning.

8.3.4 Disinfection/Sterilization

The following procedures should be established:

1. Proper disinfectant is used routinely to disinfect table tops and carts before and after laboratory work.
2. Receptacles of contaminated items are marked.
3. Performance checks of autoclave and hot air ovens are conducted with the use of spore strips, spore ampoules, indicators, etc.
4. Safety glasses are provided to employees.
5. Safety glasses are used with toxic or corrosive agents and during exposure to UV light.

8.3.5 Biohazard Control

Biohazard control procedures should be established as follows:

1. Biohazard tags and signs are posted in hazardous areas.
2. Safety cabinets of the appropriate type and class are provided.
3. Floors are wet-mopped weekly with a disinfectant solution.

8.3.6 Handling and Storage of Chemicals and Gases

The following handling and storage procedures should be initiated:

1. Containers of reagents and chemicals should be labelled properly.
2. Flammable solvents should be stored in an appropriate storage cabinet or well-ventilated area away from burners, etc.
3. Bottle carriers should be provided for hazardous substances.
4. Gas cylinders should be securely clamped to a firm support.
5. Toxic chemicals should be clearly marked toxic or poison.

8.3.7 Emergency precautions

The following emergency precautions should be followed:

1. Foam and carbon dioxide fire extinguisher should be installed and maintained within easy access to laboratory.
2. Eye stations, safety showers, respirators, and fire blankets should be available within easy access.
3. Fire exits should be marked clearly.
4. First aid kit should be available and in good condition.
5. At least one full-time employee should be trained in first aid.

9 Laboratory Quality Assurance and Quality Control

9.1 INTRODUCTION

9.1.1 QUALITY ASSURANCE (QA)

Quality Assurance (QA) is a definite plan for laboratory operation that specifies standard procedures that help to produce data with defensible quality and reported results with a high level of confidence. It is a necessary part of data production, and it serves as a guide for the operation of the laboratory for production data quality. The basic requirements of a quality assurance program are to recognize possible errors and to develop techniques and plans, and to minimize errors. It also includes evaluating what was done and reporting evaluated data which are technically sound and legally defensible.

9.1.2 QUALITY CONTROL (QC)

Quality Control (QC) is a mechanism established to control errors. It is a set of measures within a sample analysis methodology to assure that the process is in control. It consists of the use of a series of procedures that must be rigorously followed. Quality Control practices should be documented and the records should be available for inspection at any time.

9.1.3 QUALITY ASSESSMENT

Quality Assessment is the mechanism to verify that the system is operating within acceptable limits. It is the overall system of activities whose purpose is to provide assurance that the overall quality control job is being done effectively.

Special problems exist in microbiology because of analytical standards, known additions, and reference samples usually are not available. Personal judgment is required more frequently. An effective program must control all factors, from sample collection through data reporting, that can influence the results. Each laboratory has to develop its own QA/QC program, which should be delineated in a QA/QC Manual. This program should be continually reviewed and updated as needed.

9.2 REQUIREMENTS FOR FACILITIES AND PERSONNEL

9.2.1 VENTILATION

The laboratory should be well-ventilated. Ventilation should be maintained free from dust, drafts, and extreme temperature changes (see Laboratory Safety, Section 8.2).

9.2.2 LABORATORY BENCH AREAS

Laboratory bench areas should provide a minimum of a 2 m linear bench space per analyst and additional areas for other activities. The bench top should be smooth, have an impervious surface that is inert and corrosion resistant and has a minimum number of seams. Even glare-free lighting with about 1000 lux intensity should be provided at the working surface.

9.2.3 Walls and Floors

The walls have to be covered with a smooth finish that is easily cleaned and disinfected (see Laboratory Safety, Section 8.2). Floors should be covered with good quality tiles or other heavy duty material which can be maintained with skid-proof wax. Bacteriostatic agents contained in some wall or floor finishes increase the effectiveness of disinfection.

9.2.4 Laboratory Cleanliness

Maintain high standards of cleanliness in work areas. Regularly clean laboratory rooms and wash benches, shelves, floors, and windows. Wet-mop floors and treat with disinfectant solution; do not sweep or dry-mop. Wipe and disinfect bench tops before and after use (see Laboratory Safety, Section 8.3).

9.2.5 Air Monitoring

The laboratory can be monitored for cleanliness by tests regularly on a weekly or monthly or other basis to monitor the counts of the bacterial populations in the same work areas over time or to make comparisons between different work areas.

9.2.5.1 RODAC Plates

Work areas and other surfaces can be checked by Reproducible Organisms Detection And Counting (RODAC) plates which contain general growth media for total counts or selective media for coliforms, enteric pathogens, streptococci, staphylococci, or other microorganisms. The RODAC dish has a test area of about 25 cm and it is specifically designed to enumerate the microbial population of flat solid surfaces by contact techniques.

1. To sample an area, remove the plastic cover and carefully press the agar to the solid surface being sampled. It is important that the entire agar layer contacts the test surface.
2. Use three rolling motion with uniform pressure on the back of the plate for complete contact.
3. Replace the cover.
4. Incubate in an inverted position for the appropriate time and temperature.
5. Report the counted colonies per RODAC plate.

9.2.5.2 Air Density Plates

The numbers and types of airborne microorganisms can be determined by exposing petri plates for a specified time. This exposure method can be used to monitor total bacteria or to monitor specific organisms being tested.

1. Pour the petri dishes with the appropriate agar, allow to harden (store in refrigerator if not used the same day).
2. Remove the petri dish cover and place dishes top-side up on sterile towel.
3. Expose plates in selected work sites for 15 min.
4. Mark the plates with sample site identification.
5. Replace cover and incubate for proper time and temperature.
6. Count the colonies and report. The number of organisms that settle in 15 min of exposure on a petri dish is equivalent to that for 1 ft^2 per min, because the area of a standard size petri dish is 1/15 ft^2. Calculate and express in metric system:

$$1 \text{ ft}^2 = 0.09 \text{ m}^2 \text{ or}$$
$$15 \text{ colonies/ft}^2 = 15 \text{ colonies}/0.09 \text{ m}^2$$
$$15:0.09 = 166 \text{ colonies/m}^2$$

The number of colonies on the air density plate test should not exceed 166 colonies per m^2 per 15 min exposure or 15 colonies per plate per 15 min. (The area of a petri dish is 0.65 m^2, therefore, $15 \times 0.65 = 0.975 = 1$ colony per petri dish per min and 15 colonies per plate per 15 min).

9.2.6 PERSONNEL

Ideally, bacteriological work should be done by a professional microbiologist. If that is not possible, a professional microbiologist or trained analyst should be available for guidance and supervision. Additional training for laboratory personnel is necessary to advance skills and update knowledge. The microbiologist and technician should be encouraged to attend courses at centers of expertise, such as colleges, universities, commercial manufacturers, and federal and state governmental agencies.

9.3 QUALITY CONTROL FOR LABORATORY EQUIPMENT AND INSTRUMENTATION

Quality control of laboratory apparatus includes servicing and monitoring the operation of the equipment. Each item of the equipment should be tested to verify that it meets the manufacturer's claims and the user's needs for accuracy and precision.

Constant care and routine maintenance is the secret for maintaining proper functioning of laboratory instruments. Maintenance activities for each instrument are found in the manufacturer's instruction manual that accompanied each instrument. A written maintenance schedule for each instrument must be available. The laboratory must have one maintenance expert or should have contracted with the vendor specialist for regular maintenance and simple repair. Directly record all QC checks in a permanent log book. The frequency of the check depends on the manufacturer's recommendation, the workload, and the samples analyzed.

9.3.1 THERMOMETERS AND TEMPERATURE-RECORDING INSTRUMENTS

Quality control functions for the care of thermometers and temperature-recording instruments include:

- All temperature related equipment must have their temperature checked on a daily basis and the temperature recorded on log forms, as shown in Tables 9.1 to 9.4.
- Thermometers must be graduated in 0.5°C. Check mercury column for break.
- Check the accuracy of the thermometers at least semiannually against a certified National Institute of Standards and Technology (NIST), previously called National Bureau of Standards (NBS), thermometer. Record data in a QC log, as shown in Table 9.5.

9.3.2 BALANCES

Quality control functions for the care of balances include:

- Wipe balance before and after each use, and wipe spills immediately with a damp towel.
- Check weights monthly against certified weights and document as shown in Table 9.6.
- For weighing 2 g or less, use an analytical balance, and for larger quantities, use a toploading balance with the sensitivity of 0.1 g at a 150 g load.
- Protect balance and weights from laboratory atmosphere, corrosion, and moisture.
- Maintenance of the balance should be done by qualified experts under an annual contract. Maintenance activities are also recorded.

TABLE 9.1
Incubator Temperature Control Log (Temperature 37°C ± 2°C)

Instrument Description ______________________________

	Jan.	Feb.	Mar.	Apr.	May	June	July	Aug.	Sept.	Oct.	Nov.	Dec.
1.												
2.												
3.												
4.												
5.												
6.												
7.												
8.												
9.												
10.												
11.												
12.												
13.												
14.												
15.												
16.												
17.												
18.												
19.												
20.												
21.												
22.												
23.												
24.												
25.												
26.												
27.												
28.												
29.												
30.												
31.												

9.3.3 pH Meter

Quality control functions for the care of pH meters include:

- Calibrate the meter and electrode each time with pH buffer 7.00 and 4.00 or 7.00 and 10.00.
- Compensate for temperature with each use.
- Record the source, date of arrival, date of opening, and expiration date of the buffers as shown in Table 9.7 and check monthly against another pH meter if available. Do not reuse buffer solutions. Buffers are discussed in Section 4.2.2.
- Because of the wide variety of pH meters, detailed operation procedures are given for each instrument.
- Have the pH meter inspected at least yearly as part of a maintenance contract.

9.3.4 Water Deionization Unit

Proper mixed-bed deionization resin columns produce satisfactory reagent grade water. Commercial systems are available that contain prefiltration, mixed-bed resins, activated carbon,

TABLE 9.2
Incubator Temperature Control Log (Temperature 44°C ± 2°C)

Instrument Description ____________________

	Jan.	Feb.	Mar.	Apr.	May	June	July	Aug.	Sept.	Oct.	Nov.	Dec.
1.												
2.												
3.												
4.												
5.												
6.												
7.												
8.												
9.												
10.												
11.												
12.												
13.												
14.												
15.												
16.												
17.												
18.												
19.												
20.												
21.												
22.												
23.												
24.												
25.												
26.												
27.												
28.												
29.												
30.												
31.												

and final filtration to produce ultra-pure water. Deionization systems tend to produce the same quality water until resins or activated carbon are near exhaustion and quality abruptly becomes unacceptable. It is better to change cartridges at the intervals recommended by the manufacturer or as indicated by analytical results. Filter product water through a 0.22 μm pore size membrane filter to remove bacterial contamination. Quality check of the prepared water will be discussed in Section 9.5.

9.3.5 UV Sterilizer

Quality control functions for the care of the UV sterilizer include:

- Remove plug from outlet and clean UV lamp monthly with a soft cloth moistened with ethanol.
- Test UV lamp by light meter quarterly. If it emits less than 80 percent of the initial rated output, replace the lamp.
- Perform spread plate irradiation test quarterly, as follows:

TABLE 9.3
Refrigerator Temperature Control Log (Temperature 4°C ± 2°C)

Instrument Description ______________________________

	Jan.	Feb.	Mar.	Apr.	May	June	July	Aug.	Sept.	Oct.	Nov.	Dec.
1.												
2.												
3.												
4.												
5.												
6.												
7.												
8.												
9.												
10.												
11.												
12.												
13.												
14.												
15.												
16.												
17.												
18.												
19.												
20.												
21.												
22.												
23.												
24.												
25.												
26.												
27.												
28.												
29.												
30.												
31.												

a. Prepare 100 ml heterotrophic plate count agar.
b. Pour 10 to 15 ml of the melted agar into petri dishes. Keep covers opened slightly until agar has hardened and moisture and condensation has evaporated. Close dishes and keep in refrigerator until needed.
c. Prepare a coliform culture and dilute to give about 200 to 250 count per 0.5 ml.
d. Pipet 0.5 ml from the culture for selected dishes. Spread inoculum with a sterile glass-rod or sterile spreader, or rotate plates several revolutions so the inoculum will spread uniformly over the surface of the agar.
e. With cover removed, place agar spread plates under the UV lamp.
f. Place one inoculated plate under ordinary laboratory lighting as a control.
g. Expose plates for 2 min.
h. Close plates and incubate at 35°C. Count the colonies.
i. Control plates should contain 200 to 250 bacterial colonies.

UV treated plates should show 99 percent reduction in counts. If the reduction is less than 80 percent, replace the UV lamp.

TABLE 9.4
Hot Air Oven Temperature Control Log (Temperature 160°C–170°C)

Instrument Description ____________________

	Jan.	Feb.	Mar.	Apr.	May	June	July	Aug.	Sept.	Oct.	Nov.	Dec.
1.												
2.												
3.												
4.												
5.												
6.												
7.												
8.												
9.												
10.												
11.												
12.												
13.												
14.												
15.												
16.												
17.												
18.												
19.												
20.												
21.												
22.												
23.												
24.												
25.												
26.												
27.												
28.												
29.												
30.												
31.												

9.3.6 Membrane Filter Apparatus

Quality control functions for the care of membrane filter apparatus include:

- Check funnel support for leaks on a daily basis.
- Check funnel and funnel support to make certain they are smooth. Discard funnel if inside surfaces are scratched.
- Clean thoroughly after each time used, wrap in aluminum foil, and autoclave.

9.3.7 Centrifuge

Quality control functions for the care of a centrifuge include:

- Check brushes and bearings semiannually.
- Check the rheostat control against a tachometer at various loadings every six months to ensure proper gravitational fields.

TABLE 9.5
Calibration Chart for Thermometer

Date	Reading on Working Thermometer, °C	Reading on Certified Thermometer, °C	Difference between Readings, °C	Remarks	Signature

TABLE 9.6
Balance Check

Model ______________________________

Date	0.0100 g	0.1000 g	1.0000 g	10.0000 g	Signature

TABLE 9.7
Calibration Stock and Standard Solution Log Form

Name and concentration
of solution ______________________________

Test for used ______________________________

Source
Manufacturer ______________________________
House prepared ______________________________

Lot number ______________ Certification number ______________

Date received ______________ Date opened ______________

Expiration date ______________ Date of disposal ______________

Mode of disposal ______________________________

Remark ______________________________
____

______________ ______________
Signature of logger Date

9.3.8 Hot-Air Oven

Quality control functions for the care of a hot-air oven include:

- Test performance quarterly with commercially available spore strips or spore suspensions.
- Monitor temperature with a thermometer accurate in the 160 to 180°C range. Daily monitoring form is shown in Table 9.4.
- Use heat-indicating tape to identify supplies and materials that have been exposed to sterilization temperature.

9.3.9 Autoclave

Quality control functions for the care of an autoclave include:

- Autoclave must have pressure and temperature gauges on exhaust side and an operating safety valve.
- Check operating temperature with a minimum/maximum thermometer on a weekly basis.
- Depressurization must not produce air bubbles in fermentation media.
- Record items sterilized, temperature, pressure, and time for each run.
- Check performance with spore strips or suspensions weekly and record. Use heat-indicating tape to identify supplies and materials that have been sterilized. If evidence of contamination occurs, check until the cause is identified and eliminated.
- Schedule semi-annual preventive maintenance inspections.

Table 9.8 shows the proper time and temperature for sterilization of different materials.

TABLE 9.8
Time and Temperature for Autoclave Sterilization

Material	Time at 121°C
Membrane filters and pads	10 min
Carbohydrate-containing media (Lauryl Triptose, BGB broth, etc.)	12–15 min
Contaminated materials and discarded cultures	30 min
Membrane filter assemblies (wrapped)	15 min
Dilution water, 99 ml in screw-cap bottles	15 min
Rinse water volumes of 0.5 to 1 l	30 min

9.3.10 Refrigerators and Freezers

Quality control functions for the care of refrigerators and freezers include:

- Check and record temperature daily as shown in Table 9.3.
- Monthly cleaning and rearrangement of the refrigerator is recommended.
- All material should be dated and identified clearly.
- Defrost and discard outdated materials in refrigerator and freezer quarterly, and record the activities.

9.3.11 Incubators

Quality control functions for the care of incubators include:

- Check and record temperature daily on the shelf area in use as shown in Tables 9.1 and 9.2. If a glass thermometer is used, submerge bulb and stem in water or glycerine to the stem mark.
- Place incubator in area where the temperature is maintained at 16 to 27°C.

9.3.12 Water Bath

Quality control functions for the care of a water bath include:

- Keep an appropriate thermometer immersed in the water bath, monitor and record temperature daily. Bath must maintain the uniform temperature needed for the test in use.
- Use only stainless steel, plastic coated, or other corrosion-proof racks in the water bath.
- Clean bath monthly or as needed.

9.3.13 Safety Hood

Quality control functions for the care of a safety hood include:

- Check filters monthly for plugging and dirt accumulation and clean or replace as needed.
- Once a month, expose plate count agar plates to air and flow for 1 h. Incubate plates at 35°C for 24 h and examine for contamination. A properly operating safety hood should produce no growth on the plates.
- Inspect cabinet for leaks and the rate of air flow quarterly.
- Use pressure monitoring to check efficiency of hood performance.
- Maintain hood as described by the manufacturer.

9.3.14 Microscope

Quality control functions for the care of a microscope include:

- Use lens-paper to clean optics and stage after each use.
- Cover microscope when not in use.
- Permit only trained technicians to use florescence microscope and light source. Monitor fluorescence lamp with a light meter and replace when a significant loss in fluorescence is observed. Log lamp operating time, efficiency, and alignment. Periodically, check lamp alignment, particularly when the bulb has been changed. Use known positive 4+ fluorescence slides as controls.
- Establish yearly maintenance contract.

9.3.15 Spectrophotometer

Quality control functions for the care of a spectrophotometer include:

- Maintain calibration and quality control check as recommended by the manufacturer.
- Have yearly inspection by a factory maintenance man.

9.4 QUALITY CONTROL OF LABORATORY SUPPLIES

9.4.1 Glassware

Quality control procedures for the use of glassware include:

- Before each use check glassware and flasks for chipped or broken edges and etched surfaces. Discard chipped or badly etched glassware.
- Inspect after washing: if water beads are excessive on the cleaned surface, wash again.
- Check for alkaline or acid residues as follows:
  a. Preparation of 0.04 percent bromthymol blue: dissolve 0.1 g bromthymol blue indicator in 18 ml of 0.01 *M* NaOH and fill up to 250 ml with DI water (De-ionized).
  b. Add 0.04 percent bromthymol blue indicator and observe the color. Bromthymol blue is yellow in acidic, blue-green in neutral and blue in alkaline solutions.
- Check for residual detergent.
  a. Wash and rinse six petri dishes in the usual manner (Group A).
  b. After normal washing, rinse a second group of six petri dishes one time with DI (Deionized) water (Group B).
  c. Wash six petri dishes with detergent and dry without rinsing (Group C).
  d. Sterilize all three groups of glassware.
  e. Proceed heterotrophic plate count determination by using 0.5 ml of coliform inoculate.
  f. After incubation, read bacterial counts on plates.
  g. Interpretation of the result:

     Differences in bacterial counts of 15 percent or more among groups indicate the detergent has no toxicity or inhibitory effect.

     Differences in bacterial growth of 15 percent or more between groups A and B demonstrate that inhibitory residues are left on glassware after normal washing procedures.

     Disagreement in averages of less than 15 percent between groups A and B, and greater than 15 percent between groups A and C indicate that the detergent used has inhibitory properties that are eliminated during routine washing.

9.4.2 Glass, Plastic, and Metal Utensils for Media Preparation

Quality control procedures for the use of glass, plastic, and metal utensils include:

- Washing processes should provide glassware free of toxic residue as demonstrated by the inhibitory residue test discussed above.
- Glassware free of chips and cracks.
- Utensils are clean and free from foreign residues.
- Plastic items are clear with visible graduation.
- Do not use copper utensils!

9.4.3 Culture Dishes

Quality control procedures for the use of culture dishes include:

- Record the brand and type of petri dishes used. May use sterile plastic or glass dishes.
- Open packs of disposable sterile culture dishes must be resealed between uses.

9.4.4 Sterility Check on Glassware

After sterilization of bottles, flasks, or tubes, and after one of each bench have been sterilized, add aerobic broth, incubate, and check for bacterial growth. Lauryl triptose broth is excellent for this purpose.

9.4.5 Chemicals and Reagents

Quality control procedures for the use of chemicals and reagents include:

- Use only chemicals in ACS or equivalent grade, because impurities can inhibit bacterial growth, provide nutrients, or fail to produce the desired reaction.
- Date chemicals and reagents when received and when first opened.
- Make reagents to volume in volumetric flasks and transfer for storage to good quality inert plastic or borosilicate glass bottles with borosilicate, polyethylene, or other plastic stoppers or caps.
- Label reagents properly as shown in Figure 9.1.

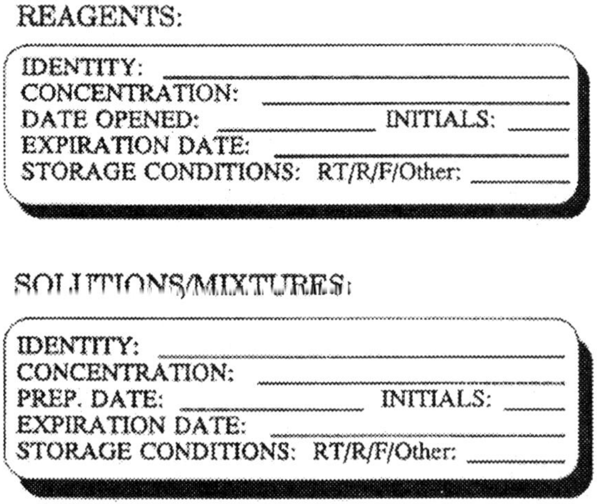

REAGENTS:

IDENTITY: ______
CONCENTRATION: ______
DATE OPENED: ______ INITIALS: ______
EXPIRATION DATE: ______
STORAGE CONDITIONS: RT/R/F/Other: ______

SOLUTIONS/MIXTURES:

IDENTITY: ______
CONCENTRATION: ______
PREP. DATE: ______ INITIALS: ______
EXPIRATION DATE: ______
STORAGE CONDITIONS: RT/R/F/Other: ______

FIGURE 9.1 Label formats used for reagents and solvents. (RT equals room temperature, R equals refrigerator, and F equals freezer.)

9.4.6 Dyes and Stains

Quality control procedures for the use of dyes and stains include:

- In microbiological analysis, organic chemicals are used as *selective agents* (e.g., brilliant green), as *indicators* (e.g., phenol red), and as *microbiological stains* (e.g., Gram stain).
- Use only dyes certified by the Biological Stain Commission.

9.4.7 Membrane Filters and Pads

The quality and performance of membrane filters vary with the manufacturer, type, brand, and lot. These variations result from differences in manufacturing methods, materials, quality control, and storage conditions.
Membrane filters and pads for water analysis should meet the following specifications:

- Filter diameter 47 mm, mean pore diameter 0.45 μm.
- The filters are floated on reagent grade water, the water diffuses uniformly through the filters in 15 s with no dry spots on the filters.
- Flow rates are at least 55 ml per min per cm^2 at 25°C.
- Filters are nontoxic, free of bacterial-growth-inhibiting substances, and free of materials that directly or indirectly interfere with bacterial indicator system in the medium; ink grid is nontoxic.
- The arithmetic mean of five counts on filters must be at least 90 percent of the arithmetic mean of the counts on five agar spread plates using the same sample volumes and agar media.
- Filters retain the organisms from a 100 ml suspension of *Serratia marcescent* containing 1×10^3 cells.
- Water extractables in filter do not exceed 2.5 percent after the membrane is boiled in 100 ml reagent grade water for 20 min, dried, cooled, and brought to constant weight.
- Absorbent pad has a diameter of 47 mm, thickness is 0.8 mm, and is capable of absorbing 2.0 ± 0.2 ml Endo broth.
- Pads release less than 1 mg acidity calculated as $CaCO_3$, when titrated to the phenolphthalein end point with 0.02 *N* NaOH.
- Some manufacturers provide information beyond that required by specification and certify that their membranes are satisfactory for water analysis. They report retention, flow rate, pore size, sterility, pH, percent recovery, and limits for specific inorganic and organic chemical extractables.
- After incubation, colonies should be well-developed with well-defined color and shape as defined in the test procedure. Colonies should be distributed evenly across the membrane surface.

9.4.8 Culture Media

Criteria for powdered culture media are as follows:

- Use the best available materials.
- Order media in quantities to last no longer than one year.
- Use media on the first-in first-out basis.
- When practical, order media in quarter pound (114 g) multiples rather than one pound (445 g) bottles, to keep the supply sealed as long as possible.
- Record kind, amount, and appearance of media received, lot number, expiration date, and dates received and opened.

- Check inventory quarterly for reordering.
- Discard media that are caked, discolored, or show other deterioration.
- Because temperature, light, and moisture conditions differ among laboratories, it is impossible to establish absolute shelflife limits for unopened bottles of media.

A conservative limit for unopened bottles of media is two years at room temperature. Use open bottles of media within six months after opening. Once bottles are opened, store them in a desiccator immediately after use.

9.4.8.1 Quality Control Criteria for Prepared Media

Maintain in a bound book a complete record of each batch of media prepared. The record should include:

- Name of preparer and the date of preparation.
- Name and lot number of medium.
- Amount of media weighed and the volume of media prepared.
- Sterilization time and temperature.
- pH measurement and adjustments.
- Sterility check.
- Positive and negative control checks.
- Ready media stored in refrigerator, with the name of the media and the date of preparation. Hold no longer than three months.
- Media stored in refrigerator are incubated overnight prior to use, and tubes with air bubbles are discarded.
- Broth media are used within 96 h.
- Agar media are used within two weeks. Seal prepared agar plates in plastic bags to retain moisture and refrigerate.
- Ampouled media is stored in refrigerator, the expiration date is designated by the manufacturer.

Preparation and sterilization of media are discussed in Chapter 11. Holding time for the ready media is given in Table 9.9.

9.5 QUALITY CONTROL OF LABORATORY PURE WATER

9.5.1 Checks and Monitoring Criteria

Checks and monitoring criteria with the acceptable limits for deionized water used in a microbiology laboratory are summarized in Table 9.10. If these limits are not met, investigate the problem and correct it. Although pH measurement of purified water is characterized by drift, extreme readings are indicative of chemical contamination.

9.5.2 Test for Bacterial Quality

The test is based on the growth of *Enterobacter aerogenes* in a chemically defined minimal growth medium. The presence of a toxic agent or a growth-promoting substance will alter the 24 h population by an increase or decrease of 20 percent or more when compared to a control. Perform the test annually, or when the source of the reagent water is changed, and when an analytical problem occurs.

The test is complex, requires skill and experience. It requires four days, ultrapure water from an independent source as a control, high purity reagents, and extreme cleanliness on glassware.

TABLE 9.9
Holding Times for Prepared Media

Medium	Holding Time
Membrane filter (MF) broth in screw-cap flasks at 4°C	96 hours
MF agar in plates with tight-fitting covers at 4°C	2 weeks
Agar or broth in loose-cap tubes at 4°C	1 week
Agar or broth in tightly-closed screw-capped tubes at 4°C	3 months
Poured agar plates with loose-fitting covers in sealed plastic bags at 4°C	2 weeks
Large volume of agar in tightly-closed screw-cap flask or bottle at 4°C	3 months

9.5.2.1 Reagents

Use only reagents and chemicals that are ACS grade, and prepare reagents in water freshly redistilled from a glass still.

Sodium citrate solution—Dissolve 0.29 g sodium citrate ($Na_3C_6O_7 \cdot 2H_2O$) and dilute to 500 ml.

Ammonium sulfate solution—Dissolve 0.26 g ammonium sulfate (NH_4SO_4) and dilute to 500 ml.

Salt-mixture solution—Dissolve 0.26 g magnesium sulfate ($MgSO_4 \cdot 7H_2O$), 0.17 g calcium chloride ($CaCl_2 \cdot 2H_2O$), 0.23 g ferrous sulfate ($FeSO_4 \cdot 7H_2O$), and 2.50 g sodium chloride (NaCl), and dilute to 500 ml. Salt-mixture solution develops a slight turbidity within 3 to 5 days caused by the oxidation of ferrous iron (Fe^{2+}) to ferric iron (Fe^{3+}). To avoid this very short life of this solution, it is best to prepare it without adding ferrous sulfate ($FeSO_4 \cdot 7H_2O$), and just before using the solution, add the appropriate amount of iron salt (e.g., 50 ml of salt solution needs 0.026 g iron sulfate).

Stock phosphate buffer solution—Dissolve 34 g potassium dihydrogen phosphate, KH_2PO_4, and dilute to 500 ml. Adjust pH to 7.2 ± 0.5 with 1 *N* sodium hydroxide (NaOH), dilute to 1 l. Keep in refrigerator.

Working phosphate buffer solution—Dilute stock phosphate (see above) solution to 1:25 in laboratory pure water. Discard solution if it becomes turbid.

TABLE 9.10
Quality Check of Laboratory Pure Water

Parameter	Monitoring Frequency	Limit
Conductivity	D	1–2 μmhos/cm
pH	D	5.5–7.5 unit
Total organic carbon	A	< 1.0 mg/l
Trace metal single Cd, Cr, Cu, Ni, Pb, Zn	A	< 0.05 mg/l
Trace metal total Cd, Cr, Cu, Ni, Pb, Zn	A	< 1.0 mg/l
Ammonia, as $NH_3 - N$	M	< 0.1 mg/l
Free chlorine, Cl_2	M	< 0.1 mg/l
Heterotrophic count		
Fresh water	M	< 1000 cnt/ml
Stored water	M	< 10,000 cnt/ml
Water suitability test	A	Ratio: 0.8–3.0

A = annually; M = monthly; D = daily; cnt = count (bacterial).
μmhos/cm = micromhos/cm

Boil all reagents 1 to 2 min to kill vegetative cells. Store solutions in sterilized glass stoppered bottles in the dark at 5°C, properly labeled as shown in Figure 9.1 for up to several months. Test solutions for sterility before each use.

9.5.2.2 Sample Preparation

Collect 150 to 200 ml of laboratory reagent grade water and control water (ultrapure water from independent source) in sterile borosilicate glass flasks and boil for 1 to 2 min. Avoid longer boiling to prevent chemical changes.

9.5.2.3 Preparation of Bacterial Suspension

Preparation of the *Enterobacter aerogenes* suspension used as inoculum is outlined as follows:

1. On the day before making the distilled water suitability test, inoculate a strain of *Enterobacter aerogenes* onto a nutrient agar slant with a slope approximately 6.3 cm long in a screw cap tube. Streak entire agar surface to develop a continuous-growth film and incubate 18 to 24 h at 35°C.
2. Pipet 1 to 2 ml sterile water from a 99 ml water blank onto the 18 to 24 h incubated culture as prepared above. Emulsify growth on slant by gently rubbing bacterial film with pipet, being careful not to tear agar. Pipet 1 ml suspension back into the original 99 ml water blank.
3. Make a 1:100 dilution of the original 99 ml water blank, as prepared above, into a second water blank, and make 1:100 dilution of second bottle into a third water blank, and a 1:10 dilution of the third bottle into the fourth water blank, shaking vigorously after each transfer. The procedure gives $1{:}10^2$, $1{:}10^4$, and $1{:}10^5$ dilutions (See Section 13.1).

9.5.2.4 Test

The outline of the test used to evaluate bacterial quality of the laboratory pure water used in microbiology work is as follows:

1. Label five flasks or tubes, A, B, C, D, and E.
2. Add water samples, media reagents, and ultrapure water to each flask as indicated in Table 9.11.
3. Pipet 1 ml of the 1:10 ($1{:}10^5$) dilution into each flask A, B, C, D, and E. This procedure should result in a final dilution of the organisms to a range of 30 to 80 viable cells per ml of the test solution.
4. Make an initial bacterial plate count by plating in triplicate 1 ml portions from each flask in plate count agar.
5. Incubate flasks A through E at 35°C for 24 ± 2 h.
6. Prepare final plate counts from each flask, using dilutions of 1, 0.1, 0.01, 0.001, and 0.0001 ml.

9.5.2.5 Calculation

1. For growth-inhibiting substance:

$$\text{Ratio} = \text{colony count/ml, Flask B/Flask A} \tag{9.1}$$

A ratio of 0.8 to 1.2 shows no toxic substances. A ratio of less than 0.8 shows growth-inhibiting substances in the water sample.

2. For nitrogen and carbon sources that promote bacterial growth:

$$\text{Ratio} = \text{colony count/ml, Flask C/Flask A} \tag{9.2}$$

3. For nitrogen sources that promote bacterial growth:

$$\text{Ratio} = \text{colony count/ml, Flask D/Flask A} \quad (9.3)$$

4. For carbon sources that promote bacterial growth:

$$\text{Ratio} = \text{colony count/ml, Flask E/Flask A} \quad (9.4)$$

For the last three ratios (nitrogen and carbon sources) a value above 1.2 indicates an available source for bacterial growth. Do not calculate the last three ratios when the first ratio indicates a toxic reaction.

9.5.2.6 Interpretation of the Results

The colony count from Flask A after 20 to 24 h at 35°C will depend on number of organisms initially planted in Flask A and the strain of *Enterobacter aerogenes* used.

For this reason run the control, Flask A, for each individual series of tests. However, for a given strain of *E. aerogenes*, under identical environmental conditions, the terminal count should be reasonably constant when the initial plant is the same. The initial colony counts on Flask A and Flask B should be approximately equal. When the ratio exceeds 1.2, assume that growth stimulating substances are present. However, this procedure is so sensitive, and ratios up to 3.0 have little significance in actual practice. The ratio could go high as 3.0 from 1.2 without any undesirable consequences.

9.6 ANALYTICAL QUALITY CONTROL PROCEDURES

9.6.1 Quality Control in Routine Analysis

Each laboratory must establish quality control checks over the microbiological analysis.

9.6.1.1 Negative (Sterile) Control

Include one negative control with each series of samples using buffered dilution water and the medium batch at the start of the series and following every tenth sample. When sterile controls

TABLE 9.11
Reagent Addition For Water Quality Test

	Control Test ml		Optional Tests ml		
Media Reagents	Control A	Test Water B	Carbon/Nitrogen Available C	Nitrogen Source D	Carbon Source E
Sodium citrate solution	2.5	2.5	—	2.5	—
Ammonium sulfate solution	2.5	2.5	—	—	2.5
Salt-mixture solution	2.5	2.5	2.5	2.5	2.5
Phosphate buffer (7.3 ± 0.1)	1.5	1.5	1.5	1.5	1.5
Unknown water	—	21.0	21.0	21.0	21.0
Redistilled water	21.0	—	5.0	2.5	2.5
Total volume	30.0	30.0	30.0	30.0	30.0

indicate contamination, data on samples affected should be rejected and a request made for immediate resampling of the samples involved.

9.6.1.2 Positive Control

Test a minimum of one pure culture of known positive reaction at the end of the analytical series. When positive control indicate faulty media, data on samples affected should be rejected. After preparation and checking a new media, make a request for resampling.

9.6.1.3 Duplicate Analysis

Run a duplicate analysis on at least one sample per test run and on 5 percent of all samples. Calculate method precision as discussed in Section 9.6.2.

9.6.2 Measurement of Method Precision

Calculate precision of duplicate analysis for each different type of sample (e.g., drinking water, surface water, wastewater, etc.) according to the following procedure:

1. Collect the results of 15 duplicate analyses. Record duplicate analyses as D_1 and D_2.
2. Calculate the logarithm of each result (L_1 and L_2). If either of a set of duplicate results is 0, add +1 to both values before calculating the logarithm.
3. Calculate the range (R) of each logarithmic pair ($L_1 - L_2$).
4. Calculate the mean (R) of the ranges. An example is given in Table 9.12.

Keep this record updated, and use the most recent values for the daily routine check of analytical precision.

Calculate the precision of duplicate analysis on 5 percent of samples or at least one sample per test run as follows:

1. Collect the result of these duplicates and record them as D_1 and D_2.
2. Calculate the logarithm of each result and record them as L_1 and L_2. If either of a set of duplicate results is 0, add +1 to both values before calculating the logarithm.
3. Calculate the range (R) of each logarithmic pair ($L_1 - L_2$).
4. Compare this calculated range of each sample to the mean of the ranges of the corresponding type of sample.
5. If the range of the duplicates is greater than $3.75R$, analyst variability is excessive.
6. If the problem related to this imprecision result is not acceptable, discard all analytical results since the last precision check.
7. Determine the cause of the unacceptable precision, and resolve the analytical problem before making further analysis. See Table 9.13.

9.6.3 Reference Sample

Laboratories should analyze reference samples quarterly when available for the parameters measured.

9.6.4 Performance Sample

Laboratories should analyze at least one unknown performance sample per year when available for the parameters measured.

TABLE 9.12
Calculation of Precision Criteria

	Duplicate of Counts		Logarithms of Counts		Range of Logarithms (R_{log})
Sample No.	D_1	D_2	L_1	L_2	$L_2 - L_2$
1	89	71	1.9494	1.8513	0.0981
2	38	34	1.5798	1.5315	0.0483
3	58	67	1.7634	1.8261	0.0627
4	7	6	0.8451	0.7782	0.0669
5	110	121	2.0414	2.0828	0.0414
6	24	25	1.3802	1.3979	0.0177
7	14	18	1.1461	1.2553	0.1092
8	3	4	0.4771	0.6020	0.1249
9	222	200	2.3464	2.3010	0.0454
10	66	61	1.8195	1.7853	0.0342
11	2	1	0.3010	0	0.3010
12	99	92	1.9956	1.9638	0.0318
13	75	63	1.8751	1.7993	0.0758
14	44	47	1.6435	1.6721	0.0286
15	32	28	1.5051	1.4472	0.0579
16	89	80	1.9494	1.9031	0.0463
17	20	31	1.3010	1.4914	0.1907
18	230	212	2.3617	2.3263	0.0354
19	29	20	1.4624	1.3010	0.1614
20	11	17	1.0414	1.2304	0.1890
				Total	1.4155
				Avg (R)	0.0708

Precision criterion = $3.27R = 3.27 \times 0.0708 = 0.2315$

TABLE 9.13
Daily Checks on Precision of Duplicate Counts

	Duplicate Counts		Logarithms of Counts		Range of Logarithms	
Date	D_1	D_2	L_1	L_2	R_{log}	Acceptance
	71	65	1.8513	1.8129	0.0383	A
	102	180	2.0086	2.2553	0.2467	U
	73	25	1.8633	1.6990	0.1663	A

The calculated precision criterion is 0.2315 (see Table 9.12). Therefore, the precision given by the first and the third duplicates values are acceptable. However, it is unacceptable in the second duplicate counts.

A = acceptable U = unacceptable

9.6.5 Membrane Filter Method (MF) Verification

Of the analyses performed, 5 percent should be verified as described in Chapters 17, 18, and 19.

9.6.5.1 Total Coliform

Pick up at least ten isolated sheen colonies from each sample. *Transfer into Lauryl Tryptose Broth*, incubate, and read. Transfer positive tubes into *Brilliant Green Bile Broth* for verification of coliforms. The laboratory should make every effort to detect coliforms from samples with excessive noncoliforms on the membrane filter. Any sheen colonies appearing in mixed confluent growth must be verified. See Sections 17.2.4 and 17.6.

9.6.5.2 Fecal Coliforms

Pick at least ten isolated colonies containing blue to blue-green pigment and transfer to *Lauryl Tryptose Broth*. Incubate and read. Transfer positive tubes to *EC Broth* where gas production verifies fecal coliform organisms. See Section 18.2.6.

9.6.5.3 Fecal Streptococci

Pick up at least ten isolated pink to red colonies from MF plates. Transfer to *BHI agar or broth*. After growth, perform catalase test. If negative (possible fecal streptococci) transfer growth to BHI (Brain Heart Infusion) broth and BHI broth with 40 percent Bile tubes, incubate at 45°C and 35°C, respectively. Growth at both temperatures verifies fecal streptococci. See Section 19.3.6.

9.7 RECORDS AND DATA REPORTING

Keep records of microbiological analyses for at least five years. Actual laboratory reports may be kept, or data may be transferred to tabular summaries, provided that all of the following information is included:

- Date, time, and place of sampling.
- Name of sample collector.
- Identification of sample.
- Date sample received by the laboratory.
- Date and time of the analysis.
- Analytical method used.
- Raw data and the calculated result.
- Verify that each result was entered correctly from the bench sheet and initialed by the analyst.

Results should be calculated and entered on the sample report form to be forwarded. A careful check should be made to verify that each result was entered correctly from the bench sheet and initialed by the analyst.

Positive results are reported as preliminary without waiting for MF (Membrane Filter) verification or MPN (Most Probable Number) completion. After verification and completion, the adjusted counts should be reported.

When action response is a designated laboratory responsibility, the proper authorities should be notified promptly on unsatisfactory sample results, and the request should be made for resampling from the same sampling point.

9.8 INTERLABORATORY QUALITY CONTROL

An interlaboratory quality control program is a system of agreed upon requirements and laboratory practices necessary to maintain minimal quality standards among a group of participant laboratories.

After methods and laboratory operational standards are established, an independent agency inspects laboratory facilities and conducts method validation and performance evaluation studies. The goal of this program is the identification of problems and technical assistance. An example of interlaboratory quality assurance is the federal/state program for the certification of water supply laboratories, developed under the Safe Drinking Water Act and its amendments. In the certification program, public water supply laboratories must be certified according to the minimal criteria and procedures and quality assurance described in the EPA manual on certification. Annually, laboratories are required to perform acceptably on unknown samples, as samples are available. The responsible authority follows up on problems identified in the on-site inspection and performance evaluation and requires corrections within a set period of time. Primary causes for discrepancies in drinking water laboratories have been inadequate equipment, improperly prepared media, incorrect analytical procedures, and insufficiently trained personnel.

To participate in the program the following preparations are required:

- A written description for the current laboratory quality control program must be available for review. Management, supervisors, and analysts participate in setting up the quality control program.
- Each participant should have a copy of the quality control program and a detailed guide of his or her own portion.
- A record on analytical quality control tests and quality control checks on media, materials, and equipment must be prepared and retained.

10 Collecting and Handling Environmental Samples for Microbiological Examination

10.1 SAMPLING

The sample is the most critical element of the analytical process. The quality of any data produced primarily depends on the sample analyzed. A sample must be representative of the whole so that the final result represents the entire system that it is intended to represent.

10.1.1 Sampling Program

A sampling program must be planned to satisfy the objectives of the study and yet remain within the limitations of available manpower, time, and money. The survey should use the minimum number of samples that adequately represent the body of water. The number of samples and location of sample sites should be determined prior to the survey and must satisfy the requirements needed to establish water quality standard or effluent permit violations.

Sampling programs should contain the following objectives:

- Objective of the study.
- Sample source.
- Number and matrix of the samples.
- Location of sampling sites.
- Sample collector(s).

10.1.2 Type of Samples

Samples collected at a particular time and place are called grab or individual samples. This type of sample represents conditions at the time it was collected. Therefore, a grab sample should not be used as a basis for a decision about pollution abatement.

A composite sample refers to a mixture of grab samples collected at the same point at different times. A series of smaller samples are collected in a single container and blended for analysis.

Samples for microbiological testing should never be composited and should be only grab samples!

Duplicate samples are collected for checking the preciseness of the sampling process.

10.1.3 Sample Containers

The material of which the sample container is made should be chosen so that it will not react with the sample. It should be resistant to leakage and breakage and should have the proper volume necessary for the analyte(s) of interest. For microbiological testing, one may use plastic or glass sterile containers with a minimum of a 125 ml capacity. Sterilization of the containers may be done in the laboratory, but commercially available sterile plastic cups, or individually wrapped or sterile whirl-pack plastic bags are commonly used, see Figure 10.1.

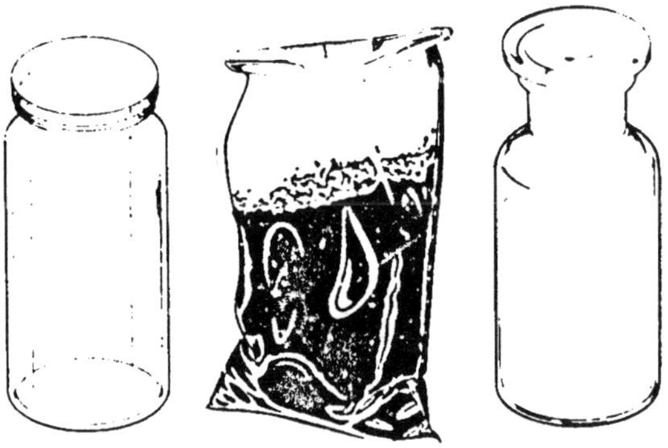

FIGURE 10.1 Suggested sample containers: a. Screw cap sterile glass or plastic bottle. b. Sterile plastic bag (whirl pack). c. Sterile glass stoppered bottle.

10.1.4 Dechlorinating Agent

Add a reducing agent to containers intended for the collection of water samples having residual chlorine or other halogens. *Sodium thiosulfate* ($Na_2S_2O_3$) is a satisfactory dechlorinating agent that neutralizes any residual halogen and prevents continuation of bactericidal action during sample transit. The examination then will indicate more accurately the true microbial content of the water at the time of sampling. The chemical must be placed in the container prior to sterilization.

A 0.1 ml of 10 percent solution of sodium thiosulfate per 125 ml sample will neutralize approximately 15 mg per l of residual chlorine, and a 0.1 ml of 3 percent solution of sodium thiosulfate per 125 ml sample will neutralize 5 mg per l of residual chlorine. Alternatively, 0.02 and 0.05 g of powdered sodium thiosulfate should be added to sample containers before sterilization. The amount need not be weighed. An estimated amount on the tip of a spatula is sufficiently accurate.

10.1.5 Chelating Agents

When the collected water samples are high in copper or zinc and wastewater samples are high in heavy metals, the sample containers should contain a chelating agent to reduce metal toxicity.

Add 0.3 ml of 15 percent *EDTA (ethylene diamine tetra acetic acid disodium salt)* solution per 125 ml sample bottle before sterilization.

10.1.6 Sampling Procedures

Samples are collected by hand or with a sampling device if depth samples are required or the sampling site has a difficult access such as a manhole, dock, bridge, or bank adjacent to the water.

When the sample such as treated waters, chlorinated wastewaters, or recreational waters are collected, the sample bottle must contain a dechlorinating agent (see Section 10.1.4).

When the sample is collected, leave ample air space in the container (at least 2.5 cm) to facilitate mixing by shaking prior to examination.

Collect samples that are representative of the water being tested, flush or disinfect sample ports, and use aseptic techniques to avoid sample contamination. Keep sample bottle closed until the moment it is to be filled.

1. Remove the stopper or cap, taking care to avoid soiling; do not contaminate inner surface of stopper or cap and neck of the bottle.
2. Fill container without rinsing.
   The volume of the sample should be sufficient to carry out all tests required, preferably not less than 100 ml.
3. Replace stopper or cap immediately.

10.2 SAMPLE COLLECTION FROM DIFFERENT SOURCES

10.2.1 Collecting Potable Water Samples

10.2.1.1 Sampling from Distribution System

Select a tap that is supplying water from a service pipe connected with the main. Open tap fully and let run for 2 to 3 min or for a time sufficient to permit clearing the service line. Reduce water flow to permit filling the sample container without splashing. If the tap cleanliness is questionable, disinfect tap by washing with sodium hypochlorite solution (100 mg/l NaOCl) and after let water run for an additional 2 to 3 min.

In drinking water evaluation, collect samples of finished, treated water from the distribution sites selected to assure systematic coverage during each month. Carefully select sampling locations to include dead-end sections to ensure that localized contamination does not occur through cross-connections, breaks, or reduction in positive pressure.

Sampling should be representative of the distribution system and include sites such as municipal buildings, public schools, airports, restaurants, theaters, gas stations, industrial plants, and private residences.

The minimum number of samples that must be collected and examined each month is based upon the population density served by the distribution system. See Table 10.1. In the event of an unsatisfactory sample, repetitive samples must be collected until two consecutive samples yield satisfactory quality water. Repetitive samples from any single point or special purpose samples must not be counted in the overall total of monthly samples.

TABLE 10.1
Sampling Frequency for Drinking Waters Based on Population

Population Served	Minimum Number of Samples per Month	Population Served	Minimum Number of Samples per Month
25 to 1,000	1	90,001 to 96,000	95
1,001 to 2,500	2	96,001 to 111,000	100
2,501 to 3,300	3	111,001 to 130,000	110
3,301 to 4,100	4	130,001 to 160,000	120
4,101 to 4,900	5	160,001 to 190,000	130
4,901 to 5,800	6	190,001 to 220,000	140
5,801 to 6,700	7	220,001 to 250,000	150
6,701 to 7,600	8	250,001 to 290,000	160
7,601 to 8,500	9	290,001 to 320,000	170
8,501 to 9,400	10	320,001 to 360,000	180
9,401 to 10,300	11	360,001 to 410,000	190
10,301 to 11,100	12	410,001 to 450,000	200
11,101 to 12,000	13	450,001 to 500,000	210
12,001 to 12,900	14	500,001 to 550,000	220
12,901 to 13,700	15	550,001 to 600,000	230
13,701 to 14,600	16	600,001 to 660,000	240
14,601 to 15,500	17	660,001 to 720,000	250
15,501 to 16,300	18	720,001 to 780,000	260
16,301 to 17,200	19	780,001 to 840,000	270
17,201 to 18,100	20	840,001 to 910,000	280
18,101 to 18,900	21	910,001 to 970,000	290
18,901 to 19,800	22	970,001 to 1,050,000	300
19,801 to 20,700	23	1,050,001 to 1,140,000	310
20,701 to 21,500	24	1,140,001 to 1,230,000	320
21,501 to 22,300	25	1,230,001 to 1,320,000	330

TABLE 10.1 continued
Sampling Frequency for Drinking Waters Based on Population

Population Served	Minimum Number of Samples per Month	Population Served	Minimum Number of Samples per Month
22,301 to 23,200	26	1,320,001 to 1,420,000	340
23,201 to 24,000	27	1,420,001 to 1,520,000	350
24,001 to 24,900	28	1,520,001 to 1,630,000	360
24,901 to 25,000	29	1,630,001 to 1,730,000	370
25,001 to 28,000	30	1,730,001 to 1,850,000	380
28,001 to 33,000	35	1,850,001 to 1,970,000	390
33,001 to 37,000	40	1,970,001 to 2,060,000	400
37,001 to 41,000	45	2,060,001 to 2,270,000	410
41,001 to 46,000	50	2,270,001 to 2,510,000	420
46,001 to 50,000	55	2,510,001 to 2,750,000	430
50,001 to 54,000	60	2,750,001 to 3,020,000	440
54,001 to 53,000	65	3,020,001 to 3,320,000	450
59,001 to 64,000	70	3,320,001 to 3,620,000	460
64,001 to 70,000	75	3,620,001 to 3,960,000	470
70,001 to 76,000	80	3,960,001 to 4,310,000	480
76,001 to 83,000	85	4,310,001 to 4,690,000	490
83,001 to 90,000	90	4,690,001 or more	500

10.2.1.2 Sampling from Wells

If the sample is taken from a well fitted with a pump, pump the water for about 5 min before the actual sampling. If the well does not have pumping machinery, collect the sample using a weighted sterilized sample bottle or frame sampler, see Figure 10.2.

Caution must be taken to prevent contaminating the sample with fingers, gloves, or other materials.

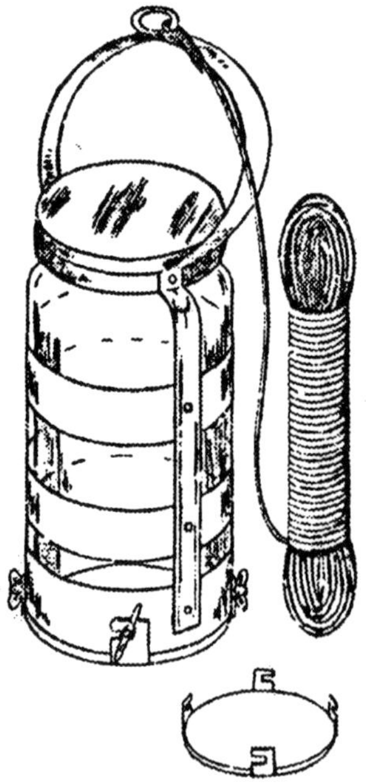

FIGURE 10.2 Weighted bottle frame and sample bottle for grab sampling.

10.2.2 Collect Samples from River, Stream, Lake, Spring, or Shallow Well

Take samples by holding the container near its base in the hand and plunging it, neck downward, below the surface. Turn bottle until neck points upward and mouth is directed toward current as shown in Figure 10.3.

For surface water sampling use the *frame sampler*, the bottle in a weighted frame that holds the bottle securely as shown in Figure 10.2.

For depth sampling use different samplers. It is undesirable to take samples too near to the bank, too far from the point of draw-off, or at a depth above or below the point of draw-off.

The *Kemmerer sampler* is a depth sampler that has been used without sterilization to collect bacteriological water samples in high pollution areas, see Figure 10.4.

Samples may be collected from a boat or bridges.

Choose sampling frequency to be reflective of stream or water body conditions. Sampling frequency may be seasonal for recreational waters, daily for water supplies intake, hourly where wastewater control is erratic and effluents are discharged into shellfish harvesting areas, or even continuous.

Sampling site selections from a lake or impoundment is shown in Figure 10.5. Sampling site selections from a large stream is shown in Figure 10.6 and sampling points in a water supply reservoir is shown in Figure 10.7.

10.2.3 Sample Collection from Bathing Beaches

Sampling sites at bathing beaches or other recreational areas should include upstream or peripheral areas and locations adjacent to natural drains that would discharge stormwater, or run-off areas draining septic wastes from restaurants, boat marinas, or garbage collection areas.

Collect samples in the swimming area from a uniform depth of approximately 1 m. Consider sediment sampling of the water-beach interface because of exposure of young children at the water's edge. Relate sampling frequency to the peak bathing period, generally in the afternoon, but preferably collect a sample in the morning and the afternoon. Weekends and holidays must be included in the sampling program. Correlate bacteriological data with turbidity levels or rainfall over the watershed.

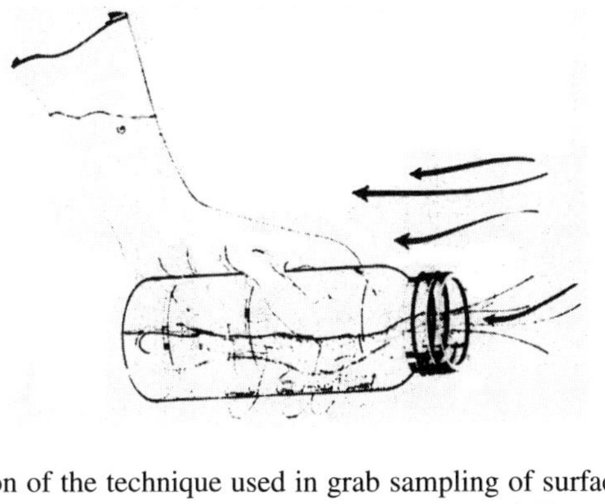

FIGURE 10.3 Demonstration of the technique used in grab sampling of surface waters. Grasp the bottle at the base with one hand, plunge the bottle mouth down into the water to avoid introducing surface scum. Position the mouth of the bottle into the current away from the hand of the collector and away from the side of the sampling platform. The sampling depth should be 15 to 30 cm (6 to 12 in.) below the water surface. Fill the bottle. After removing the bottle from the stream, pour out a small portion of the sample to allow an air space (2.5 to 5.0 cm) above the sample. Tightly stopper and label the bottle.

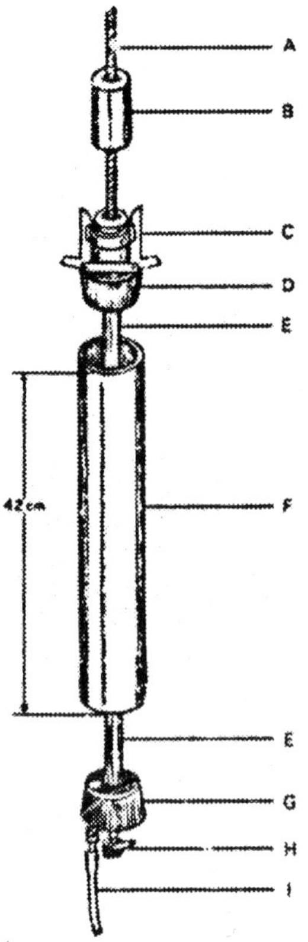

FIGURE 10.4 Kemmerer depth sampler: a. Nylon line. b. Messenger. c. Catch set so that the sampler is open. d. Top rubber valve. e. Connecting rod between the valves. f. Tube body. g. Bottom rubber valve. h. Knot at the bottom of the suspension line. i. Rubber tubing attached to the spring loaded check valve.

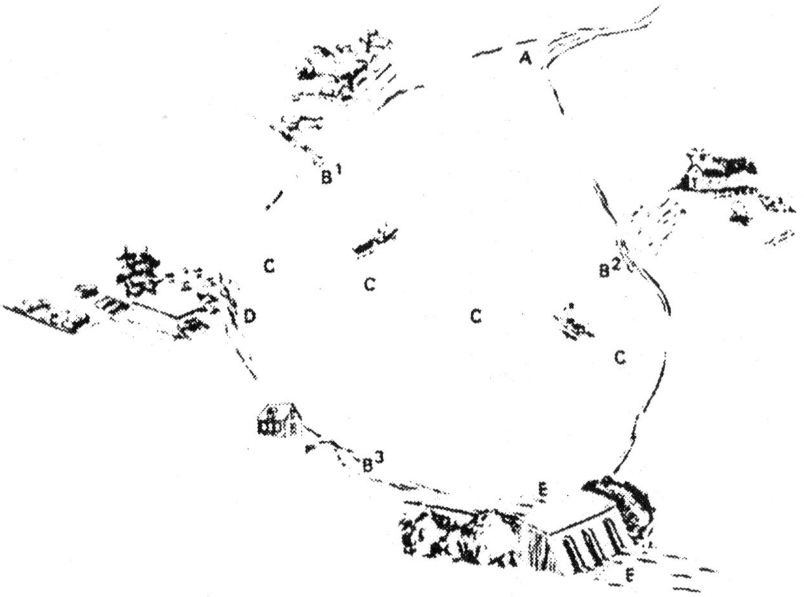

FIGURE 10.5 Sampling a lake or impoundment: a. Inlets. b. Potential source of pollution. b1. Village. b2. Agricultural run-off. b3. Home septic tank. c. Multipoint transcent. d. Bathing beach. e. Outlet above and below.

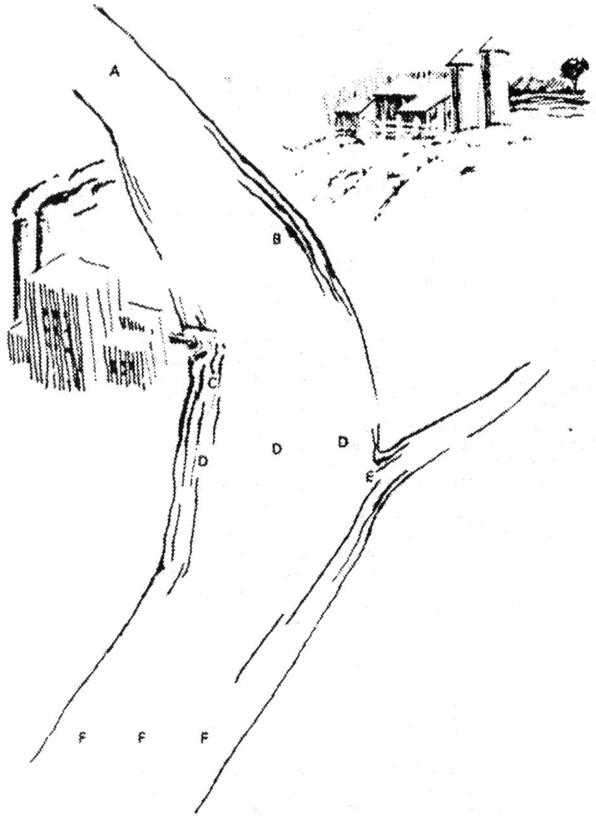

FIGURE 10.6 Sampling a large stream: a. Control station. b. Agricultural pollution. c. Industrial discharge. d. Quarter point transect. e. Tributary. f. Downstream monitoring.

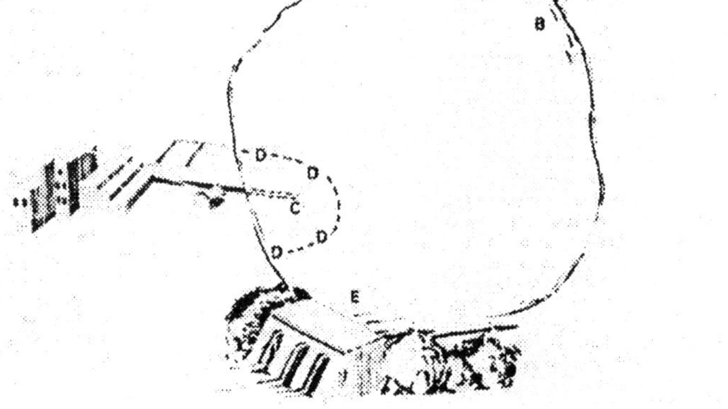

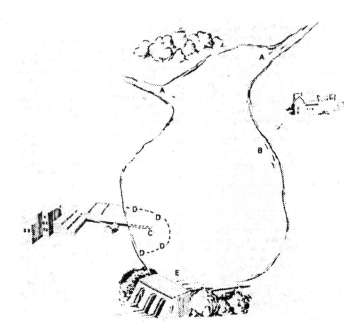

FIGURE 10.7 Sampling a water supply reservoir: a. Influent stream. b. Possible agricultural contamination. c. Water plant intake. d. Multipoint sampling around intake. e. Reservoir outlet.

Swimming pool water should monitored daily during maximum use periods, preferably at the overflow. It is important to test water for neutralization of residual chlorine at the pool side.

10.2.4 Marine and Estuarine Sampling

Sampling marine and estuarine waters requires the consideration of other factors in addition to those usually recognized in fresh water sampling. They include tidal cycles, current patterns, bottom currents, and countercurrents, stratification, seasonal fluctuations, dispersions of discharges, and multidepth sampling. The frequency of sampling varies with the objectives.

10.2.5 Sample Collection from Domestic and Industrial Discharges

In situations where the plant treatment efficiency varies considerably, grab samples are collected around the clock at selected intervals for three or five day periods. In no case should a composite sample be collected for microbiological examination!

The NPDES (National Pollution Discharge Elimination System) has established wastewater treatment plant limits for all dischargers. These are often based on maximum and mean values. A sufficient number of samples must be collected to satisfy the permit and to provide statistically sound data and give a fair representation of the bacteriological quality of the discharge.

10.2.6 Collecting Samples from Sediments and Sludges

10.2.6.1 Bottom Sediments

The bacteriology of bottom sediments is important in water supply reservoirs, in lakes, rivers, and coastal waters used for recreational purposes, and in shellfish-growing waters. Seasonal changes in water temperatures and storm water runoffs influence microbiological quality. Sediment samples are usually taken as a part of surface water samples. All surface water samples should be taken prior to the sediment collection. The most commonly used equipment for sediment sampling are the *Peterson*, the *Eckman*, and *Ponar sampler*. The Peterson and Ponar samplers are suitable for hard rocky sediments, whereas the Eckman sampler is mostly used for sand, silt, or mud type sediments. The Eckman bottom grab sampler is shown in Figure 10.8.

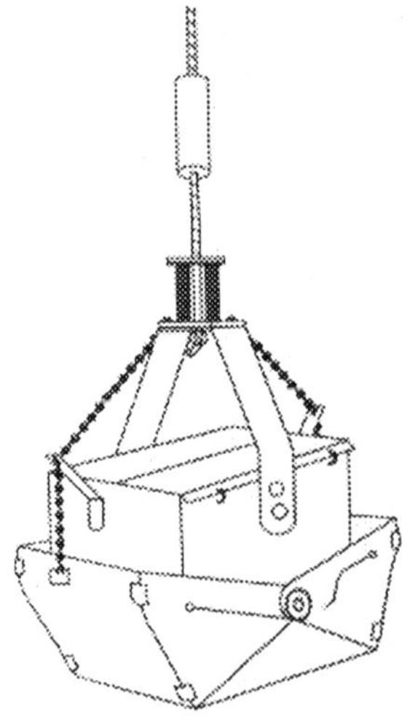

FIGURE 10.8 Eckman bottom grab sampler is used for sand, silt, and mud-type sediments.

10.2.6.2 Sludges

Sludges from water and wastewater treatment plants are checked for microbiological quality to determine the impact of their disposal into receiving waters, ocean dumping, or burial in landfill operations. Microbiological testing of sludges also indicates the effectiveness of the treatment processes.

Collect sediment, mud or sludge samples into sterile whirl pack plastic bag.

10.2.7 Soil Sampling

Selection of the sampling site is based on knowledge of the area and the purposes of the analyses, that is, surface sampling for natural background, surface contamination, or below surface sampling to monitor treatment effect, such as irrigation, or stormwater runoff. The actual sites for sampling and the number of points to be sampled must be predetermined by the survey objectives. Soil sampling has the advantage of permitting the survey planners to lay out a stable grid network for sampling and resampling over a given time period.

If a surface sample is desired, use a sterile scoop or spatula to remove the top surface of one inch or more from a one foot square area. Use a second sterile scoop or spoon to take the sample.

For subsurface soil sampling, dig a hole with a sterile shovel to the desired depth or, as an alternative, use a stainless steel bucket auger. Typical locations of subsamples after sampling by an auger are shown in Figure 10.9.

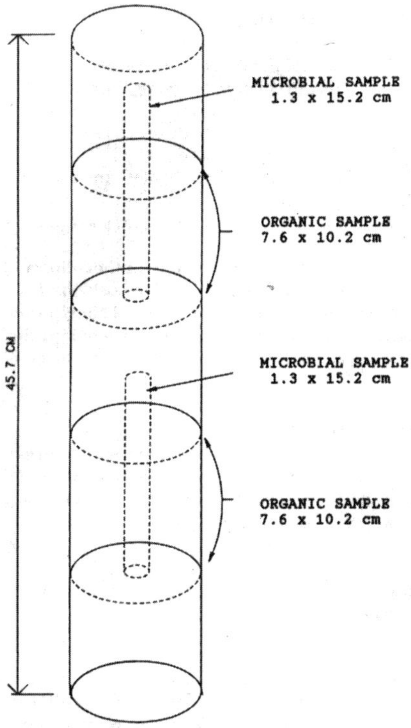

FIGURE 10.9 Typical location of subsamples; successful sampling of subsurface soils requires acquisition of subsurface solids at desired depth. After a subcore for microbiological analysis has been removed from core, a 10 cm (4 in.) length of core material for chemical analysis is obtained. Polyethylene bags, which allow the passage of air, but not water vapor, are good sample containers because the samples have access to air and are kept from drying.

Place sample in a sterile one quart wide mouth screw cap bottle until it is full. Depending on the moisture content of the sample, a one quart container holds 300 to 800 g of soil.

10.3 SAMPLE IDENTIFICATION

Samples should be accompanied by complete and accurate identification. Do not accept inadequately identified samples for examination. Documentation during sample collection and transportation includes the *chain of custody* form (see Table 10.2) and the *sample log* sheet (see Table 10.3) and contains all the information about the field activities. An agency must demonstrate the reliability of its evidence in pollution cases by proving the chain of possession and custody of samples which are offered for evidence or which form the basis for analytical results introduced into evidence. It is imperative that the office and the laboratory prepare written procedures to be followed when samples are collected, transferred, stored, analyzed, or destroyed. The chain of custody record includes the name of the study, the collector's signature, sample location, date and time of the sample collection, type of sample, sequence number, number of containers, and the analyses required. When turning over the possession of the samples, the transferor and transferee signature, date, and time are noted on the sheet.

A field record should be completed on each sample to record the full details on sampling and other pertinent remarks such as flooding, rain, or extreme temperature which are relevant to the interpretation of the results. The field notebook should be specially designed for field work, containing

TABLE 10.2
Chain of Custody

Field I.D. ______________________ Site Name ______________________
Date Sample Received ______________________ Address ______________________
Sampler(s) ______________________ Laboratory ______________________

		Sample Container Description									
Sample Identity	Date Sampled									Total	Remarks
						Total Number of Containers					

Relinquished By: __________ Organization: __________ Received By: __________ Organization: __________
Date: __________ Time: __________ Date: __________ Time: __________
Relinquished By: __________ Organization: __________ Received By: __________ Organization: __________
Date: __________ Time: __________ Date: __________ Time: __________
Delivery Method: ______________________ (attach shipping bills, if any)
Use extra sheets if necessary

TABLE 10.3
Sample Log Sheet

Purpose of Analysis ______________________ Sample Field ID ______________________

Type of Sample ______________ Sampler ______________ Date/Time ______________

Sample Site Number	Sample Source Description	Bottle Type	Bottle No	Preservative	Analysis Required							

Remarks:

*** Field Measurements**

waterproof paper, a hard cover, and entries into all field records should be written in waterproof ink to maintain a permanent legible mark. Any errors in all documents should be deleted by a single line through the incorrect information with the date and initial of the person making the correction.

Specific details on sample identification are entered on a permanent sample label. A sample label will be affixed to all sample containers. It should be waterproof, and all of the information should be written in waterproof ink to maintain a permanent legible mark, and its size should be sufficient for the necessary information. A typical sample label form is shown in Table 10.4. The tag or label should contain the location, date and time of sampling, type of sample, sequence number (first sample of the day—sequence no. 1, second sample—sequence no. 2, etc.). Labels must be securely attached to the sample bottle, but removable, when necessary. Do not accept insufficiently or improperly labeled samples for examination.

10.4 SAMPLE TRANSPORTATION, PRESERVATION AND HOLDING TIME

The adherence to sample preservation and holding time limits is critical to the production of valid data. Samples exceeding these limits should not be analyzed.

10.4.1 Sample Transportation and Preservation

If a sample custodian has not been assigned, the field custodian or field sampler has the responsibility for packaging and dispatching samples to the laboratory for analysis. The transferee must sign

TABLE 10.4
Sample Label

Field sequence no. ______________________________

Field sample No. ____________ Date ____________ Time ____________

Sample location ______________________________

Sample source ______________________________

Preservative used ______________________________

Analyses required ______________________________

____

Collected by ______________________________

Remarks ______________________________

____

____

Final pH checked ______________________________

Additional preservative used (if applicable) ______________________________

____

and record all data and the time of transportation on the chain of custody form. Samples must be carefully packed in shipment containers, such as an ice chest, to avoid breakage. Packages must be accompanied by the chain of custody record showing identification of the contents. The original must be sent with the shipment and a copy is retained by the survey coordinator or project manager.

10.4.2 Laboratory Custody Procedure

The laboratory has to designate a sample custodian and an alternate to act in the absence of the custodian.

In addition, the laboratory should designate a sample storage security area, an isolated room with sufficient refrigerator space or just a separate refrigerator in the laboratory.

Samples should be handled by the smallest possible number of persons. Incoming samples can be received only by the custodian, who will indicate receipt by signing the chain of custody. Immediately upon receipt, the custodian places the samples in the sample room or into the designated refrigerator. The custodian must be sure that microbiological samples are properly stored and maintained at 1 to 4°C, and only the custodian will distribute samples to personnel who are to perform tests.

10.4.3 Holding Time

Start microbiological examination of the sample promptly after collection. If samples cannot be processed within 1 h after collection, use an ice cooler for storage during transportation. Maximum transportation time is 6 h. Refrigerate these samples upon receipt in the laboratory and process within 2 h. Unfortunately, these requirements seldom are realistic in the case of individual potable water samples shipped directly to the laboratory by mail, bus, etc.

The permitted holding time for these samples is 24 h. When refrigeration of individual samples sent by mail is not possible, a thermos-type insulated sample bottle (or equivalent that can be sterilized) may be used.

Record the time and temperature of storage of all samples and consider information in the interpretation of the results.

In situations where it is impossible to meet the 6 h maximum holding time, the use of temporary field laboratories located near the collection site should be considered.

If sampling and transit conditions require more than 6 h, and the use of field laboratories is impossible, the delayed incubation procedure for total and fecal coliforms and fecal streptococci should be considered.

The adherence to sample preservation and holding time limits is critical to the production of valid data. Samples exceeding these limits should not be analyzed.

The analyst records information onto a laboratory notebook or analytical worksheet, describing the sample, the procedures performed, and the results of the testing. The notes should be dated and initialed by the analyst. The notes should be retained in the laboratory as a permanent record.

10.4.4 Discard Samples

Once the sample testing is completed, microbiological samples can be discarded, but identification labels or tags and laboratory records should be returned to the custodian. Tags and laboratory records of tests may be destroyed only upon the order of the laboratory director.

11 Laboratory Equipment and Supplies in the Environmental Microbiology Laboratory

11.1 LABORATORY EQUIPMENT

11.1.1 INCUBATORS

Incubators are constant temperature air chambers or water baths that provide controlled temperature environments. Incubators must maintain a uniform and constant temperature at all times in all areas, that is, they must not vary more than ± 0.5°C in the areas used. Obtain such accuracy by using a water-jacketed or anhydric type incubator with thermostatically controlled low-temperature electric heating units properly insulated and located in or adjacent to the walls or floor of the chamber and preferably equipped with mechanical means of circulating the air. Incubators equipped with high-temperature heating units are unsatisfactory, because such sources of heat, when improperly placed, frequently cause localized overheating and excessive drying of media, with consequent inhibition of bacterial growth. It is desirable where, ordinary room temperatures vary excessively, to keep laboratory incubators in special rooms maintained at a few degrees below the recommended incubator temperature. Alternatively, use special incubating rooms that are well insulated and equipped with properly distributed heating units, forced air circulation, and air exchange ports, provided that they conform to desired temperature limits.

Provide incubators with open metal wire or perforated sheet shelves so spaced as to assure temperature uniformity throughout the chamber. Leave a 2.5 cm space between walls and stacks of dishes or baskets of tubes.

Maintain an accurate thermometer (regularly checked by a thermometer certified by the National Institute of Standards and Technology, NIST) with the bulb immersed in liquid (glycerine, water, or mineral oil) on each shelf in use in the incubator and record daily temperature readings (preferably morning and afternoon).

A water bath with a gabled cover to reduce water and heat loss, or a solid heat sink incubator may be used to maintain a temperature of 44.5 ± 0.2°C. Keep water depth in the incubator sufficient to immerse fermentation tubes to the upper level of the media. Typical laboratory incubator is shown in Figure 11.1.

11.1.2 HOT-AIR STERILIZING OVEN

Hot-air sterilizing ovens are constructed to give a uniform and adequate sterilizing temperature of 170 ± 0.2°C and are equipped with suitable thermometers.

11.1.3 AUTOCLAVES

Autoclaves are used for moist heat sterilization. They are normally operated at 15 lb/in^2 in steam pressure for 15 min, producing a temperature inside the autoclave of 121.6°C (250°F) at sea level.

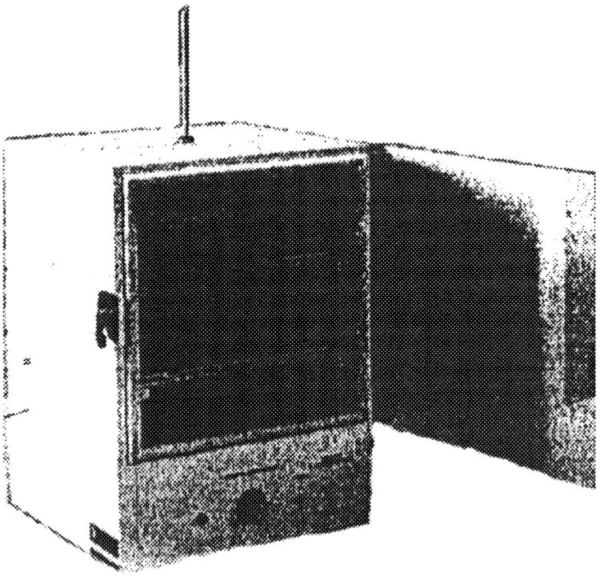

FIGURE 11.1 Typical laboratory incubator.

Steam under pressure provides effective sterilization because it has good penetrating power and coagulates microbial protoplasm.

Use of a pressure cooker is not recommended because of the difficulty in adjusting and maintaining the sterilization temperature and it is a potential hazard. If a pressure cooker is used in emergency or special circumstances, equip it with an efficient pressure gauge and a thermometer where the bulb is 2.5 cm above the water level.

Performance checks of autoclaves and hot air ovens are conducted with the use of spore strips, spore ampuls, indicators, etc.

11.1.4 pH Meter

A pH meter must be accurate to at least 0.1 pH unit to determine the pH of the prepared media. At a given temperature, the intensity of the acidic and basic character of the solution is indicated by pH.

A pH meter functions by measuring the electric potential between two electrodes that are immersed into the solution of interest. The basic principle is to determine the activity of the hydrogen ions by potentiometric measurement using a glass and a reference electrode. The most popular, called a combination electrode, incorporates the glass and the reference electrode into a single probe. The glass electrode is sensitive to the hydrogen ions, and changes its electrical potential with the change of the hydrogen ion concentration. The reference electrode has a constant electric potential. The difference in potential between these electrodes, measured in millivolts (mV), is a linear function of the pH of the solution. The scale of pH meters is designed so that the voltage can be read directly in terms of pH.

pH is the negative logarithm of the hydrogen ion concentration,

$$pH = -\log [H^+] \tag{11.1}$$

The term concentration of the hydrogen ion is written $[H^+]$. The brackets mean mole concentration and H^+ means hydrogen ion. The concentration of the H^+ is expressed in moles per liter.

Water dissociates by a very slight excess into H^+ and OH^- ions. It has been experimentally determined that in pure water

$$[H^+] = 1 \times 10^{-7} \quad (11.2)$$

and

$$[OH^-] = 1 \times 10^{-7} \quad (11.3)$$

therefore,

$$[H^+] = [OH^-] \quad (11.4)$$

The negative logarithm of 1×10^{-7} is 7.00. Therefore, the pH of pure water is 7.00. Such a solution is neutral and has no excess of H^+ or OH^-. An excess of $[H^+]$ is indicated by a pH below 7.00, and the solution is said to be acidic. pH values above 7.00 indicate an excess in $[OH^-]$, and the solution is alkaline.

A glass electrode usually consists of a silver and a silver chloride electrode in contact with dilute aqueous HCl, surrounded by a glass bulb that acts as a conducting membrane. The hydrogen ion concentration of the HCl solution inside the electrode is constant; therefore, the potential of the glass electrode depends on the hydrogen ion concentration outside of the glass membrane.

The reference, also called the calomel electrode, contains elemental mercury (Hg), calomel (Hg_2Cl_2) paste, and Hg metal. This paste is contacted with an aqueous solution of potassium chloride (KCl) solution. KCl serves as a salt bridge between the electrode and the measured solution.

> The electrode must be visually inspected every month. The level of the solution should be checked every day and refilled as needed.

Because of the wide variety of pH meters, detailed operation procedures are given for each instrument. The analyst must be familiar with the operation of the system and with the instrument functions. General rules in pH measurement are given in Appendix F.

11.1.5 Balances

Microbiology laboratories use two types of balances:

Top loading balance—Sensitivity should be at least 0.1 g at a load of 150 g.
Analytical balance—Sensitivity should be 0.0001 g under a load of 10 g. Analytical balance used for weighing small quantities (under 2 g) of material. Single-pan rapid-weigh balances are most convenient.

11.1.6 Optical Counting Equipment

Microbiology laboratories use the following types of optical counting equipment.

Pour and spread plates use the *Quebec-type colony counter*, dark-field model preferred, or one providing magnification of 1.5 diam, and satisfactory visibility.

Membrane filters use a binocular microscope with magnification of 10 to 15×. Provide daylight fluorescent light source at an angle of 60 to 80° above the colonies and use low-angle lighting for nonpigmented colonies.

11.1.7 Refrigerator

Use a refrigerator maintaining a temperature of 1.0 to 4.4°C to store samples, media, reagents, etc. Do not store volatile solvents, food, or beverages in this refrigerator. Frost-free refrigerators may cause excessive media dehydration when storing samples longer than 2 weeks.

11.1.8 Membrane Filtration Equipment

Membrane filtration equipment consists of a filter base and funnels. The filter funnel and membrane holder are made of seamless stainless steel, glass, or autoclavable plastic that does not leak and is not subject to corrosion. These are wrapped in aluminum foil and sterilized before using them. Different membrane filter funnels are seen in Figure 11.2.

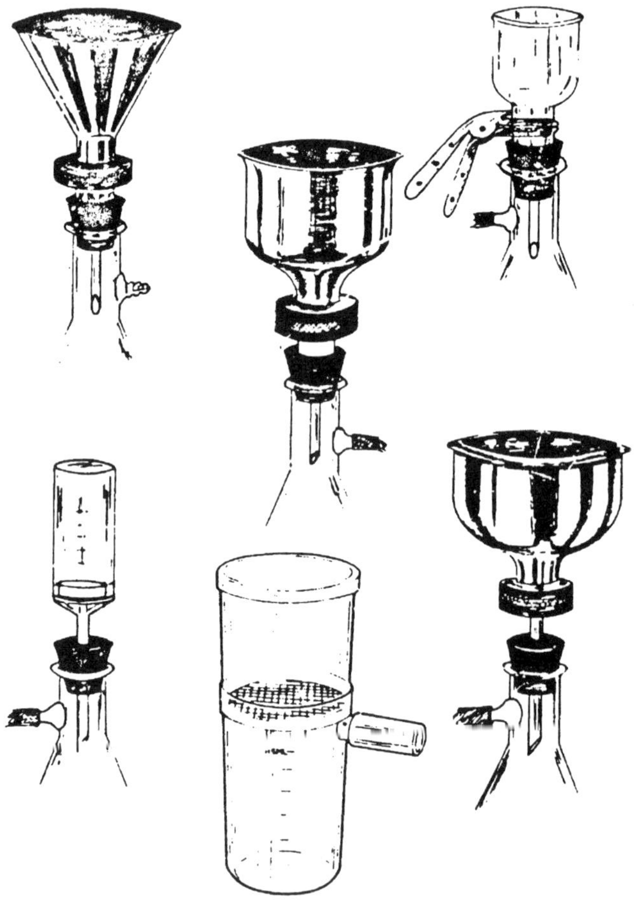

FIGURE 11.2 Membrane filtration funnels made by various manufacturers for detection of bacteria in aqueous solutions.

11.1.9 LINE VACUUM OR ELECTRIC VACUUM PUMP

An aspirator is used as a vacuum source.

11.1.10 INOCULATING NEEDLES OR LOOPS

Needles and loops of nichrome, platinum, or platinum-iridium wire are used to transfer microbes aseptically from one growth medium to another. Use loops at least 3 mm in diameter. The loops are sterilized by heating to redness in a gas flame (see Figure 11.3), or by dry heat or steam.

Single service hardwood applicators also may be used. Make these 0.2 to 0.3 cm in diameter and at least 2.5 cm longer than the fermentation tube. Sterilize and store in glass or other nontoxic containers.

11.1.11 MICROSCOPE

Because microbial objects are very small, modern technology has made available a broad range of instruments for viewing microbial objects. All microscopes operate on the same basic principle: energy is projected toward an object, such as a microorganism. The energy bounces off the object and creates an impression on a sensing device. This device may be a television screen, a photographic film, or the human eye. The image reveals the form, shape, size, and other structural features of the object.

11.1.11.1 Parts of the Microscope

Parts of a microscope include:

1. *Base and arm*—serve to hold in position the essential optical parts. They are wide and sturdy to minimize vibration.

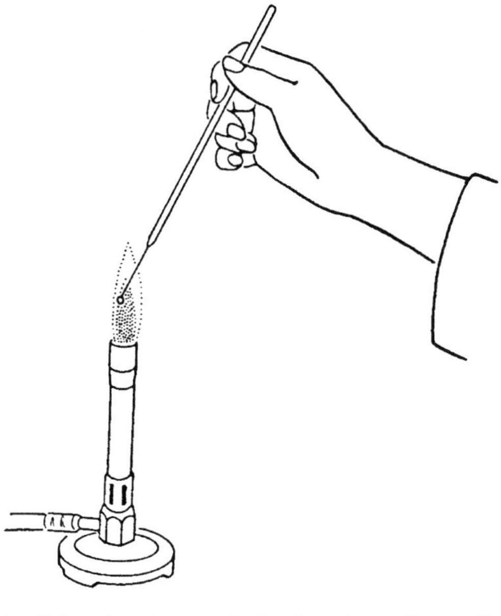

FIGURE 11.3 Sterilizing the wire inoculating loop in the flame of a Bunsen burner.

2. *Inclination joint*—Permit the upper part of the microscope to be tipped back to any degree desired.
3. *Body tube*—The body tube is the part to which the principal lenses are attached.
4. *Nose piece with the objectives*—Nose pieces limit the size of the image we see and they are also largely responsible for the quality of the image. Most microscopes are equipped with three objectives of different magnifying powers.
5. *Ocular or eyepiece*—The eyepiece includes a short tube with two lenses which fit into the upper end of the body tube. The function is to magnify the image of the object formed by the objective. The total magnification of the system is determined by multiplying the magnification of the objective by the magnification of the ocular or the eyepiece as discussed in the next section.
6. *Mirror*—The mirror collects and reflects light up into the microscope. Before the light reaches the object on the stage, it is condensed and focused by passage through the large condensing lens of the substage.
7. *Iris diaphragm*—Often a subject is too brilliantly illuminated if all the light from the mirror passes into the condenser. The size of the opening in the diaphragm may be reduced to that of a pinhead or to any intermediate size by moving a hand level. The amount of light admitted to the condenser can be accurately controlled.
8. *Stage*—The place where an object is placed.
9. *Coarse and fine focus adjustor*—With the course adjustor, the stage with the condenser move up and down. With the fine adjustment, the tube and the stage are raised and lowered by very slight degrees.

Parts of a compound microscope are illustrated in Figure 11.4.

11.1.11.2 Light Microscope

Light microscopes use light rays that are magnified and focused by means of lenses. It is also called the bright-field microscope because visible light passes directly through its lenses until it reaches the eye.

1. *Dissecting microscope or stereo microscope*—is designed to study objects in three dimensions at low magnification.
2. *Compound light microscope*—it has a two lens system with the objective lens nearer the object and the ocular lens nearer the eye. It is used for the examination of small or thinly sliced cross sections or longitudinal sections of subjects under magnification higher than that of dissecting microscopes. Illumination is from below, and the light passes through clear sections, but does not pass through opaque sections. To improve contrast, the microscopist uses stains or dyes that bind to cellular structure.

A light microscope usually has three objective lenses, the low-power, high-power, and oil immersion lenses. Generally these lenses magnify an object 10× (low power), 40× (high power), and 100× times (oil immersion lens), respectively.

The *magnification* is represented by the multiplication symbol "×." Total magnifications are determined by multiplying the objective magnification by the ocular magnification. For example, 10× ocular and 10× objective give a magnification of 100× or 10× ocular and 40× objective give 400× magnification. The actual lens magnification depends on the type of the microscope and the manufacturer.

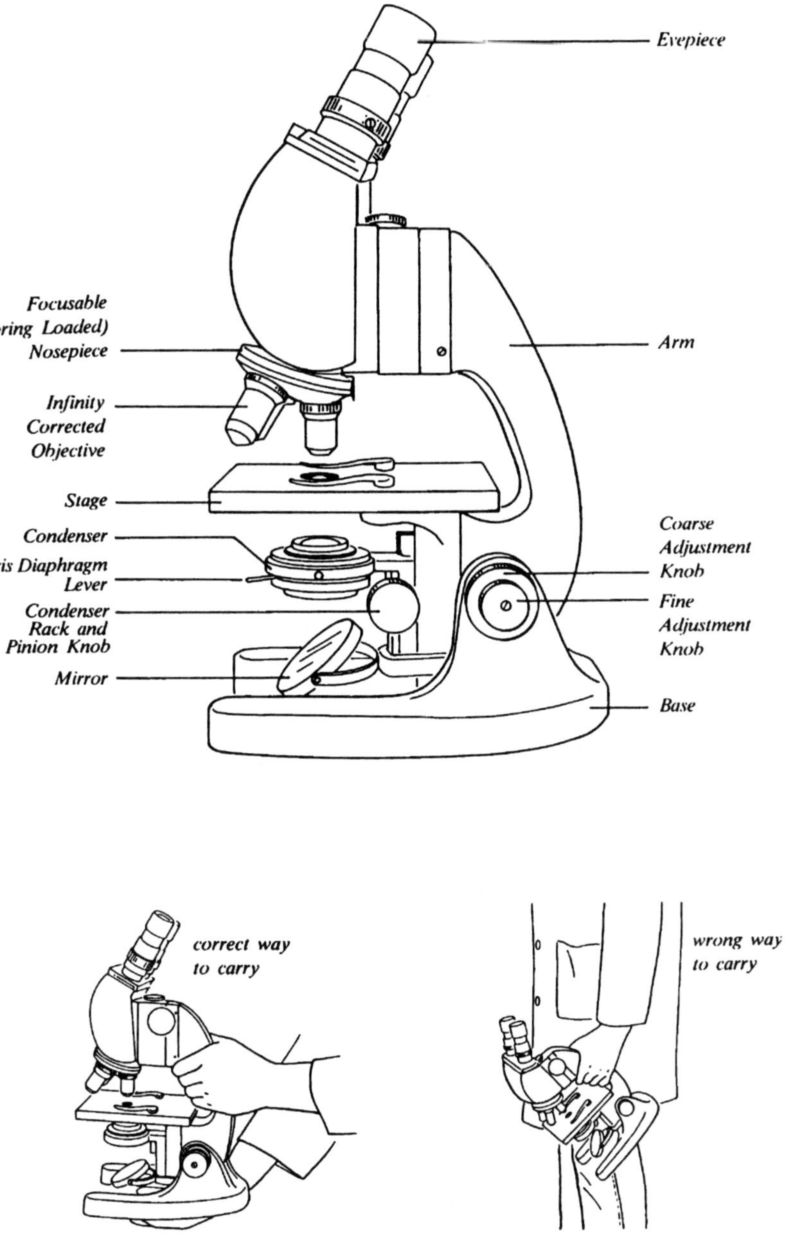

FIGURE 11.4 Parts of a compound light microscope.

For an object to be seen distinctly, the lens system must have good *resolving power* (RP), that is, it must transmit light without variation and allow closely spaced objects to be clearly distinguished. For example, a car seen in the distance at night may appear to have a single headlight because the eyes lack resolving power. However, when the car comes closer, the two headlights can be seen clearly as the resolving power of the eye increases.

Numerical aperture (NA) is the mathematical expression for the size of the beam of light that an objective can utilize. The maximum NA of the low- and high-power objectives is always less than 1.0, and for most oil immersion, the objective is 1.25.

$$RP = \frac{\lambda}{2 \times NA} \tag{11.5}$$

In this formula, the greek letter λ (lambda) represents the wavelength of light and is usually set at 550 nm (nanometer), the half way point between the limits of visible light.

For example, for a low-power objective with an NA 0.25, the resolving power (RP) is calculated as:

$$RP = \frac{550}{2 \times 0.25} = \frac{550}{0.5} = 1100\ \text{nm} \quad \text{or} \quad 1.1\ \mu\text{m}$$

It means, any object smaller than 1.1 μm could not be seen, but an object larger than 1.1 μm would be visible. For an oil immersion lens, the NA is 1.25, therefore,

$$RP = \frac{550}{2 \times 1.25} = \frac{550}{2.5} = 220\ \text{nm} \quad \text{or} \quad 0.22\ \mu\text{m}$$

In this situation, objects as small as 0.22 μm may be visualized.

Oil is needed for *oil-immersion microscopy*, because light bends abruptly as it leaves the glass slide and enters the air. Immersion oil has a refraction index of 1.5, which is identical to the refraction index of the glass. The refraction index is a measure of the light-bending ability of a medium. The refraction index of air is 1.0, which accounts for the abrupt bending as light enters it. The oil, thus, provides a homogeneous pathway for light from the slide to the objective, and the resolution of the object increases.

11.1.11.3 Focusing the Microscope

11.1.11.3.1 Low power (4X or 10X)

For low power microscopes:

1. With the coarse adjustment knob, raise the nose-piece until it stops.
2. Place the slide on the stage.
3. Make sure the lowest objective is in place. Then, as you look from the side, decrease the distance between the stage and the nose-piece until the nose-piece comes to an automatic stop or is no closer than 1/8 in. above the slide.
4. Looking into the ocular, rotate the diaphragm to give the maximum amount of light.
5. Slowly increase the distance between the stage and the nose-piece, turn the course adjustment knob until the object comes into view or focus.
6. Once the object is seen, it may be necessary to adjust the amount of light. To increase or decrease contrast, rotate the diaphragm slowly.
7. Use the fine adjustment knob to sharpen focus, if necessary.

11.1.11.3.2 Higher power

Compound light microscopes are parfocal, so once the object comes into focus with the lowest power, should be almost in focus with a higher power.

1. Bring the object into focus under low power by following the preceding instructions.
2. Bring the high power objective into place by turning the nose-piece.
3. If any adjustment is needed, use only the fine adjustment knob.
4. After the adjustment, rotate the nose-piece until the lowest-power objective clicks in place and then remove the slide from the stage.

11.1.11.4 Rules for Using Microscopes

Follow the following rules for using microscopes:

1. Have both eyes open when looking through the eyepiece.
2. Use only lens paper (not a paper towel or tissue paper) for cleaning the lenses.
3. Always remove the slide and clean the lenses when finished.
4. Clean stage and condenser if oily, keep clean to prevent rust and corrosion.
5. Always raise the objective (or lower the stage) while focusing so that the specimen and lens move apart from each other.
6. Use two hands when carrying the microscope. Grasp the microscope arm with one hand and place the other hand under the base. Do not tilt the microscope when carrying it.
7. Do not force parts.
8. Do not remove parts.
9. Store the microscope with the low-power objective down and there should be no oil on lenses or stage.
10. Keep the microscope dust-free by covering it during storage.

11.1.11.5 Dark Field Microscopy

In dark-field microscopy, the background remains dark and only the object is illuminated.

11.1.11.6 Phase-Contrast Microscopy

In phase contrast microscopy, small differences in the densities of objects show up as different degrees of brightness and contrast. The Dutch scientist *Fritz Zernicke* received the 1935 Nobel Prize in Physics for his development of this system. The same organism seen under three different microscopes: light, dark field, and phase contrast microscopes are shown in Figure 11.5.

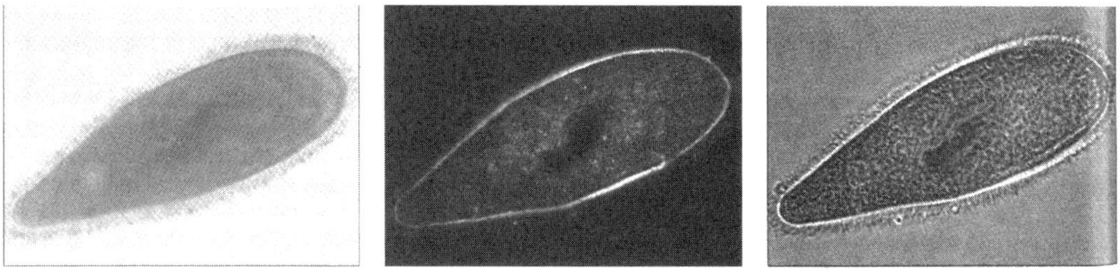

(a) (b) (c)

FIGURE 11.5 The same organism as seen under different microscopes. a. Brightfield illumination shows internal structures and the outline of the transparent sheath. b. Against the dark background seen with darkfield microscopy, edges of the cell are bright, some internal structures seem to sparkle, and the sheath is almost visible. c. Phase contrast microscopy shows greater differentiation among internal structures and also shows the sheath wall. The wide light band around the cell is an artifact resulting from this type of microscopy—a phase halo.

11.1.11.7 Fluorescent Microscopy

In fluorescent microscopy, microorganisms are coated with a fluorescent dye, such as fluorescein, and are illuminated with ultraviolet light energy. This energy excites electrons on the dye, and they move to higher energy levels. However, they quickly drop back to their original energy levels and give off the excess energy as visible light. The coated microorganisms appear to fluoresce.

11.1.11.8 Electron Microscopy

The electron microscopic technique uses a beam of electrons that is magnified and focused on a photographic plate by means of electromagnets. The key to electron microscopy is the extraordinary short wavelength of the beam of electrons. Measured at 0.005 nm (as compared to 550 nm visible light), the short wavelength dramatically increases the resolving power of the system and makes possible the visualization of viruses, fine cellular structures, and large molecules such as DNA. The electron microscope grew out of an engineering design made up in 1933 by the German physicist *Ernst Ruska* (1986 Noble Prize winner in physics).

1. *Scanning electron microscope*—Analogous to the dissecting microscope, the electron beam sweeps across the surface of an object and gives an image of the surface as is.
2. *Transmission electron microscope*—Analogous to the compound light microscope, the object is thinly sliced with a diamond knife and treated with heavy metal (such as gold or palladium) to make certain parts dense and improve contrasts. It achieves a total magnification over 20 million times. Objects as small as 2.0 nanometers can be seen. Table 11.1 shows the comparison of the various types of microscopes.

11.2 LABORATORY GLASSWARE

11.2.1 Petri Dishes

11.2.1.1 Petri Dishes with Tight Fitting Lids

50 × 12 mm plastic dishes with tight fitting lids are preferred for MF (membrane filter) procedures because they retain humidity (see Figure 11.6).

11.2.1.2 Petri Dishes with Loose Fitting Lids

60 × 15 mm with loose fitting lids can be used in incubators with controlled humidity or in plastic boxes with tight covers containing moist towels (see Figure 11.7).

11.2.2 Pipets

Pipets are designated for the transfer of known volumes of liquid from one container to another. Pipets that deliver a fixed volume are called volumetric or transfer pipets.

Other pipets, known as measuring pipets, are calibrated in convenient units so that any volume up to maximum capacity can be delivered.

Volumetric pipets have a single calibration line. They deliver even volumes from 0.5 to 100 ml as labeled. Volumetric pipets labeled as Class A have a certain time imprinted near the top, which is the length of time necessary to wait before terminating the delivery of the liquid. The correct use of a volumetric pipet is outlined in Figure 11.8.

TABLE 11.1
Summary of Various Types of Microscopes

Microscope Type	Distinguishing Feature	Principal Use
Light		
Brightfield	Uses visible light, cannot resolve structures smaller than about 0.2 μm. Specimen appears against a dark background.	Observe various stained specimens, and to count microbes; inexpensive, simple to use.
Darkfield	Uses special condenser that blocks light from entering the objective directly; light reflects off specimen.	Bright specimen on dark background; observation of living or difficult to stain organisms.
Phase contrast	Uses special condenser and diffraction plate to diffract light rays so that they are out of phase with one another.	Observation of internal structure of living organisms; specimen has different degrees of brightness and darkness.
Fluorescence	Uses an ultraviolet source of illumination that causes fluorescent compounds in a specimen to emit light.	Diagnostic tool for detection of organisms or antibodies in clinical specimens or immunologic studies.
Electron		
Transmission	Uses a beam of electrons instead of light; structures smaller than 0.2 μm can be resolved; the image is two dimensional.	Used to examine viruses or the internal ultrastructure in thin sections of cells; Magnification is 10,000× to 100,000×.
Scanning	Same as in transmission, but the produced image is three dimensional.	Used to study the surface features of cells and viruses; usual magnification is 1000× to 10,000×.

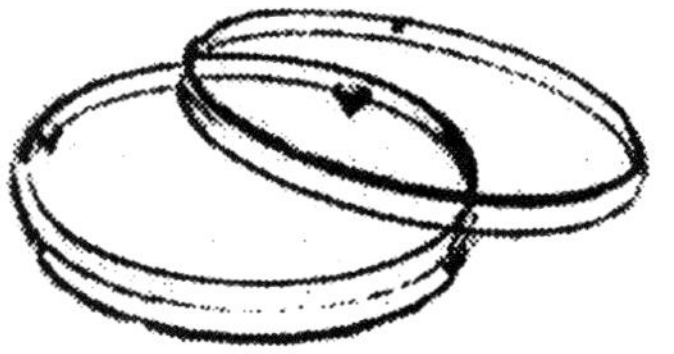

FIGURE 11.6 Tight fitting petri dish lids: 50 × 12 mm plastic with tight fitting lids are preferred for membrane filter (MF) procedure because they retain humidity.

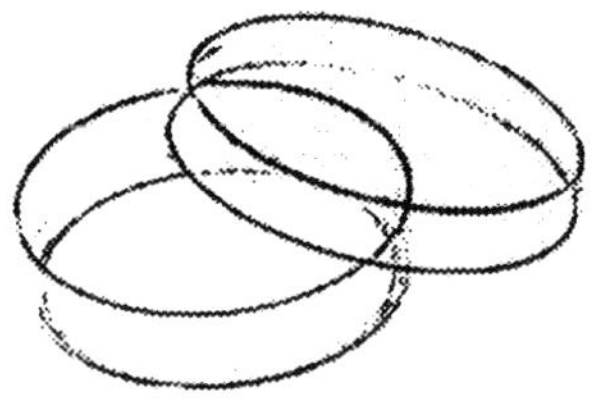

FIGURE 11.7 Loose fitting petri dish lids: 60 × 15 mm plastic with loose fitting lids can be used in incubators with controlled humidity or in plastic boxes with tight covers containing moist towels.

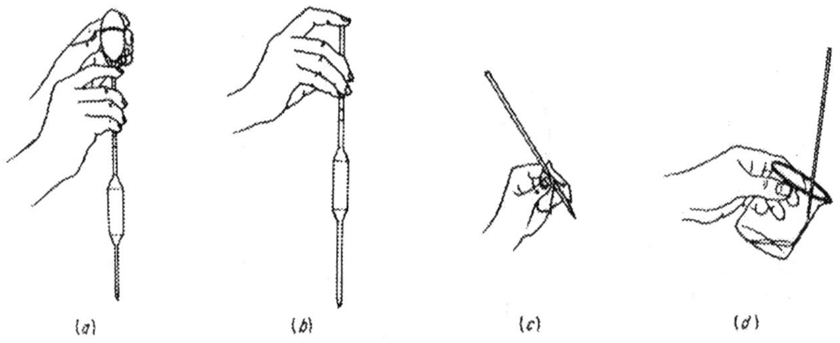

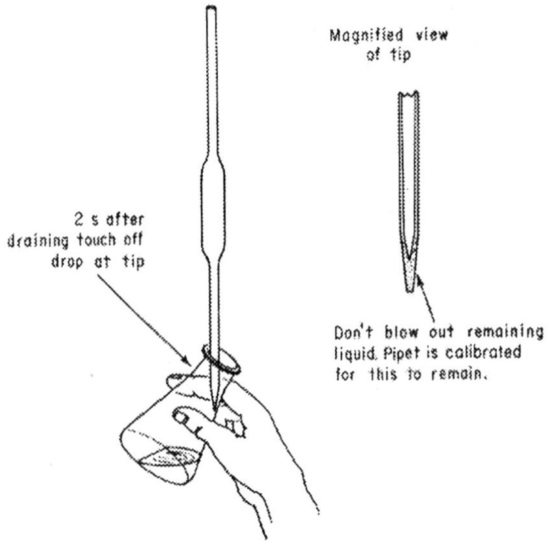

FIGURE 11.8 Technique for using a volumetric pipet: a. Draw the liquid up in the pipet using a pipet bulb until past the graduation mark. b. Use forefinger to maintain liquid level above the graduation mark. Release pressure on the index finger to allow the meniscus to approach. c. Tilt pipet slightly and wipe away any drops on the outside surface. d. Allow pipet to drain freely. Remove the last drop by touching the wall of the container. Do not blow out pipet.

Measuring pipets have graduation lines. They are two types, the *Mohr pipet* and the *serological pipet*. The difference is whether or not the calibration lines stop short of the tip (Mohr pipet) or go all the way to the tip (serological pipet). The serological pipet is better in the sense that the meniscus need be read only once, because the solution can be allowed to drain completely down.

Disposable pipets are most often serological the blow-out type. Disposable pipets are less expensive, because the calibration lines are not permanently affixed to the outside wall of the pipet.

Pipets marked on the neck of the pipet as *TC (to contain)* are calibrated to hold an exact amount, specified by calibration. The pipet must be completely emptied to provide the stated volume.

Pipets marked on the neck of the pipet as *TD (to deliver)* are designed to release the exact calibrated amount when the pipet tip is held vertically against the receiving vessel wall until draining stops.

Blow-out pipets are for rapid use. Pipets deliver the calibrated amount when they are completely emptied. Blow-out pipets are marked with a double band etching on the neck of the pipets. Different type of pipets are shown in Figure 11.9.

For microbiological analysis, only bacteriological, serological, or Mohr pipets are recommended for use in the TD (to deliver) mode.

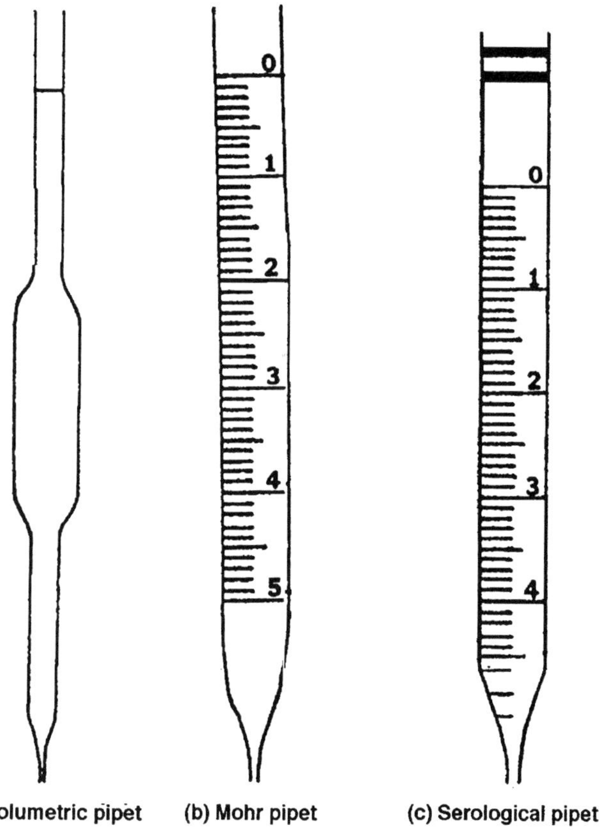

FIGURE 11.9 Different types of pipets. a. Volumetric pipet delivers a fixed volume. Like a volumetric flask, it has a single calibration line. Volumetric pipets are not calibrated to blow out. b. and c. Measuring pipets have graduation lines much like a buret and are used for measuring volumes of solutions more accurately than could be done with graduated cylinders. There are two types of measuring pipets—the Mohr pipet b and the serological pipet c. The difference is whether or not the calibration lines stop short of the tip (Mohr pipet) or go all the way to the tip (serological pipet). The serological pipet is better in the sense that the meniscus need to be read only once, since the solution can be allowed to drain completely out. A double or single frosted ring circumscribing the top of the pipet indicates the pipet is calibrated to blow out. Disposable pipets are most often of the serological blow-out type.

Use *sterile pipets* of any convenient size, provided that they deliver the required volume accurately and quickly. Use pipets having graduations distinctly marked and with unbroken tips. The error of calibration for a given manufacturer's lot must not exceed 2.5 percent.

For extremely small volumes, the so-called *pipettors* are used in various designs. These typically handle volumes from 1 μl to 500 μl.

Remember: Always use a rubber pipet bulb to fill the pipet; do not use the mouth for suction.

11.2.3 Graduated Cylinders

A graduated cylinder is a volumetric glassware designed to measure volumes greater than 10 ml. Graduated cylinders are calibrated to deliver (TD). Glassware designed TD will do so with accuracy only when the inner surface is so scrupulously clean that water wets it immediately and forms a uniform film upon emptying. The calibration must obviously take this thin film into account in the sense that it will not be a part of the delivered volume.

When a liquid is confined in a narrow tube, the surface is found to exhibit a marked curvature, called the *meniscus*. It is a common practice to use the bottom of the meniscus in calibrating and using volumetric glassware. A useful technique for reading the meniscus is shown in Figure 11.10. Location of the eye in reading any graduated tube is important. The eye must be level with the meniscus of the liquid to eliminate parallax errors, as shown in Figure 11.11.

11.2.4 Vacuum Filter Flask

A vacuum filter flask is usually a 1 l flask with appropriate tubing. Filter manifolds to hold a number of filter bases are optional. Vacuum filtration is shown in Figure 11.12.

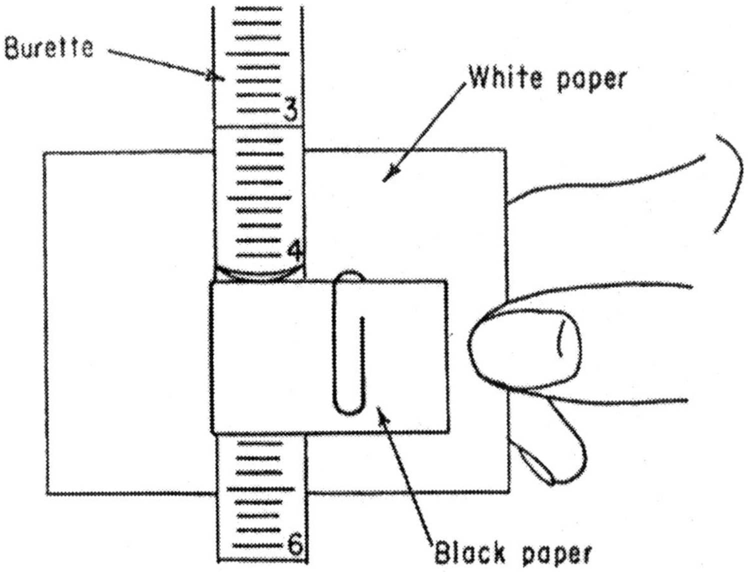

FIGURE 11.10 Useful technique for reading the meniscus render the bottom of the meniscus, which is transparent, more distinct by positioning a black-stripped white card behind the glass.

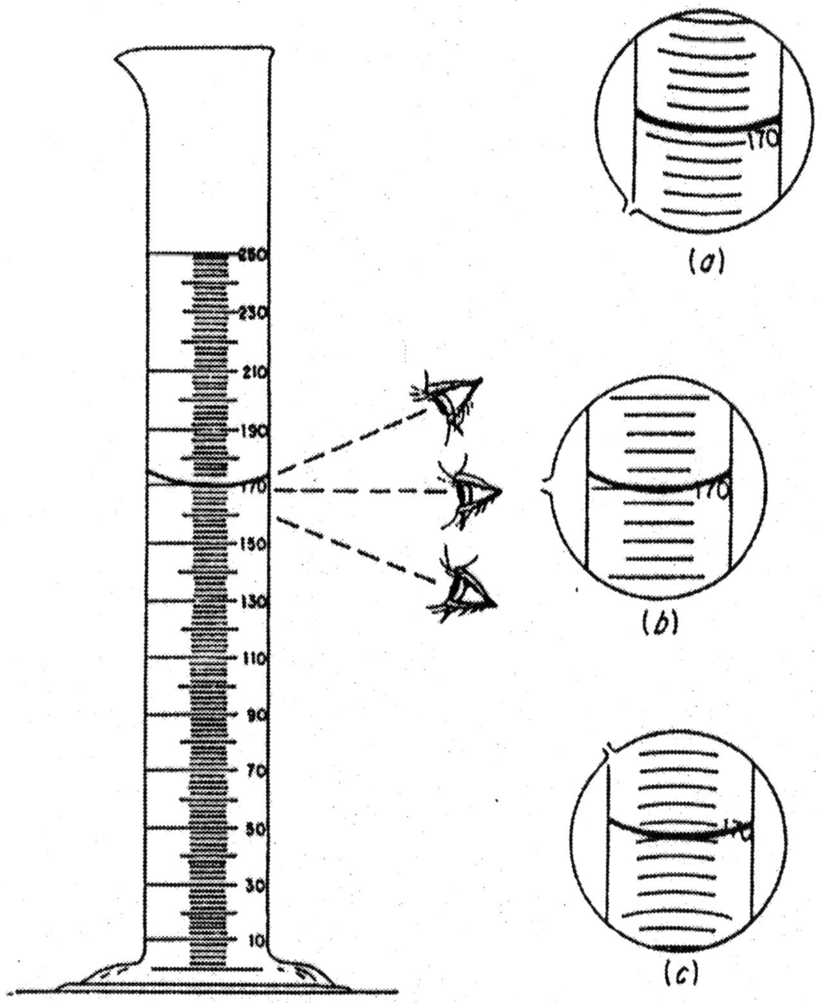

FIGURE 11.11 Avoiding parallax error in reading the meniscus. a. Eye level too high, volume too high. b. Eye level correct, volume correct. c. Eye level too low, volume too low. The eye must be level with the meniscus of the liquid to eliminate parallax errors.

11.2.5 Safety Trap Flask

The safety trap flask is placed between the filter flask and the vacuum source in the vaccum filtration apparatus as shown in Figure 11.12.

11.2.6 Dilution (Milk Dilution) Bottles

Use bottles of resistant glass, preferably borosilicate glass, closed with glass stoppers or screw caps equipped with liners that do not produce toxic or bacteriostatic compounds on sterilization. Mark graduation levels indelibly on side of the dilution bottle. Plastic bottles of nontoxic material and acceptable size may be substituted for glass provided that they can be sterilized properly. A milk dilution bottle is shown in Figure 11.13.

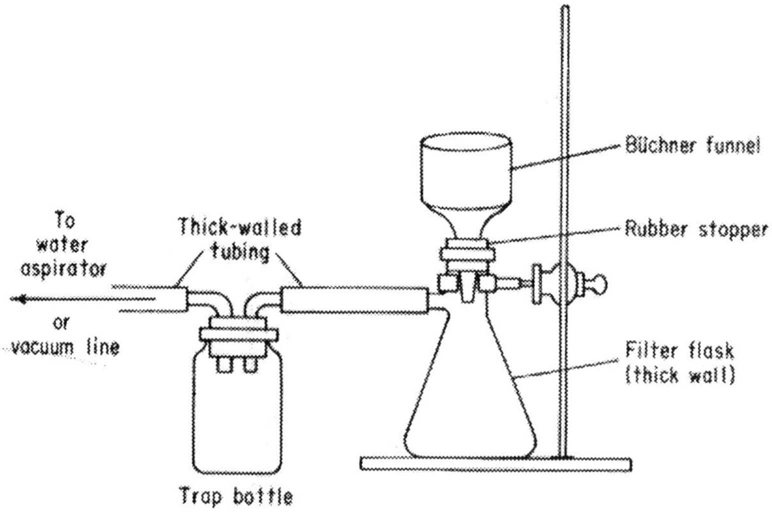

FIGURE 11.12 Vacuum filtration.

11.2.7 Fermentation Tubes and Vials

The fermentation tubes should be large enough to contain the media and inocula in no more than half of the tube depth.

Where tubes are used for a test of gas production, enclose a small vial, inverted. Use tube and vial of such size that the vial will be filled completely with the medium, at least partly submerged in the tube, and large enough to make gas bubbles easily visible.

11.2.8 Thermometers

Use glass or metal thermometers graduated to 0.5°C to monitor most incubators and refrigerators. Use thermometers graduated to 0.1°C for incubators operated above 40°C. Verify and document the accuracy of the thermometer by regular comparison with a NIST-certified thermometer.

FIGURE 11.13 Milk dilution bottle.

11.2.9 Cleaning Laboratory Glassware

In microbiology, clean glassware is crucial to ensure valid results. Previously used, and new glassware must be thoroughly cleaned with a phosphate-free detergent and hot water, then rinsed several times with hot water to remove all traces of residual washing compound, and finally rinsed with laboratory pure water. Traces of some cleaning solutions are difficult to remove completely. Before using a batch of clean glassware test several pieces for alkaline or acid residue by the addition of *bromthymol-blue indicator*. It should show color change of yellow to blue in the pH range of 6.5 to 7.3.

Preparation of bromthymol-blue indicator:
0.1 g of bromthymol-blue indicator.
16 ml of 0.01 *N* NaOH (0.4 g/l NaOH).
Dilute to 250 ml with laboratory pure water.

11.2.10 Glassware Sterilization

Sterilize glassware, except when in metal containers, for not less than 60 min at a temperature of 170°C. Glassware in metal containers heat to 170°C for not less than 2 h.

11.3 CHEMICALS AND REAGENTS

11.3.1 Chemicals and Reagents

See Section 9.4, Quality Control of Laboratory Supplies and Section 9.4.5, Chemicals and Reagents.

11.3.2 Laboratory Pure Water

See Section 9.5, Quality Control of Laboratory Pure Water.

12 Culture Media

12.1 CULTURE MEDIA AND CULTURE

Just as humans do, microorganisms need to obtain nutrients to live and grow as discussed previously in Chapter 4. A nutrient material prepared for the growth of microorganisms in a laboratory is called culture medium (plural: media). The microbes that grow and multiply in or on a culture medium are referred to as a culture. The medium must contain the right nutrients for the particular microorganism we want to grow. It should also contain sufficient moisture, oxygen, and a properly adjusted pH. The medium must initially be sterile, so that the culture will contain only the microorganisms we add to the medium. Finally, the growing culture should be incubated at the proper temperature.

Culture media are used for the isolation and identification of bacteria that are of interest in such areas as food, water, and clinical microbiology. A wide variety of media are available for the growth of microorganisms. Most of these are available from commercial sources, have premixed components, and require only the addition of water and need to be sterilized.

When a medium is in liquid form, it is called *broth.*

When it is desirable to grow bacteria on a solid medium, a solidifying agent such as agar is added to the medium. Gelatin was the agent used initially, but obviously it could only be used at low temperatures. Agar is a complex polysaccharide derived from the marine alga, *Gelidium,* and it has long been used as a thickener in foods such as jellies, soups, and ice creams. Agar has some properties that make it valuable for microbiology. While many bacteria have the capability to break down gelatin and, thus, leave a useless liquid, very few microorganisms can break agar, so it remains a solid. Also important is the fact that agar melts at about the boiling point of the water but remains liquid until the temperature drops to 40°C.

Agar media are usually contained in test tubes or petri dishes. The test tubes are called *slants* when they are allowed to solidify with the tube held at an angle so a large surface for growth is available.

When the agar solidify in a petri dish, it is called the *petri plate.*

12.1.1 Chemically Defined Media

A chemically defined medium is one whose exact chemical composition is known.

12.1.2 Complex Media

Most heterotrophic bacteria and fungi are grown on complex media. A complex medium is one in which the exact chemical composition is not known. These types of media are made up of nutrients such as extracts from yeast, beef, or plants, or digests of proteins from these or other sources. The energy, carbon, nitrogen, and sulfur requirements of the growing microbe are met largely by protein. The vitamins and other organic growth factors are provided by meat and yeast extracts. If this type of medium is in liquid form, it is called a nutrient broth. When agar is added, it is called nutrient agar.

12.1.3 Selective and Differential Media

12.1.3.1 Selective Media

Selective media are designed to suppress the growth of unwanted bacteria and to encourage the growth of the desired microbes. For example, bismuth sulfite agar is used to isolate the Gram-negative *Salmonella typhi* from feces. *Bismuth sulfite* inhibits Gram-positive bacteria and numerous Gram-negative intestinal bacteria as well.

Dyes such as *Brilliant green* selectively inhibit Gram-positive bacteria and this dye is the basis of a medium called Brilliant green agar that is used to isolate the Gram-negative salmonella.

12.1.3.2 Differential Media

Differential media make it easier to distinguish colonies of the desired organisms from other colonies growing on the same plate. Sometimes selective and differential media are used together. For example, *Staphylococcus aureus* bacteria has the ability to tolerate high concentrations of sodium chloride; another characteristic is its ability to ferment the carbohydrate mannitol to form an acid. Mannitol salt agar medium contains 7.5 percent sodium chloride and also contains a pH indicator that changes its color if the mannitol is fermented to acid. Bacteria that grow with high salt concentration and ferment mannitol by acid production can be readily identified by color change.

12.1.4 Storage of Dehydrated Culture Media

The preparation of culture media is a critical aspect of water quality testing. Commercially available dehydrated media are used and require weighing and dissolving of the powder in laboratory pure water.

Commercially prepared media in liquid form (sterile ampul or other) also may be used if known to give equivalent results.

Store dehydrated media (powders) in tightly closed bottles in the dark at less than 30°C in an atmosphere of low humidity. Do not use them if they discolor or become caked and lose the character of a free-flowing powder. Purchase dehydrated media in small quantities that will be used within six months after opening. Additionally, use stocks of dehydrated media containing selective agents such as sodium azide, bile salts or derivatives, antibiotics, sulfur containing amino acids, etc., of relatively current lot number (within a year of purchase) so as to maintain optimum selectivity. Record kind, amount, appearance of media received, lot number, expiration date, date received, and opened. Check inventory quarterly for recording. Discard media that are caked, discolored, or show other deterioration. A conservative limit for unopened bottles is two years at room temperature. Use open bottles of media within six months after opening. Complete listing of the quality requirements of the dehydrated media are detailed in Section 9.4.8.

12.1.5 Preparation of Media

To prepare the media, follow the instructions on the bottle. Care must be taken to completely dissolve and mix the ingredients before dispensing the medium into tubes, flasks, or dishes. If heat is necessary, apply with caution. Direct heat, boiling water bath, and flowing steam are used selectively. Agar and large volumes of broth require direct heating to the first bubble of boil. Such heating must be applied with stirring and constant attention until agar is dissolved. Preferably use plate-magnetic stirrer combinations. Prepare all media in deionized or distilled water of proven quality, as specified in Section 9.5. Media must be dissolved before autoclaving to ensure timing for complete sterilization.

A freshly prepared medium contains numerous microorganisms found in the ingredients, the water used to prepare it, and from the utensil surfaces and glassware. Therefore, it must be sterilized. Prior to sterilization, the container is usually plugged with cotton or loosely capped. This prevents the entry of new contaminants but permits free interchange of air or gases.

12.1.5.1 Preparation and Sterilization of Media

The following procedures should be followed during the preparation and sterilization of media.

1. To avoid boiling over during heating, a vessel holding a liquid solution to be sterilized in an autoclave should never be filled more than two-thirds full. Prepare media in containers that are at least twice the volume of the medium being prepared.
2. The proper amount of powder is weighed on nonabsorbent paper or weighing dish and poured into the container in which it is to be prepared.
3. Deionized or distilled water is added with vigorous swirling to achieve an even suspension. If other liquid ingredients are to be added, they should be incorporated at this point.
4. If the solution is clear, as most broth usually are, it requires no further manipulation before autoclaving. However, most agar solutions require heating almost to the boiling point, with constant agitation, to achieve an even solution. The use of a stirrer-hot plate and a magnetic stirring-bar will greatly increase the efficiency of this stage of media making. The hot solution must be watched extremely carefully as soon as tiny bubbles begin to appear because these media tend to boil over very easily. Some media are ready to dispense at this point.
5. If the medium is to be sterilized in an autoclave, it is capped with either a plastic screw cap or a plug. A good plug can be made from a large wad of nonabsorbable cotton wrapped in a square of gauze one layer thick.
6. For sterilization of flasks of agar that are to be poured into plates by hand, a large square of aluminum foil molded the same way as the shape of the flask is an excellent cover that can be removed and replaced many times. This cover also allows the flask to remain sterile until the foil is lifted off, see Figure 12.1. The solutions are placed in the autoclave and sterilized; the timing of the sterilization should start from the moment the temperature reaches 121°C. Very large quantities of media may require a longer sterilization time than is recommended on the package label. Once the sterilization cycle is completed, the autoclave chamber is slowly returned to atmospheric pressure to prevent the liquid from bubbling over.
7. Sterilized media should not be kept in the autoclave once pressure has equalized, because prolonged heat may alter some of the ingredients. Sterilization in an autoclave can be dangerous; all the safety precautions outlined in Section 8.3.3 and the quality control practices outlined in Section 9.3.9 should be carefully adhered to by operators.
8. Media removed from the autoclave should be placed into a 55°C water bath before plates are poured.
9. Liquid media or media that are not to be dispensed may be allowed to cool on the bench top.
10. Agar will tend to settle to the bottom of the flask during sterilization, so all flasks should be swirled in a large circle on the bench top (to avoid making bubbles) before pouring.
11. Carbohydrate solutions and other liquids that may be denatured by heat can be filter sterilized by injecting the liquid through a syringe attached to a membrane filter with pores no larger than 0.2 or 0.45μ in diameter as shown in Figure 12.2.

FIGURE 12.1 Gauze covered nonabsorbable cotton plug and aluminum foil cover for flasks used in media preparation.

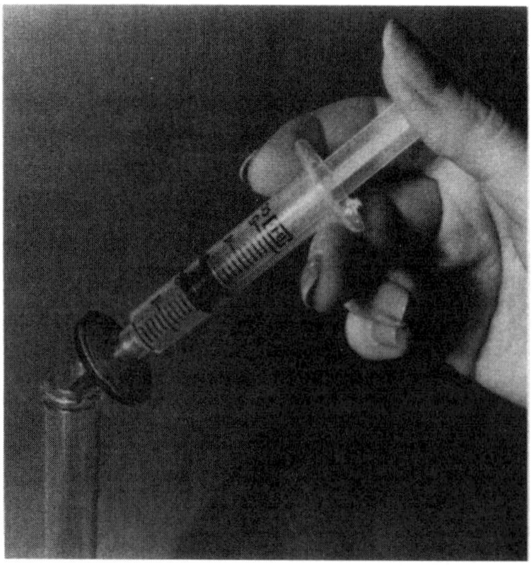

FIGURE 12.2 Use of a membrane filter system to sterilize solutions in an autoclave.

12.1.6 Sterilization of the Media

After rehydrating the medium, dispense promptly to culture vessels and sterilize within 2 h. Do not store nonsterile media. Sterilize all media, except sugar broths or broths with other specifications, in an autoclave at 121°C for 15 min after the temperature reaches the 121°C. As soon as the autoclave pressure has fallen to zero, the sterilized media should be removed from the autoclave for cooling before use or storage. To permit uniform heating and quick cooling, pack materials loosely in small containers. Never reautoclave media.

Check effectiveness of the sterilization with each run by using *Bacillus stearothermophilus* spore suspension or strips (commercially available) inside glassware. Sterilization at 121°C for 15 min kills the spores. If growth of the autoclaved spores occurs after incubation at 55°C, sterilization is inadequate.

Sterilize nonautoclavable media by filtration through a 0.22μ pore size filter in a sterile filtration and sterile receiving apparatus as mentioned previously. Filter and dispense medium in a safety cabinet or biohazard hood, if available.

12.1.7 pH Check of the Media

Check pH of a portion of each medium after sterilization and cooling. Check pH of solid medium with surface probe. Record results. Make minor adjustments in pH (less than 0.5 pH units) with NaOH or HCl solution to the pH specified in formulation. If the pH difference is larger than 0.5 units, discard the batch and resolve the problem. Incorrect pH values may indicate a problem with reagent water quality, medium deterioration, or improper preparation. Review instruction for preparation and check deionized water pH. If the pH of the water is unsatisfactory, prepare a new batch of medium using water from a new source. If the reagent grade water is satisfactory, prepare the medium from another bottle of medium and discard the failed one. Record pH measurement and problem in the media record book and report to the manufacturer if the medium is indicated as the source of error.

Examine prepared media for unusual color, darkening, or precipitation, and record observations. If any of the preceding problems occur, discard the medium.

12.1.8 Storage of Culture Media

Prepare culture media in batches that will be used in less than one week. Holding time prepared media are listed in Table 12.1. If fermentation tube media are refrigerated, incubate overnight before use and check for false air bubbles. Prepare media that are to be stored for more than one week in screw-capped or tightly capped tubes and flasks to prevent the loss of moisture. To check the loss of moisture in broth tubes, mark the original liquid level in several tubes of each batch and monitor the loss of the moisture. If estimated loss exceeds 10 percent, discard tubes.

Seal prepared agar plates in plastic bags and refrigerate to retain moisture. Plates should be stored in inverted position. Each batch should be clearly labelled by the name and preparation date of the media.

Protect media containing dyes from light. If color changes are observed, discard the medium.

TABLE 12.1
Holding Times for Prepared Media

Medium	Holding Time
Membrane filter (MF) broth in screw-cap flasks at 4°C	96 h
Membrane filter (MF) agar in plates with tight fitting covers at 4°C	2 weeks
Agar or broth in loose cap tubes at 4°C	1 week
Agar or broth in tightly closed screw-cap tubes at 4°C	3 months
Poured agar plates with loose fitting covers in sealed plastic bags at 4°C	2 weeks
Large volume of agar in tightly closed screw-cap flask or bottle at 4°C	3 months

12.1.9 Sterile Media from Commercial Sources

Prepared sterile broths and agars available from commercial sources may offer advantage when analysis are done intermittently, when staff is not available for preparation work, or when cost can be balanced against other factors of laboratory operation. Check and record date of arrival and date of expiration regularly.

12.2 BACTERIAL GROWTH IN MEDIA

To grow a microbial culture in a sterilized medium, cells (the inoculum) are transferred (inoculated) into the medium using special precautions to maintain the purity of the culture being transferred. Following inoculation, a culture is incubated in an environment providing suitable growth conditions. Growth here means the development of a population of cells from one or a few cells. The mass of daughter cells become visible to the naked eye either as a cloudiness (turbidity) in liquid broth or as an isolated population (colony) on solid media. The visible appearance of growth is sometimes an aid in differentiating microbial species.

12.2.1 Bacterial Growth in Liquid (Broth) Media

The following procedures should be followed during the preparation for bacterial growth in a liquid medium:

1. When a needle or loop inserted into a holder is used to transfer microorganisms in the inoculation procedure, it should be heated to redness by flaming immediately before and after making the transfer. Hold the wire portion in the flame in a manner to heat first the entire length of the wire and then the lower part of the holder. The inoculating procedure can see in Figure 12.3.
2. During the transfer of cells, hold the tube with the bacterial culture in the left hand as nearly horizontal as feasible and grasp the cap between the fingers of the right hand. *Caution: never lay a cap down.* The mouth of the tube must be passed through the burner flame before inserting the inoculating needle or loop. Reheat the mouth of the tube and replace the cap.
3. Pick up the tube to be inoculated, remove the cap, flame the lip and insert the loop or needle to transfer the bacterial culture. Move the wire back and forth once or twice, remove the loop or needle, and flame the loop or needle as described previously. Flame the lip of the inoculated tube and replace the cap.
4. With a little practice, both tubes can be held in one hand and done at the same time. The mouths of both tubes, from which cultures are taken and into which they are transferred, must be passed through the burner flame immediately before and after the inoculating loop or needle is introduced and removed. Do not overheat the glass and do not leave the tubes open any longer than necessary.
5. Following inoculation, the culture is incubated in an environment providing suitable growth conditions.

12.2.1.1 Observing Growth Patterns on Broth Media

Use of broth media is a convenient way to handle bacteria in stock culture and organisms often grow in a characteristic manner in the broth medium.

Cells may grow dispersed showing interruption of the light path or cloudiness called turbidity. This may be slight to moderate to heavy.

Cells may settle to the bottom to form a *sediment or button.* Cells may be attracted to each other to form clumps that settle to the bottom. These clumps may be adherent or slimy when dislodged

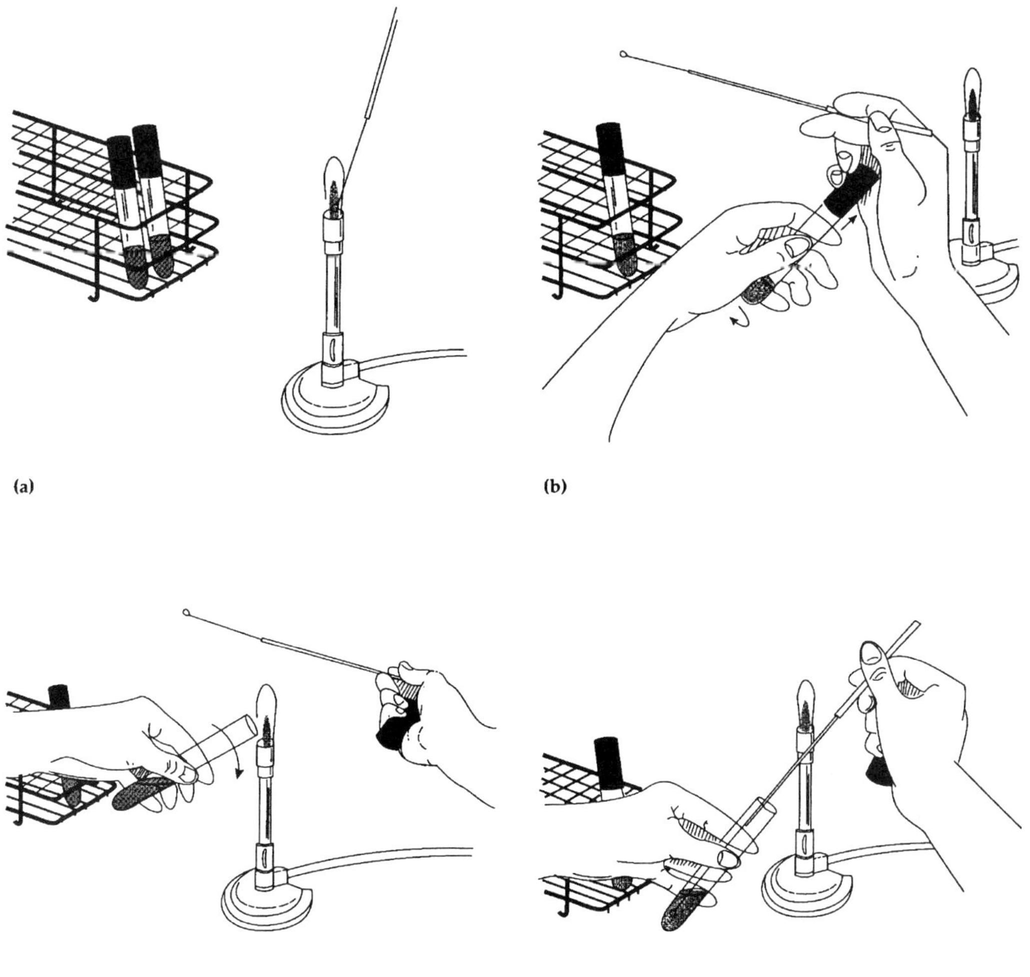

(a) (b) (c) (d)

FIGURE 12.3 Inoculating procedures. a. Sterilize the loop by holding the wire in the flame until it is red hot. b. While holding the sterile loop and the bacterial culture, remove the cap as shown. c. Briefly heat the mouth of the tube in the flame before inserting the loop for an inoculum. d. Get a loopful of the culture, heat the mouth of the tube, and replace the cap.

from the bottom. A small button often may be the only sign of growth. Examine the tube first by looking up from the bottom. Then gently tap the tube near the bottom. A slight sediment will swirl upward. A button does not always show growth but simply may be inoculum cells which have settled out. If there is no increase with time, this is probably the case.

Many organisms grow in a film across the surface called the *pellicle* which may be heavy or light and membranous.

Some organisms form a *ring of growth* around the glass-broth-air interface, seen by tilting the tube slightly. Handle tubes carefully when first observing them because heavy pellicles often fall to the bottom when disturbed.

These growth patterns often give information about an organism's relation to air and its surface activity. Cultural characteristics of broth cultures is shown in Figure 12.4.

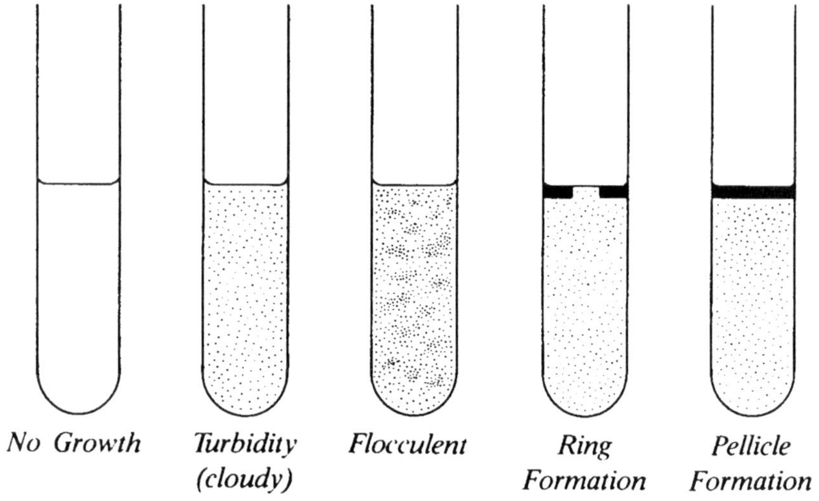

FIGURE 12.4 Cultural characteristics of broth cultures. a. No growth. b. Turbidity. c. Flocculent. d. Ring formation. e. Pellicle formation.

12.2.2 Bacterial Growth in Agar Slant Cultures

A common means of providing a solid medium is to add a solidifying agent to a broth medium during preparation which hardens as it cools after autoclaving. Agar is rehydrated by mixing into broth, melting it by heating to boiling with constant stirring, distributing it into tubes or flasks, and then autoclaving it. After cooling, it solidifies again. Agar is usually used at a concentration of 1.5 percent (see Section 12.1).

12.2.2.1 Preparation of Agar Slants

The following procedures are used in the preparation of agar slants.

1. Agar slants are prepared by melting a small amount of a solid medium in a tube and allowing it to solidify in a slanted position. The slant should begin below the cap and end about 1 to 2 cm above the curved bottom of the tube.
2. Flame inoculating loop or needle as described in Section 12.2.1.
3. Transfer bacterial culture from the culture tube to the inoculating tube as described in Section 12.2.1.
4. Streak the surface in one straight line from bottom to top while holding the loop vertical to the slant as illustrated in Figure 12.5. Holding the loop horizontally will tear the agar. Rest the loop very lightly on the slant so as not to gouge. If the entire surface is to be covered, use the loop parallel to the agar surface and move it rapidly back and forth while drawing the loop up the slant. Great care must be taken to avoid gouging the agar.
5. After proper incubation, examine the cultures.
6. Touch the growth with a sterile loop and observe the consistency of the growth, whether it is brittle, soft, gummy, buttery, or any other description you might use. Cultural characteristics of agar slant cultures are shown in Figure 12.6.

12.2.3 Bacterial Selection by Sugar Fermentation

Lactose fermentation with gas formation is a special ability of some microbes. Coliform bacteria are indicator organisms used to detect water pollution. Coliforms are defined as aerobic or facultatively

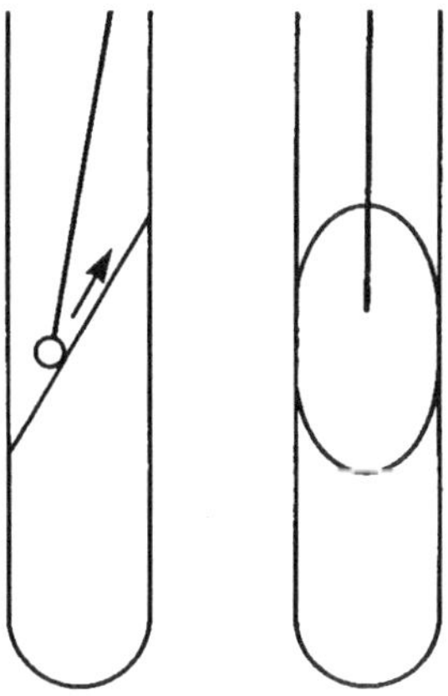

FIGURE 12.5 Inoculate agar slant by streaking back and forth across the surface of the agar being careful not to gauge the agar.

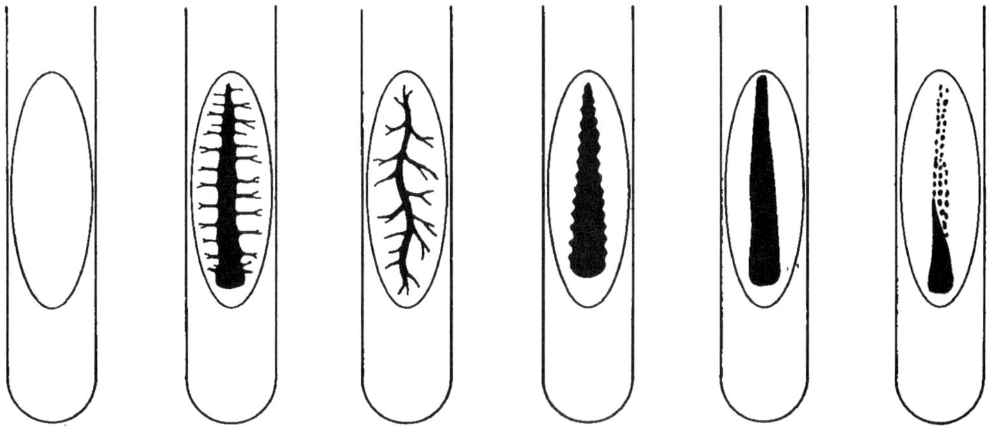

FIGURE 12.6 Cultural characteristics of agar slant culture.

anaerobic, gram negative, non-endospore-forming, rod-shaped bacteria that ferment lactose to form gas within 48 hours of being placed in lactose broth at 35°C.

Fermentation is the term ordinarily applied to the anaerobic breakdown of carbohydrates. The purpose of the fermentation process is to make energy available for utilization by the microorganism because carbohydrates are quite rich in stored energy. Whether or not a given carbohydrate is fermented depends upon the transport proteins (permeases) and the endoenzymes (or carbohydrates) possessed by the organism. The end products of fermentation often will vary from one organism to another depending on the pathways involved. However, the end products are usually acids of various types or acid and gas.

The ability to ferment can be determined by inoculating the organism in a fermentation tube containing the nutrient broth to support the growth of the organism, a single chemically-defined carbohydrate, a pH indicator, and an inverted *Durham tube* to collect gases. A Durham tube is a small

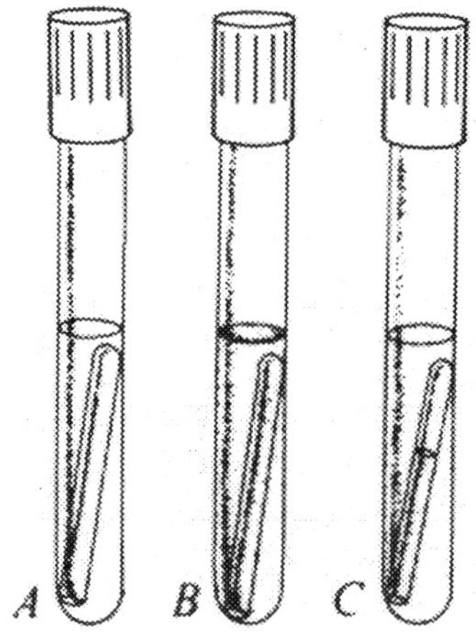

FIGURE 12.7 Broth culture tubes with Durham tubes to measure gas production. a. Before inoculation. b. Growth but no gas production. c. Growth resulting in gas production. Gases are trapped in the inverted Durham tube.

inverted vial placed inside the culture tube. The purpose of the Durham tube is to trap gas if it has been produced as an end product of metabolism. Figure 12.7 shows the culture tubes with Durham tubes to measure gas production.

pH indicators determine whether or not an acid has been produced as an end product of the metabolism. Phenol red and bromcresol purple are the two pH indicators most frequently used in microbiological work. Phenol red appears red in basic and yellow in acidic pH, bromcresol purple is purple in basic and yellow in acidic solutions.

1. Inoculate carbohydrate broths with the sample.
2. Keep one tube of the carbohydrate broth uninoculated for control.
3. Incubate the tubes at 37°C. Make observations at 24, 48, and 72 h.
4. After each incubation period, compare each of the inoculated tubes with the control tube to determine whether growth occurred and whether acid or acid and gas were produced.
5. If at any time during the incubation period should a fermentation tube contain both acid and gas, it is not necessary to continue incubation. The tube may be discarded after recording the result.

12.2.4 Obtaining Pure Culture: Streak Plate Method

Most infectious materials contain several kinds of bacteria. So do samples of soil, water, and foods. If these materials are plated out onto the surface of a solid medium, colonies will form. These are copies of the same organism theoretically, a visible colony arises from a single spore or vegetative cell or from a group of the same microorganism (i.e. a clone of cells) attached to one another in clumps or chains. Microbial colonies often have a distinctive appearance that distinguishes one microbe from another. The bacteria must be distributed widely enough so that the colonies are visibly separated from each other. Most bacteriological work requires pure cultures, or clones of bacteria.

The isolation method most commonly used to get pure cultures is the streak plate method. A sterile inoculating loop is dipped into a mixed culture that contains more than one type of microbe and is streaked in a pattern over the surface of the nutrient medium. As the pattern is traced, bacteria are rubbed off the loop onto the medium in paths of fewer and fewer cells. The last cells to be rubbed off the loop are far enough apart to grow into isolated colonies. These colonies can be picked up with an inoculating loop and transferred to a test tube of nutrient medium to form a pure culture containing only one type of bacterium.

12.2.4.1 How to Prepare Agar Plates

The following procedures are used in the preparation of agar plates:

1. After melting an agar medium, it is then cooled to 45 to 47°C.
2. Take out the melted agar from the holding water bath and wipe the outside of the container with a cloth or paper towel. Otherwise, the water will run onto the plate and introduce contaminants.
3. After removing the plug or cap from the container to pour the agar, flame the mouth of the container to kill microorganisms on the outside lips.
4. In pouring the agar from the container to the plate, raise the cover of the plate only on one side just enough to easily admit the mouth of the container.
5. Pour 15 to 20 ml of agar into the sterile petri dish. Take care not to scrape the container on the petri plate.

The procedure is shown in Figures 12.8 and 12.9.

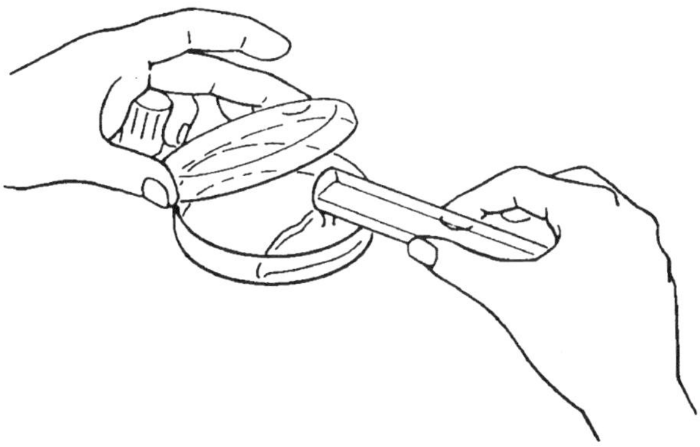

FIGURE 12.8 Using aseptic technique, lift the cover of the petri dish high enough to insert the mouth of the tube to pour the melted medium into it. Do not touch the plate with the tube.

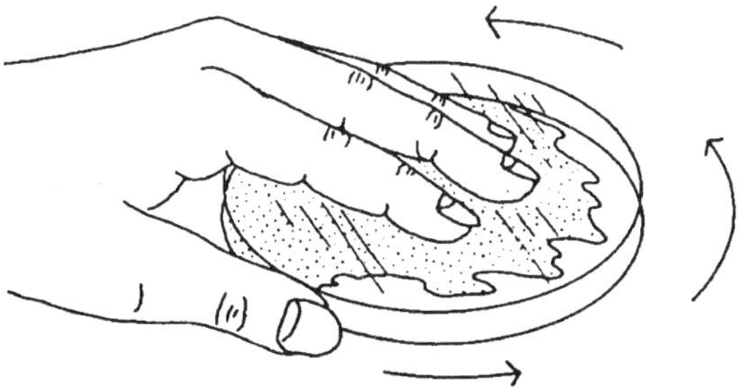

FIGURE 12.9 Rotate the petri plate so that the medium covers the bottom. Do not move the plate again until the medium has solidified.

12.2.4.2 How to Inoculate Agar Plate

The following procedures are used in the inoculation of agar plates:

1. Flame the inoculating loop and allow it to cool for a few seconds.
2. Lift the lid of the petri plate only far enough to allow the insertion of the loop. Take a loopful of the mixed broth suspension and streak back and forth over a small area near the side of the agar plate. Be careful not to cut the agar or contact the side of the plate.
3. The inoculating loop should be held with the open face parallel to the agar surface rather than vertically as illustrated in Figure 12.10.
4. Hold the handle so that the loop rests lightly on the agar surface. Slide the loop rapidly from side to side in arcs of 2 to 3 cm while drawing it toward you but without pressure. This reduces the likelihood of gouging the agar.
5. Flame the loop and allow it to cool briefly.
6. Pass the loop completely through the previous streaked area once and then back and forth in a restricted area as shown in Figure 12.11. Repeat this step until there is no further room on the plate.
7. Incubate at 37°C for 24 h.
8. Make observation on isolated colonies.

12.2.4.3 How to Incubate Agar Plates

After streaking, incubate petri dishes at 37°C for 24 h in inverted position (bottom-side up) to prevent condensation of water on the agar surface. Moisture interferes with development of isolated colonies by spreading bacterial growth over the agar surface.

After incubation, examine plates for single, well-isolated colonies.

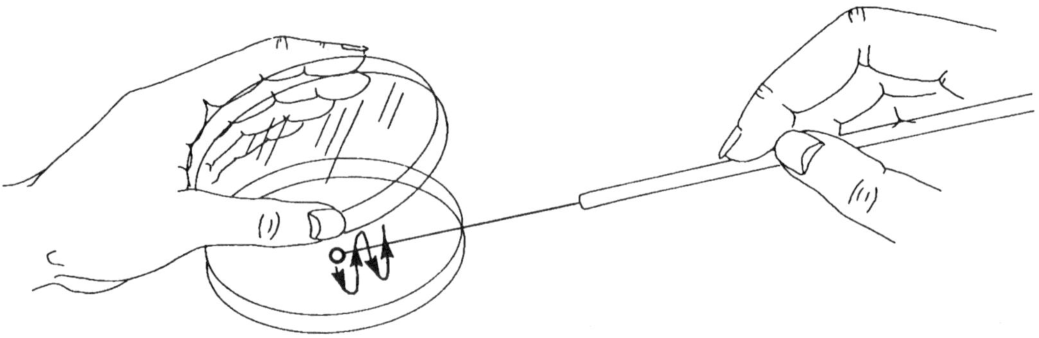

FIGURE 12.10 Proper handling of the cover of the petri dish while preparing a streak plate. Hold the loop parallel to the agar to avoid gouging.

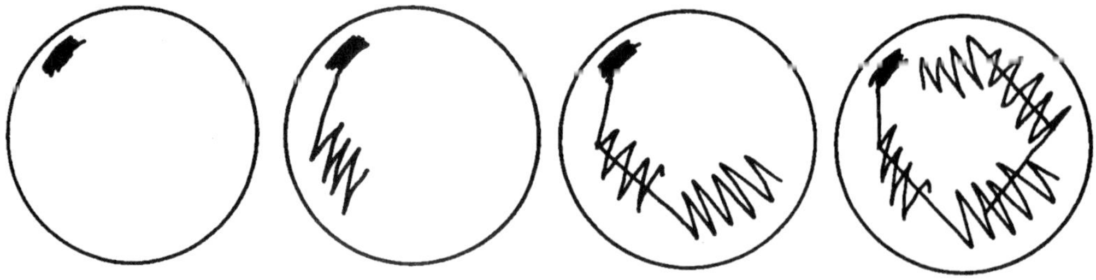

FIGURE 12.11 Steps in the preparation of a streak plate. The inoculating loop is sterilized between each set of streaks.

13 Direct Measurement of Microbial Growth

The growth of microbial populations can be measured in a number of ways. Some methods measure cell numbers, other methods measure the population's total mass, which is directly proportional to cell numbers. Population numbers are usually recorded as the number of cells in 1 ml of liquid or in a g of solid material. Because bacterial populations are usually very large, most methods of counting them are based on direct or indirect counts of very small samples; calculations then determine the size of the total population. Assume, for example, that one millionth of 1 ml (10^{-6} ml) of sample is found to contain 70 bacterial cells. Then there must be 70 times 1 million, or 70 million cells per ml. However, it is not practical to measure out one millionth of a ml of liquid or a millionth of a g of solid material.

Dilutions of the original are often necessary to reduce the number of bacterial cells to measurable levels or to isolate single cells. This requires mixing a small accurately measured sample with a large volume of dilution water called the diluent, or dilution blank.

13.1 DILUTIONS

13.1.1 Single Dilution

A single dilution is calculated as follows:

$$\text{Dilution} = \frac{\text{volume of the sample}}{\text{total volume of the sample and the diluent}} \tag{13.1}$$

For example, the dilution of 1 ml sample into 9 ml diluents equals

$$\frac{1}{1+9} = \frac{1}{10} \text{ and is written } 1{:}10 \text{ or } 10\times$$

13.1.2 Serial Dilution

Experience has shown that better accuracy is obtained with very large dilutions if the total dilution is made out of a series of smaller dilutions rather than one large dilution. This series is called a serial dilution, and the total dilution is the product of each dilution in the series.

For example, if 1 ml is diluted with 9 ml, and then 1 ml of that dilution is put into a second 9 ml diluent, the final dilution will be

$$\text{Dilution} = \frac{1}{10} \times \frac{1}{10} = \frac{1}{100} \text{ or } 1{:}100 \text{ or } 100\times$$

To facilitate calculations, the dilution is written in exponential notation (see Appendix A). In the previous example, the final dilution 1:100 would be written 10^{-2}.

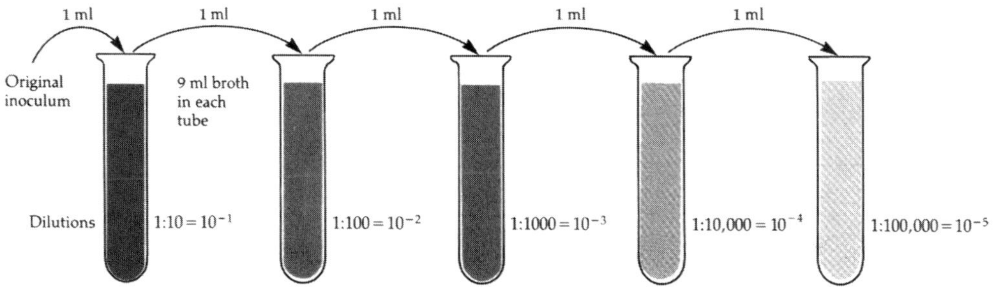

FIGURE 13.1 Serial dilution technique. First dilution represents a 100× dilution, so the sample size is 10^{-2} ml. The second dilution represents a 100 × 100 = 10,000 times dilution, the sample size is 10^{-4} ml. The third dilution is 10,000 × 10 = 100,000 times dilution, with original sample volume of 10^{-5} ml. The fourth, last dilution is 100,000 × 10 = 1,000,000 times dilution and the original sample size is 10^{-6} ml.

When a 1 ml sample is added to 99 ml of dilution water (first dilution), from that dilution add 1 ml into a new 99 ml of dilution water (second dilution), from that dilution add 1 ml to 9 ml of dilution water (third dilution), and from that dilution add 1 ml to 9 ml of dilution water (fourth dilution), and calculate the final dilution, serial dilution technique shown in Figure 13.1.

$$\frac{1}{100} \times \frac{1}{100} \times \frac{1}{10} \times \frac{1}{10} = \frac{1}{1{,}000{,}000} \text{ or } 1:1{,}000{,}000 \text{ or } 10^{-6}$$

If the reading on the final dilution was 22 colonies, the final result will be 22 × 1,000,000 = 22,000,000 colonies per ml sample.

13.1.3 Prompt Use of Dilutions

The potential toxicity of phosphate dilution water increases rapidly with time. Therefore, dilutions of samples should be tested as soon as possible after make up and should be held no longer than 30 min after preparation.

13.1.4 Dilution Water

The ideal dilution water is neutral in effect. It maintains bacterial populations without stimulating cell growth and reproduction, damaging cells or reducing their ability to survive, grow, or reproduce. Its basic purpose is to stimulate the conditions of the natural environment which are favorable to cell stability.

13.1.4.1 Phosphate Buffered Dilution Water

13.1.4.1.1 Stock phosphate buffer solution

The following procedure is used to prepare stock phosphate buffer solution.

1. Measure 34 g potassium dihydrogen phosphate (KH_2PO_4).
2. Dissolve in laboratory pure water and fill up to 500 ml.
3. Adjust the pH to 7.2 with 1 *N* sodium hydroxide (NaOH). (Preparation of 1 *N* NaOH dissolve 40 g NaOH in laboratory pure water and dilute to 1 l. *Caution:* the reaction between NaOH and water liberates extreme heat. Prepare under laboratory hood. Very slowly add water to the NaOH and wait between water additions until the solution cools down. Wear safety glasses and safety face shield to safeguard your eyes and face from the splattering

hot, strong alkaline solution. Wear protective laboratory coat, laboratory apron, and asbestos gloves, too).

4. Bring the volume to 1000 ml with laboratory pure water.
5. Sterilize by autoclaving for 15 min at 121°C and 15 lb pressure.
6. After sterilization, store in a refrigerator and handle aseptically. If evidence of mold or other contamination appears, the stock solution should be discarded and fresh solution prepared.

13.1.4.1.2 Working phosphate buffer solution

The following procedure is used to prepare a working phosphate buffer solution.

1. Measure 1.25 ml of stock phosphate buffer solution.
2. Add 5.00 ml of magnesium chloride ($MgCl_2$) solution. [Preparation of magnesium chloride solution: add 81.1 g of magnesium chloride hexahydrate ($MgCl_2 \cdot 6H_2O$) per 1 l of laboratory pure water. The solution is good for 6 months.]
3. Fill up to 1000 ml with laboratory pure water.
4. Final pH should be 7.2 ± 0.1.

Dispense 102 ml volume of working phosphate buffer solution into borosilicate glass, screw-cap dilution bottles scribed at 99 ml. Loosen screw-caps and sterilize by autoclaving for 15 min. Final volume after sterilization should be 99 ± 2 ml. Cool, tighten screw caps, and store in cool place.

13.2 PLATE COUNTS

The most frequently used method for measured bacterial populations is the plate count. An important advantage of this method is that it measures the number of viable cells. One disadvantage may be that it takes some time, usually 24 h, or more, for visible colonies to form. The plate count is based on three assumptions: that each bacterium grows and divides to produce a single colony, that the original inoculum is homogeneous, and no aggregates of cells are present. When a plate count is performed, it is important that only a limited number of colonies develop in the plate. When too many colonies are present, some cells are overcrowded and do not develop; these conditions cause inaccuracies in the count. Generally, only plates with 25 to 250 colonies are counted. To ensure that some colony counts will be within this range, the original inoculum is diluted several times in a process called serial dilution as discussed in Section 13.1.

The media used support the growth of most heterotrophic bacteria; therefore, the routinely performed test is called a heterotrophic plate count, previously named by standard plate count. Because colonies can originate from more than one cell, results may be reported as *colony forming units* (CFUs). Each colony may arise from one individual cell.

A plate count is done by either pour plate or the spread plate method. Plate count is also performed as streak plate count, but it is only an indirect, qualitative estimation of the bacterial population.

13.2.1 POUR PLATE METHOD

The agar medium is inoculated while it is still liquid, and so colonies develop throughout the medium not only on the surface. Usually 1 ml and 0.1 ml samples or appropriate dilutions of the sample are introduced into sterile petri dishes. The nutrient medium, in which the agar is kept liquid at 44 to 46°C, is poured over the sample, which is then mixed into the medium by gentle agitation of the plate. When the agar solidifies, the plates are inverted and incubated for a predetermined time.

The pour plate technique has some drawbacks because some relatively heat-sensitive microorganisms may be damaged by the melted agar and will, therefore, be unable to form colonies. This technique is used in the heterotrophic plate count method.

Detailed discussion of the method, counting, and reporting the results is presented in Section 16.2.

13.2.2 Spread Plate Method

To avoid problems associated with the pour plate method, the spread plate technique is frequently used instead.

After melting the agar medium, it is then cooled to 45 to 47°C and about 15 to 20 ml of the cooled, but still liquid agar is poured into a sterile, covered petri plate. Prior cooling prevents excess condensation of moisture in the petri plates when the liquid agar is cooling.

A 0.1 ml of sample (inoculum) or 0.1 ml of appropriate dilution is added to the surface of a prepared, solidified agar medium. The inoculum is then spread uniformly over the surface of the medium with a sterile glass rod. The inoculum is spread uniformly by holding the stick at a set angle on the agar and rotating the agar plate or rotating the stick until the inoculum is distributed. This method positions all the colonies on the surface and avoids cell contact with melted agar.

Cover plates partially, leaving open slightly to evaporate excess moisture for about 15 to 30 min. When agar surfaces are dry, close dishes, invert them, and incubate as required for the specified test. After incubation at the proper time and temperature, isolated surface colonies should develop.

Detailed outline of the method, counting, and reporting the results are presented in Section 16.3.

13.2.3 Streak Plate Method

The purpose of the streak plate is to produce well-separated colonies of bacteria from concentrated suspensions of cells. The streak plate method is a qualitative, indirect detection of microorganisms. However, the pour plate and spread plate methods are quantitative, direct measurements of the bacterial growth. The technique is discussed in Chapter 16.

13.3 MEMBRANE FILTER (MF) TECHNIQUE

When the quantity of bacteria is very small, bacteria can be counted by filtration method. A sample size of 100 ml, or a smaller sample size diluted to 100 ml, are passed through a thin membrane whose pores are too small to allow bacteria to pass. Filters with a variety of pore sizes are available. Nitrocellulose or polyvinyl membrane filters with 0.45 μ pores are commonly used to trap bacteria. Bacteria are retained in the filter and then placed on a suitable medium. Nutrients that diffuse through the filter can be metabolized by bacteria trapped on the filter. Each bacterium that is trapped will develop into a colony. Bacterial colonies growing on the medium can be counted. When a selective or differential medium (Section 12.1.3) is used, desired colonies will have a distinctive appearance.

Thus, the bacteria are sieved out and retained on the surface of the filter. This filter is then transferred to a petri dish containing a pad soaked in a liquid medium, or a solidified medium, where colonies arise from the bacteria on the filter surface. This method is applied frequently to coliform bacteria, which are indicators of fecal pollution in food or water samples. The colonies formed by these bacteria are distinctive when a different nutrient medium is used.

13.3.1 Advantages of the Membrane Filter Technique

Membrane filter methods are preferred over most probable number (MPN), see Section 13.4, or other techniques, where applicable because of the following advantages.

One of the primary advantages of this method is its speed. Definitive results can be obtained in 22 to 26 h, whereas 48 to 96 h are required for the multiple-tube fermentation method.

Considerably larger, more representative water samples can be examined than with MPN.

The precision is greater with the MF because it makes a direct count of colonies per unit volume.

The method represents savings in time, labor, space, supplies, and equipment.

It is also practical in field studies.

13.3.2 Limitations of the Membrane Filter Technique

Some samples contain large quantities of colloidal materials or suspended solids, such as iron, manganese, alum flocs, or clay. Other samples may contain algae. These substances can clog the filter pores and prevent filtration or can cause the development of spreading bacterial colonies. When the bacterial counts of such samples are high, a smaller volume or a higher sample dilution can be used to minimize the effect of sample turbidity. However, the membrane filter method is not applicable for waters with high turbidity but low bacterial count.

Industrial wastewaters may contain zinc, copper, or other heavy metallic compounds which absorb onto the membrane surface and interfere with subsequent bacterial development.

Inhibition may result in seawater or toxic materials such as chlorine or phenols.

13.3.3 Outline of the Membrane Filter Technique

The following list outlines the membrane filter technique.

1. Set up filtration apparatus.
2. Filter.
3. Inoculate and incubate.
4. Verify—a verified membrane filter test establishes the validity of colony differentiation on a selected medium and provides support evidence of colony interpretation. Verification is required for all positive samples from potable waters. Verification is also recommended for nontypical colonies to determine that false negative colonies do not occur. Verification is also used for establishing quality control with new test waters, new procedures, or new technicians, for identifying unusual colony types, and as a support for data used in legal actions.
5. Count colonies and record—examine, count, and calculate bacterial population for a 100 ml sample as described in individual methods. The grid lines are used to count colonies. The acceptable number of colonies that is countable on a membrane is the function of the parameter tested. The general formula used is

$$\frac{\text{counts}}{100\text{ ml}} = \frac{\text{number of colonies}}{\text{volume of sample filtered (in ml)}} \times 100 \qquad (13.2)$$

Detailed steps used during test procedures are discussed in individual methods in Chapters 17, 18, and 19.

13.4 MOST PROBABLE NUMBER (MPN) METHOD

Another method for counting the number of bacteria in a sample is the most probable number (MPN) method. This statistical estimating method is based on the fact that the greater the number of bacteria in a sample, the more dilution is needed to reduce the density to the point at which no bacteria are left to grow in the tubes in a dilution series. The MPN method is most useful when the microbes being counted will not grow on solid media. The MPN is only a statement that there is a 95 percent chance that the bacterial population falls within a certain range and that the MPN is statistically the most probable number.

The MPN test is used to detect the coliform group in water samples. As discussed previously, coliforms occur in large numbers in the intestinal tract of man and animals. While not normally pathogenic themselves, they do indicate the presence of sewage and, thereby, pathogens because they come from the same site of the body.

The MPN test for coliforms consists of three steps: a presumptive test, a confirmed test, and a completed test. It attempts to determine the number of organisms in the water which are Gram negative and ferment the carbohydrate lactose with the production of gas at 35°C. They must be facultative anaerobes and nonspore formers.

Decimal dilutions of the samples are inoculated in a series into liquid tube media. Positive tests are indicated by growth and/or fermentative gas production. Bacterial densities are based on combinations of positive and negative tube results read from the MPN table. The detailed method for the MPN test is discussed in Sections 17.4, 18.4, and 19.5.

13.4.1 Presumptive Test

The presumptive test provides a preliminary estimate of bacterial density based on enrichment in minimally-restrictive tube media. The result of this test is never used without further analyses. A detailed method is discussed in Chapter 17.

13.4.2 Confirmed Test

The confirmed test is performed by verifying positive tubes from the presumptive test by using one appropriate medium and 24 and 48 h incubation, as discussed above in the presumptive test. After 24 and 48 h incubation, examine the tubes for gas and/or growth and record negative and positive results. In routine practice, most sample examination is terminated at the end of the confirmed test.

However, for quality control, at least 5 percent of the confirmed tests samples and a minimum of one sample per test run should be carried through to the completed test.

13.4.3 Completed Test

In some cases, the organisms must be isolated and stained to provide the completed test.

1. Streak Lewin's EMB (eosin-methylene blue) agar plates from each positive confirmatory tube and incubate at 35°C for 24 h. Pick typical colonies or atypical colonies and inoculate into Lauryl tryptose broth and incubate for 24 to 48 h at 35°C. The formation of gas in the fermentation tube constitutes a positive completed test for total coliform.
   - Typical colonies—colonies with a golden green metallic sheen or reddish purple color with nucleation.
   - Atypical colonies—red, pink, or colorless, unnucleated and mucoid.
2. Instead of checking gas formation in the Lauryl tryptose broth, the Gram stain test has also been used in the completed MPN test. Pick up typical or atypical colonies from the EMB agar plate. Inoculate into nutrient agar slants, incubate for 24 h at 35°C and follow the staining procedure as described below in Section 13.5.

13.4.4 Calculation and Reporting of MPN Values

Detailed calculation and reporting of the results are discussed in Sections 17.4.7, 18.4.5, and 19.5.3.

13.5 STAINING PROCEDURES

Microbe are very small and very similar in water or in refracting light and are very difficult to see. Although modern microscopy has helped, the development of phase contrast and interferences microscopes, staining of the cells to make them more visible, still remains the standard tool.

A stain or dye uses a molecule that can bind to a cellular structure and give it color. Staining techniques make the microorganisms stand out against the background. A stain or dye is also used to help investigators group major categories of microorganisms, examine the structural and chemical differences in cellular structures, and look at the parts of the cell.

Stains are either simple stains, consisting of the addition of one dye that serves to delineate morphology but renders all structures the same hue. Differential stains consists of more than one dye added in several steps and the stained structures are differentiated by color as well as by shape.

13.5.1 Preparation of Bacterial Smears

In order to stain microorganisms, a thin layer of cells, called smear, must be made first. This is a simple process of spreading an aqueous suspension of cells on a glass slide and allowing it to air dry. This is followed by fixation (causing the cells to adhere to the slide) and the application of the staining solutions.

Stains are generally made on smears prepared from colonies or slants because the mass of cells is very great. The main problem with this is gathering too many cells. Making smears of the right density can be learned only by experience.

Clean, grease-free slides are essential for obtaining good stained preparations.

1. Place a small drop of laboratory pure water on a clean slide.
2. Using a sterilized inoculating loop, pick a small amount of growth from the agar slant. Mix the bacteria with the drop of water on the slide and spread evenly over an area the size of a quarter.
3. Allow the smear to air dry for at least 30 min.
4. The smear must be heat fixed to be certain that the cells will adhere to the slide and not be washed off during the staining process. This is accomplished by quickly passing the slide (smear side up) several times through a portion of the flame. Do not overheat the slide. You should be able to touch the underside of the slide comfortably to the back of the hand.

The steps of the preparation of a smear from a solid culture and the heat fixation of an air dried specimen are shown in Figure 13.2 and Figure 13.3, respectively.

13.5.2 Gram Stain

Gram stain is a general test for characterization of bacteria and for examination of culture purity. The Gram differentiation is based upon the application of a series of four chemical reagents: primary dye, mordant, decolorizer, and counterstain.

1. *Primary dye*—The purpose of the primary dye, crystal violet, is to impart a purple or blue color to all organisms regardless of their designated Gram reaction.

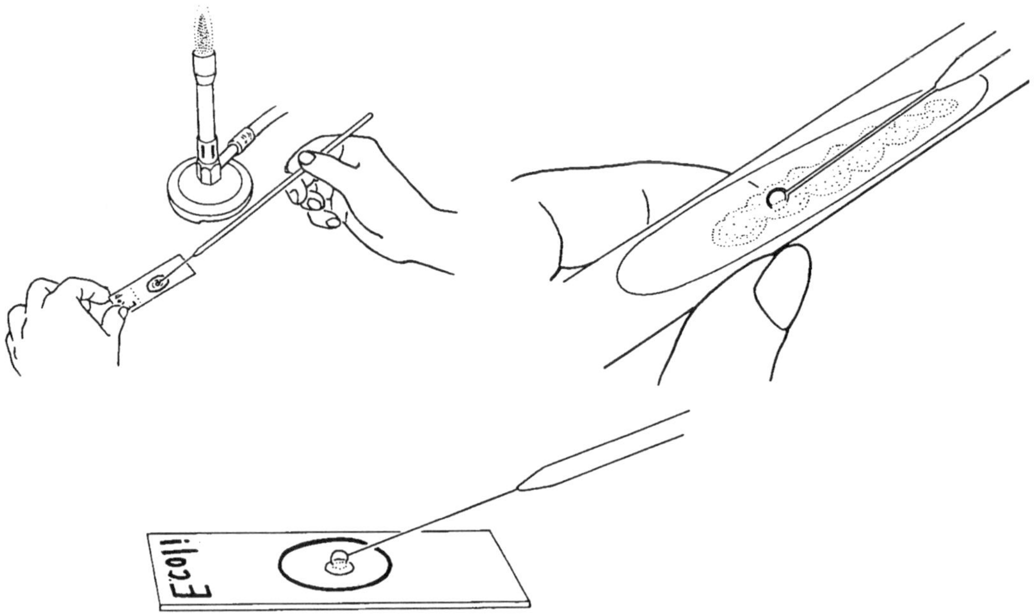

FIGURE 13.2 Preparation of a smear from a solid culture.

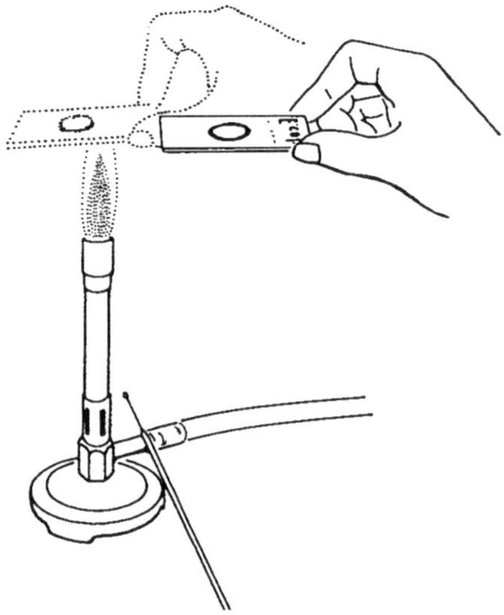

FIGURE 13.3 Heat fixation of an air dried specimen.

2. *Mordant*—The crystal violet treatment followed by the application of Gram's iodine, which acts as a mordant by enhancing the union between the crystal violet dye and its substrate by forming a complex.
3. *Decolorizer*—The decolorizing solution of acetone–alcohol extracts the complex from certain cells more readily than from others.

4. *Counterstain*—A safranin counterstain is applied in order to see those organisms previously decolorized by the removal of the complex.

Gram positive—purple or blue	Gram negative—pink or red
Organisms retaining the complex.	Organisms losing the complex.

The Gram-staining procedure is shown in Figures 13.4 and 13.5.

13.6 DIRECT MICROSCOPIC COUNTS

Bacterial growth can be measured by direct microscopic counts. In this method, a known volume of medium is introduced into a specially calibrated etched glass slide called the *Petroff-Hausser counting chamber.*

A bacterial suspension is introduced onto the chamber with a calibrated pipet. After the bacteria settle and the liquid currents have slowed down, the microorganisms are counted in specific calibrated areas. Their number per unit volume of the original suspension is calculated by using an appropriate formula.

The number of bacteria per ml of medium can be estimated with a reasonable degree of accuracy. The accuracy of the direct microscopic counts depends on the presence of more than 10 million bacteria per ml of culture. This is because counting chambers are designed to allow accurate counts only when large numbers of cells are present.

This technique has the disadvantage of generally not distinguishing between living and dead cells.

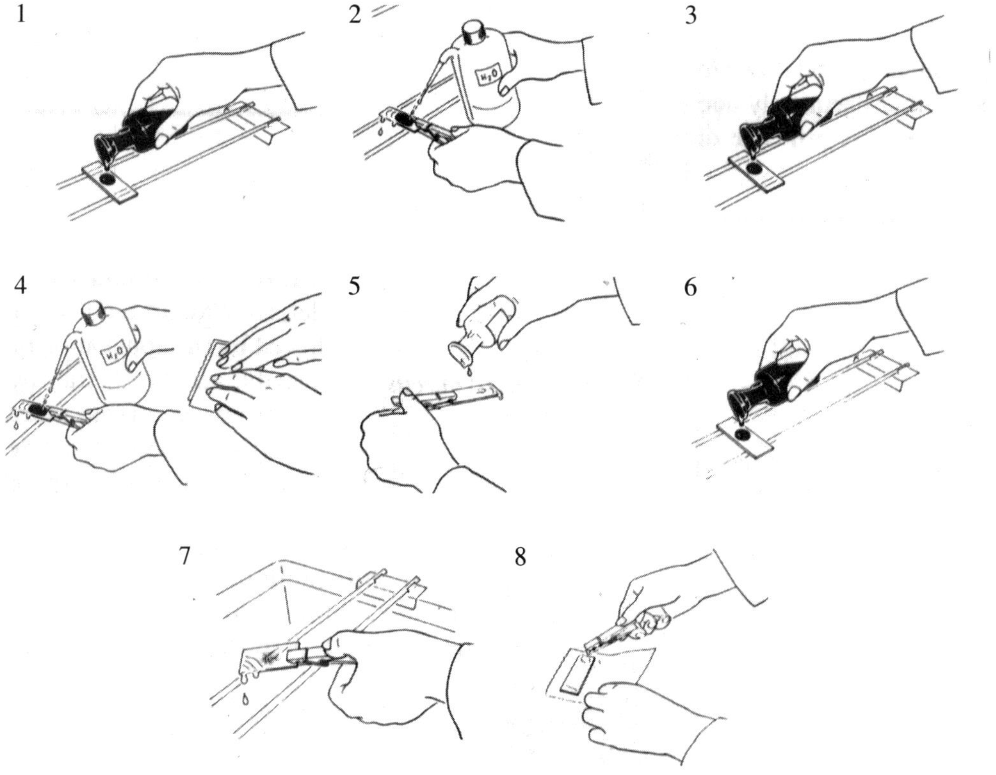

FIGURE 13.4 Steps in the Gram-stain procedure.

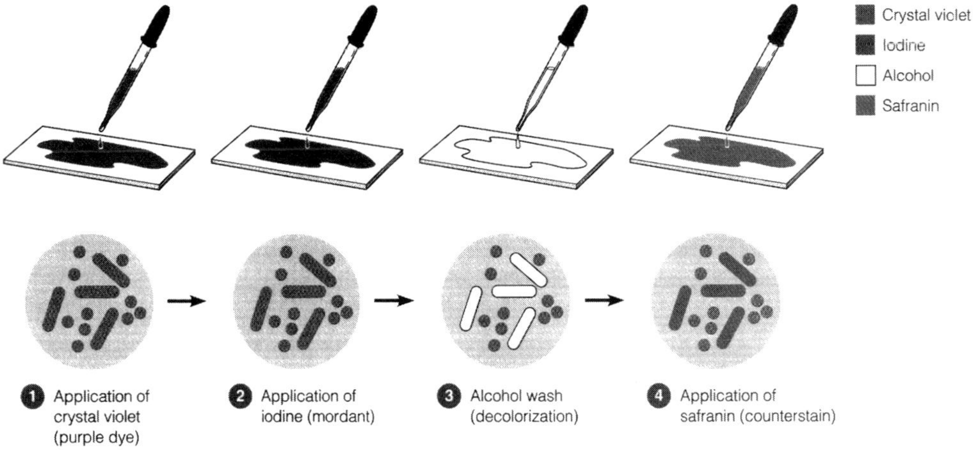

FIGURE 13.5 Gram-staining procedure. 1. A heat fixed smear of cocci and rods is first covered with a basic purple dye (primary stain) such as crystal violet, and then the dye is washed off. 2. Then the smear is covered with iodine (a mordant) and washed off. At this time, both Gram-positive and Gram-negative bacteria are purple. 3. The slide is washed with ethanol and washed with water. Now Gram-positive cells are purple and Gram-negative cells are colorless. 4. In the final step, safranin is added as a counterstain, and the slide is washed, dried, and examined microscopically. Gram-positive bacteria retain the purple dye. Gram-negative bacteria appear pink because they pick up the safranin counterstain.

14 Estimation of Bacterial Numbers by Indirect Methods

It is not always necessary to count bacterial cells to estimate their numbers. In science, microbial numbers and activity are determined by some of indirect means discussed below.

14.1 TURBIDITY

Estimating turbidity is a practical way to monitor bacterial growth. As bacteria multiply in a liquid medium, the medium becomes turbid or cloudy with cells. The instrument used to measure turbidity is a spectrophotometer. In the spectrophotometer, a beam of light is transmitted through a bacterial suspension to a photometric cell. *The amount of light lost or scattered is inversely proportional to the cell concentration or directly proportional to the absorbance* (also called *optical density*, OP). The light loss can be determined by measuring the amount scattered or reflected (*nephelometry*) or the amount of light transmitted (*turbidimetry*).

Scales of spectrophotometers (*galvanometers*) are measured in percent transmittance (% T) and/or absorbance. Bacterial numbers are directly proportional to absorbance and indirectly proportional to % T. The two scales are related as follows:

$$\text{Absorbance} = \log 100 - \log \% \text{T} \tag{14.1}$$

Unless a given absorbance reading can be quantitatively related to a number of bacteria per ml, the density scale must be in arbitrary units. In order to quantify the absorbance reading, a quantitative enumeration must be made of the number of bacteria in unit volume. A unit volume of sample is usually 1 ml or 1 g. If absorbance readings are matched with plate counts of the same culture, this correlation can be used in future estimations of bacterial numbers obtained by measuring turbidity. A graph of absorbance versus bacterial count of known concentration will form an approximately straight line.

More than one million cells per ml must be present for the first traces of turbidity to be visible. About 10 million to 100 million cells per ml are needed to make a suspension turbid enough to be read on a spectrophotometer. Therefore, turbidity is not a useful measure of contamination by a relatively small number of bacteria.

Preparation of the curve from *Escherichia coli* culture includes:

1. Prepare serial dilution from a pure *E. coli* culture.
2. Use 1 ml from each dilution to prepare pour plate counts as described in Sections 16.2.4 and 16.2.5. After 24 to 48 h incubation at 37°C read and calculate bacterial counts per 1 ml.
3. Prepare serial dilution from the same *E. coli* culture as in step 1 for absorbance determination by using a nutrient broth. With the series, always prepare a tube with the broth without the addition of a bacterial culture and mark the tube as the control.
4. Using a spectrophotometer (operation of the meter depends on the model and type) adjust the wavelength to 520 nm.
5. Adjust the meter to 0 absorbance reading by the control tube.

6. Read the absorbance for the prepared serial dilutions in step 3 starting with the undiluted *E. coli* culture.
7. From the count per ml of the original *E. coli* culture in step 2, now calculate the count of the 1:2 series of dilutions prepared above in step 3.
8. Using the absorbance and the count per ml of the 1:2 series of dilutions, plot the relationship between absorbance and bacterial count on graph paper.

14.2 BIOCHEMICAL REACTIONS AND ENZYMATIC TESTS

Enzymatic and biochemical reactions are used to identify microorganisms once they have been isolated. Some rapid methods and basic concepts of the commercial systems are currently available for performing these tests. The tests mentioned here are only a small fraction of those available, but these are some of the most universally used.

Chemical reactions that occur within all living organisms are referred to as metabolism. Metabolic processes involve enzymes, which are proteins that catalyze biologic reactions. The majority of enzymes function inside a cell; that is, they are *endoenzymes*. Many bacteria make some enzymes, called *exoenzymes*, which are released from the cell to catalyze reactions outside of the cell.

On the basis of which substrate a particular bacterium uses and which metabolic products it forms, laboratory tests have been designed to determine which enzymes the bacterium has.

These tests can produce positive reactions in less than 1 h, and most take only a few minutes, although incubation for up to 4 h may be required.

14.2.1 CATALASE TEST

The enzyme catalase catalyzes the liberation of water and oxygen from hydrogen peroxide, a metabolic end product toxic to bacteria.

All members of the *Staphylococci* are catalase positive whereas members of the genus *Streptococcus* are negative.

Catalase can also help distinguish *Bacillus species* (catalase positive) from *Clostridium species* (catalase negative). The test can be performed with a very small amount of growth removed from the agar surface.

The principle of the method, the breakdown of hydrogen peroxide into oxygen and water, is mediated by the enzyme catalase. When a small amount of an organism that produces catalase is introduced into hydrogen peroxide, rapid elaboration of bubbles of oxygen, the gaseous product of the enzyme's activity, is produced.

1. With a loop or sterile wooden stick, transfer a small amount of pure growth from the agar onto the surface of a clean, dry glass slide.
2. Immediately place a drop of 3 percent hydrogen peroxide (H_2O_2) onto a portion of a colony on the slide.
3. Observe for the evolution of gas bubbles, indicating a positive test as shown in Figure 14.1.

14.2.2 CLUMPING FACTOR TEST (SLIDE COAGULASE TEST)

Gram-positive cocci that are catalase positive belong to the family Micrococcaceae, which includes the *Staphylococci.* The clumping factor test is used to screen quickly for isolates of *S. aureus,* which are almost always coagulase positive. The clumping factor is a cell-associated substance that binds plasma fibrinogen, causing agglutination of the organisms by binding them together with aggregated

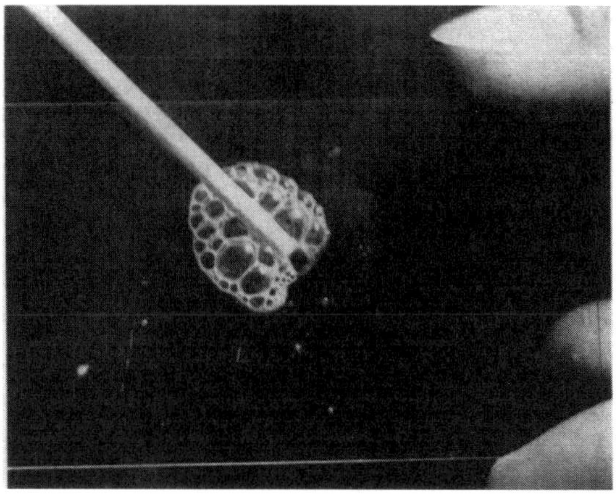

FIGURE 14.1 Catalase test. Bubbles of gas are released when a catalase-positive organism is emulsified in a drop of hydrogen peroxide on a slide.

fibrinogen. Those organisms that produce the clumping factor also elaborate the coagulase enzyme and can be identified presumptively as *S. aureus*.

1. Place a coagulase plasma (rabbit plasma with EDTA or citrate, commercially available) on a clean, dry slide.
2. Place a drop of distilled water next to the drop of plasma as a control.
3. With a loop or wooden stick, emulsify an amount of the isolated colony being tested in each drop, and inoculate the water first. Try to create a smooth suspension.
4. Watch for clumping in the coagulase plasma drop and a smooth, homogenous suspension in the control. Clumping in both drops indicates that the organism autoagglutinates and is suitable for this test. Positive organisms, such as *S. aureus*, exhibit immediate aggregation visible to the naked eye.

14.2.3 Nitrate Reduction Test

Nitrate serves as the source of nitrogen for many bacteria and fungi, but it must be broken down. Many organisms possess the enzyme nitrate reductase which is capable of converting nitrate (NO_3^-) to nitrite (NO_2^-). This process involves the reduction of the nitrogen atom in the nitrate molecule forming nitrite which accumulates in the medium. The enterobacteriaceae and many other Gram-negative bacilli, mycobacteria, and fungi reduce nitrate to nitrite.

Certain microorganisms are able to reduce nitrite further to nitrogen by replacing the remaining oxygen by hydrogens. Some pseudomonas species and other fermentative Gram-negative bacilli possess this ability.

In the nitrate reduction test, the nitrate reduced to nitrite, which combines with an acidified naphtylamine substrate to form a red colored end product. Metallic zinc catalyzes the reduction of nitrate to nitrite; thus, with the addition of zinc, a negative test will yield a red color, indicating the presence of unreacted nitrate.

1. Grow the organism in 5 ml of nitrate broth (commercially available) for 24 to 48 h. A small inverted tube (Durham tube) may be placed into the broth to trap bubbles of nitrogen gas that may be formed by nitrite-reducing organisms.

2. Prepare the reagents:
   Reagent A: Dissolve 4 g sulfanilic acid in 5 *M* acetic acid and dilute to 500 ml.
   Reagent B: 3 ml of n,N-dimethyl-1-naphtylamine and fill up to 500 ml with 5 *M* acetic acid.
3. Add 3 drops of reagent A and then 3 drops of reagent B to the suspension of the organism in the broth used in step 1.
4. Wait 30 min for the production of the red color, indicating the presence of nitrite (NO_2^-).
5. The presence of unreacted nitrate can be detected by adding a pinch of commercially available zinc powder to the broth if the red color does not develop after the initial regents were added.

14.2.4 Oxidase Test (Kovacs Method)

The oxidase test indicates the presence of the enzyme *cytochrome oxidase*. This iron-containing porphyrin enzyme participates in the electron transport mechanism and in the nitrate metabolic pathways of some bacteria. The simple reaction uses the *Kovacs method.*

The principle of the method that the cytochrome oxidase enzyme is able to oxidase the substrate *tetramethyl-p-phenylenediamine dihydrochloride*, forming a colored end product, *indophenol*. The dark-purple end product will be visible if a small amount of growth from a strain that produces the enzyme is rubbed on a substrate impregnated filter paper.

Positive organisms turn the filter paper dark purple within 10 s. Negative organisms, such as *E. coli,* remain colorless and have a negative oxidase test.

1. Prepare a solution of 1 percent tetramethyl-p-phenylenediamine dihydrochloride (the chemical is available from Kodak or Sigma or other chemical companies) by dissolve 1 g of the chemical and dilute to 100 ml with sterile water. Prepare fresh on the day of use. The reagent is also commercially available in individually sealed glass ampules for daily use.
2. Place a filter paper circle into a sterile, plastic, disposable petri dish, and moisten the paper with several drops of the fresh reagent.
3. Remove a small portion of the colony to be tested (preferably not more than 24 h old) with a platinum wire or wooden stick and rub the growth on the moistened filter paper. Do not use iron or other reactive wire because it will cause false positive reactions.
4. Observe for a color change to blue or purple within 10 s. Time is critical. See Figure 14.2.

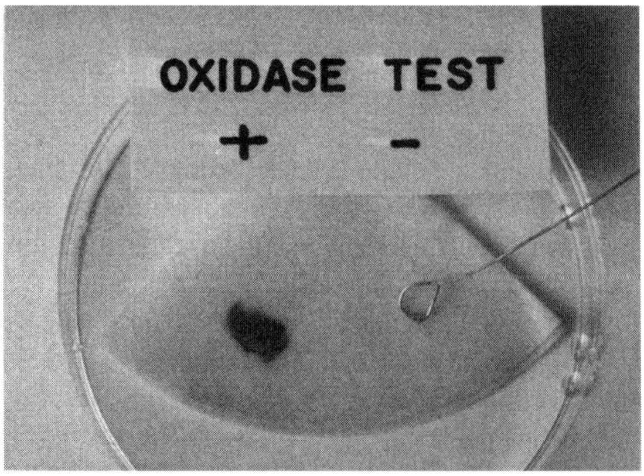

FIGURE 14.2 Kovacs oxidase test. A purple color results when a small portion from a colony of an oxidase-positive organism is rubbed onto filter paper saturated with the oxidase reagent.

14.2.5 Indole Test

The ability of an organism to degrade the amino acid tryptophane can be detected by testing for *indole*, the product of *tryptophanase*. Reaction of indole with an aldehyde results in a colored end product. Organisms that produce the enzyme tryptophanase are able to degrade the amino acid tryptophane into pyruvic acid, ammonia, and the product indole. Indole is detected by its combination with the indicator aldehyde (available commercially) to form a colored end product.

The test is excellent for the differentiation of *E. coli* and Enterobacter-Klebsiella groups. Typical strains of *E. coli* produce indole, while the Enterobacter-Klebsiella groups do not.

14.2.5.1 The Rapid Indole Test

The rapid indole test is performed with *dimethylamino-cinnamaldehyde* reagent, which turns blue or blue-green in the presence of indole. The blue-green compound formed by indole and cinnamaldehyde is visualized by rubbing bacteria that produce tryptophanase on filter paper impregnated with the substrate.

The classic test calls for p-dimethylamino-benzaldehyde as the indicator (Kovacs reagent). Broth for the indole test is commercially available. Positive result is visible as the red to pink ring around the top of the tube where the reagent floats after incubation at least for 24 h. In the negative test, no change in the reagent's color is seen.

1. Prepare 1 percent indole solution—1 g para-dimethyl-amino-cinnamaldehyde, dissolved in 10 percent v/v hydrochloric acid (10 ml conc HCl and 90 ml laboratory pure water). Store in a dark bottle in the refrigerator.
2. Saturate a qualitative filter paper (Whatman No. 1 is fine) in the bottom of a petri dish with the reagent prepared in step 1.
3. Using a wooden stick or loop, rub a portion of the colony on the filter paper. Rapid development of a blue color indicates a positive test. Most indole positive organisms turn blue within 30 s. An indole test is shown in Figure 14.3.

14.2.5.2 Regular Indole Test

The following steps are followed for the regular indole test.

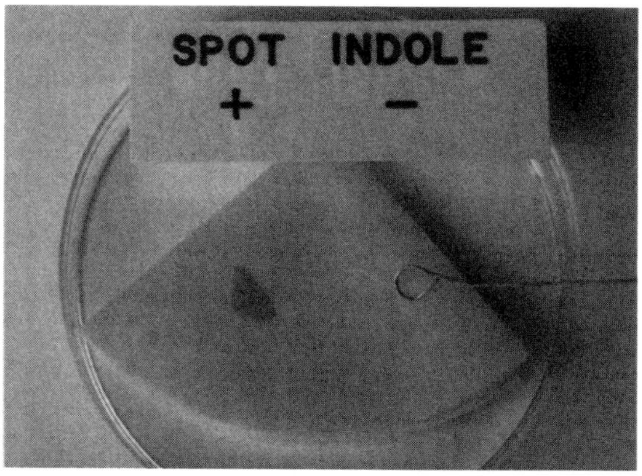

FIGURE 14.3 Spot indole test. A blue-green color results when a small portion from the colony of a indole-positive organism is rubbed onto filter paper saturated with the indole reagent.

1. The medium (tryptophane broth) and test reagent (5 g paramethyl-aminobenzaldehyde are dissolved in 75 ml isoamyl alcohol 25 ml conc *HCl* is added: the reagent should be yellow. The test reagent is commercially available.
2. Inoculate 5 ml portions of medium from a pure culture and incubate at 35°C for 24 h.
3. Add 0.2 to 0.3 ml test reagent and shake. Let stand about 10 min and observe the result.
4. A dark red color in the amyl alcohol surface layer constitutes a positive indole test. The original color of the reagent indicates a negative test.

14.2.6 Rapid Urease Test

Proteus sp., Klebsiella sp., some *Citrobacter sp.,* several yeast and fungi produce the enzyme urease, which hydrolyses urea into ammonia, water, and carbon dioxide. The alkaline end products cause the indicator phenyl red to change from yellow to pink or red. This test can be used as a component of screening tests for lactose negative colonies and help to differentiate *Salmonella* and *Shigella sp.,* which are urease negative from the urease-positive nonpathogens. Commercial reagents are available for performance of the test.

1. Inoculate a tube of rapid urea broth (commercially available) with a heavy suspension of the organism to be tested.
2. Incubate the tube at 35°C and observe at 15, 30, 60 min and up to 4 h for a change in color to pink or red.

Urease positive organisms such as *Proteus sp.* yield a bright-pink or bright-red color to the broth, negative organisms such as *E. coli* do not cause a change in the color of the broth.

14.2.7 IMViC Tests

The IMViC tests were developed as a means of separating enteric bacteria, particularly the coliforms, using a standard combination of four tests. Each letter in IMViC represents a test.

I = indole production from tryptophane.
M = methyl red test for acid production from glucose.
V = Voges-Proskauer test for production of acetoin from glucose.
i = added for easier pronunciation.
C = for the utilization of citrate as the sole carbon source.

The *Simmons citrate agar* used in this exercise contains the indicator bromthymol blue. Citric acid will be the only source of carbon. Therefore, only organisms capable of utilizing citric acid as a source of carbon will grow. When the citric acid is metabolized, an excess of sodium and ammonia ions results, and the indicator turns from green to blue, indicating alkaline conditions.

The test is specially useful in the differentiation of Gram-negative enteric bacilli, especially coliforms of the *E. coli* and the *Enterobacter-Klebsiella* groups. Because enteric organisms have very similar physiological reactions and often cannot be adequately differentiated, the IMViC tests are a considerable aid in the identification of organisms within the family Enterobacteriaceae.

14.2.8 Methyl Red and Voges-Proskauer Test

Organisms that utilize glucose may do so by producing abundant acidic end products such as formate and acetate from the intermediate pyruvate metabolite, or they may metabolize the carbon compounds to acetoin and butanediol, which are more neutral in pH. The methyl red (MR) and Voges-Proskauer (V-P) detect the presence of the end products of these two divergent metabolic pathways. The same substrate broth, *MR/V-P broth*, is used for both tests. Tubes of broth (5 ml per tube) are available commercially, or prepared by the laboratory from powder base. The organisms are inoculated below the surface of the medium in the tube, and the tube is incubated with a loose cap at 35 to 37°C for at least 48 h. A longer incubation may be necessary.

14.2.8.1 Methyl Red (MR) Test

The test depends upon the ability of the organisms to produce acid from glucose in sufficient quantity to cause the methyl red indicator to change to its acidic color red and hold this low pH of 4.4 for 5 days. Typical strain of *E. coli* are mixed acid fermenters which produce a variety of acid end products which change the methyl red color to red. The *Enterobacter-Klebsiella* organisms ferment the same amount of glucose but convert it into neutral products such as acetoin (acetyl methyl carbinol) and 2,3 butanediol. Consequently, the pH which may initially reach pH 4.4, very quickly rises again and the methyl red no longer displays its acid red color.

1. The buffered glucose broth medium and the methyl red indicator are commercially available.
2. Inoculate 10 ml portion of the medium from a pure culture. Incubate at 35°C for 5 days.
3. Take out about 5 ml portion of the culture and add 5 drops methyl red indicator solution. Testing portions of the culture at 2, 3, 4, and 5 days may provide positive test results.
4. Record a distinct red color as methyl red positive and a distinct yellow color as methyl red negative.

14.2.8.2 Voges-Proskauer (V-P) Reaction

The reaction is the result of a red color complex formed by the oxidation of 2,3 butanediol and acetoin in alkali (KOH) and the presence of creatine and alpha-naphthol. *Escherichia coli* does not produce these end products and is Voges-Proskauer negative. Since the *Enterobacter-Klebsiella* groups produce 2,3 butanediol and acetoin, they produce a positive test.

1. Use buffered glucose media (used also in MR tests) or alternatively, the salt peptone glucose medium. The reagents are naphtol solution (5 g alpha-naphtol in 100 ml of absolute ethyl alcohol) and 7 *N* KOH (40 g KOH in 100 ml reagent grade water). The naphtol solution is stable for two weeks at 5 to 10°C.
2. Inoculate 5 ml portion of either medium and incubate at 35°C for 48 h.
3. To 1 ml of culture add 0.6 ml of naphtol solution and 0.2 ml of KOH solution. Development of a pink to crimson color within 5 min constitutes a positive test. Do not read after 5 min. Discard tubes developing a copper color.

14.2.9 Citrate Utilization Test

Certain organisms are able to utilize a single substrate such as a sole carbon source. The ability to use citrate can help to differentiate among members of Enterobacteriaceae. The test determines whether one organism can grow with citrate as the sole source of carbon. The test is designed to determine whether or not the bacterium has a cell membrane transport protein to carry citrate into the cell. If the organism has the particular protein, then its grows. If the organism lacks this protein, then it can not grow.

Escherichia coli does not utilize citrate, whereas members of the *Enterobacter-Klebsiella* groups do.

The medium, *Simmons' citrate agar*, contains salts, buffer, cations, citrate, and bromthymol blue as the pH indicator. The medium can be purchased prepared or as dry powder.

Because the reaction requires oxygen, the organism is inoculated on the surface of an agar slant medium. A very light inoculum is necessary to prevent false positive results because of the carryover of substrates from previous media. The tube is incubated for 24 h or for up to 4 days at 35 to 37°C with a loose cap. Growth of the organism on the slant, with a change of the color indicator from green to blue, or turquoise blue, is evidence of a positive test.

1. Use either Koser's broth or Simmons agar (commercially available).
2. Lightly inoculate liquid medium with a straight needle, never with a pipet.
3. Incubate at 35°C for 72 to 96 h.
4. Record visible growth as positive, no growth as negative.
5. Alternatively, inoculate agar medium with a straight needle, using both a stab and a streak. Incubate at 35°C for 48 h. Record growth on the medium with a blue color as a positive reaction, record absence of growth as a negative reaction.

14.2.10 Decarboxylase Test

This procedure tests the ability of bacteria to metabolize the amino acids lysine and ornithine.

1. Use a basal medium. Divide the medium into three portions: make no addition to the first portion, add enough L-lysine-dihydrochloride to the second portion to make a 1 percent solution, and add L-ornithine-dihydrochloride to the third portion to make a 1 percent solution.
2. Dispense medium in 3 to 4 ml portions in screw-capped test tubes and sterilize by autoclaving at 121°C for 10 min. A floccular precipitate in the ornithine medium does not interfere with its use.
3. Sterilize mineral oil by autoclaving at 121°C for 30 to 60 min depending on the size of the container.
4. Lightly inoculate each of the three media, add a layer of about 10 mm thickness of mineral oil, and incubate at 37°C for up to 4 days. Examine tubes daily. A color change from yellow to violet or reddish violet constitutes a positive test; a change to bluish gray indicates a weak positive, no color change represents a negative test.

14.2.11 Motility Test

Use motility test agar medium. Inoculate by stabbing into the top of the medium column to a depth of 5 mm. Incubate for 1 to 2 days at 35°C. If negative, incubate an additional 5 days at 20 to 25°C. Diffuse growth through the medium from the point of inoculation is positive.

14.3 RAPID IDENTIFICATION SYSTEMS (MULTITEST SYSTEMS)

Rapid identification systems provide a large number of results from one inoculation.

Enterotube II is divided into 12 compartments, each containing a different substrate in agar.

API 20E consists of 20 microtubes containing dehydrated substrates. The substrates are rehydrated by adding a bacterial suspension. No culturing beyond the initial isolation is necessary for these systems. Comparisons between these rapid identification methods and conventional culture methods show that they are as accurate as conventional tube methods.

These commercial identification systems are time-, cost-, and labor-saving. Detailed operation of these multiple test systems are discussed in Chapter 20.

15 Methods for Analyzing Microbiological Quality of the Environment

The following chapters describe procedures for microbiological examination of environmental samples to determine sanitary quality. The methods are selected according to the best techniques currently available; however, their limitations must be understood thoroughly.

Chapter 7 introduced microorganisms effect on water quality, including indicator organisms and pathogens. Tests for detection and enumeration of indicator organisms, rather than that of pathogens are used. The coliform group of bacteria, as herein defined, is the principal indicator of suitability of water for domestic, industrial, or other uses. The significance of the coliform group density establishes the criterion for the degree of pollution and, thus, of sanitary quality. The significance of the test and the interpretation of the results are well authenticated and have been used as a basis for the standards of bacteriological quality of water supplies.

Examination of routine microbiological samples cannot be regarded as providing complete information concerning water quality. Always consider bacteriological results in the light of information available concerning the sanitary conditions surrounding the sample source. For a water supply, precise evaluation of quality can be made only when the results of laboratory examinations are interpreted in the light of sanitary survey data. Consider the results of the examination of a single sample from a given source inadequate. When possible, base evaluation of water quality on the examination of a series of samples collected over a known and protracted period of time.

Most commonly used methods with method numbers and references are given in Table 15.1.

15.1 METHODS AND TECHNIQUES

15.1.1 Methods for the Total Coliform Group

The membrane filter (MF) technique, which involves a direct plating for detection and estimation of coliform densities, is as effective as the multiple-tube fermentation test. In both procedures coliform density is reported conventionally as membrane filter count per 100 ml and as the most probable number (MPN) index. The results of both procedures help to establish the sanitary quality of the water and increase the effectiveness of the treatment process. Both methods may also be used for solid matrices, such as soil, sediment, sludge, etc. In this case, the report is given as the bacterial count per gram of sample. Detailed methods for the detection and enumeration of the total coliform group are discussed in Chapters 17 and 18.

15.1.2 Methods for Fecal Coliforms

Coliform group bacteria are present in the digestive system of warm-blooded animals and, generally, include organisms capable of producing gas from lactose in a suitable culture medium at 44.5°C. However, there are coliform organisms that are not present in the digestive system of warm-blooded animals and that do not produce gas in the above culture medium and temperature. According to this difference, there are total coliform bacteria that ferment lactose at 35°C and a group of the coliform bacteria called fecal coliform that ferment lactose at 44.5°C.

TABLE 15.1
Methods for Microbiological Tests

Parameter	Technique	Method No.	Reference
Heterotrophic plate count (HPC)	Pour plate method	9215	R-1
		Part III.A	R-2
Total coliform	MF	9222B	R-1
		Part III.B	R-2
	MF, delayed inc.	9222C	R-1
	MPN	0221B	R-1
		Part III.B	R-2
Fecal coliform	MF	9222D	R-1
		Part III.C	R-2
	MF, delayed inc.	9222E	R-1
	MPN	9221E	R-1
		Part III.C	R-2
Klebsiella	MF	9222F	R-1
F. Streptococcus	MF	9230C	R-1
		Part III.D	R-2
	MPN	9230B	R-1
		Part III.D	R-2
Iron bacteria	Microscopic	9240B	R-1
Sulfur bacteria	Microscopic	9240C	R-1
Actynomycetes	Plate count	9250B	R-1
		Part III.F	R-2

MF = membrane filtration, MPN = most probable number, Delayed inc. = delayed incubation method, R-1 = Standard methods for the examination of water and wastewater, APHA.AWWA.WPCF, 18th ed., 1992, R-2 = Microbiological methods for monitoring the environment, EPA, 8-78-017, 1978.

Both the membrane filter (MF) method and the multiple-tube fermentation method (MPN test) have been modified to incorporate incubation in confirmatory tests at 44.5°C to provide estimates of the density of the fecal coliform.

Procedures for fecal coliforms also include a 24 h multiple-tube test using A-1 medium and a 7 h rapid test (see Chapter 18).

This differentiation yields valuable information concerning the possible sources of pollution in water, especially its remoteness, because the nonfecal members of the coliform group may be expected to survive longer than the fecal members in the unfavorable environment provided by the water.

15.1.3 Heterotrophic Plate Count

The test provides an approximate enumeration of total numbers of viable bacteria that may yield important information about microbiological quality of the sample and may provide supporting data for the coliform test results. The count determined by spread, or pour plate count technique is discussed in Chapter 7 and detailed methodology is discussed in Chapter 16.

15.1.4 Method for Fecal Streptococcus

Fecal streptococcus is also an indicator of fecal pollution and the methods for its detection are available in Chapter 18.

15.1.5 Detection of Stressed Organisms

Indicator bacteria, such as coliform and fecal streptococci, may become stressed or injured by other disruption. The stress may be caused by temperature changes or chemical treatment, such as chorine or toxic wastes, especially toxic metals and phenols.

15.1.5.1 Ambient Temperature Effect

Extreme ambient temperatures stress microorganisms and reduce the recovery of microbiological indicators. For example, in extreme cold areas, the severe change from cold stream temperature to the 44.5°C temperature of incubation reduces recovery of fecal coliforms. The two-step MF test for fecal coliform increases recoveries by using a 2 h acclimation on an enrichment medium at 35°C before normal incubation at 44.5°C.

Water samples from natural waters at high temperatures may include large number of noncoliform organisms which interfere with sheen-production on MFs and with positive gas production in MPN analysis. An improved MF medium that provides greater selectivity is desirable but may not be possible without sacrificing recovery.

15.1.5.2 Chlorinated Effluents and Toxic Wastes

The false negative bacteriological findings could result in an inaccurate definition of water quality, or even worse, lead to the erroneous acceptance of a potentially hazardous condition resulting from the presence of resistant pathogens. Recent publications support the health significance of injured coliform bacteria.

Enteropathogenic bacteria are less susceptible than coliforms to injury under conditions similar to those in drinking water and injured pathogens retain the potential for virulence. Viruses and waterborne pathogens that form cysts are also more resistant to environmental stressors than indicator bacteria. Since the MF method yields for low and variable recovery from chlorinated water samples, the MPN method is preferable.

When MPN procedure results consistently higher than those obtained from parallel membrane filter tests, consider injury problems.

Recovery of injured total coliform and fecal coliform bacteria by MF method needs modified media and an agar overlay technique. The MF modified methods are discussed in Chapters 17 and 18.

15.2 RAPID DETECTION METHODS

Applications of rapid methods may range from analysis of wastewater to potable water quality assessment. There are emergencies involving water treatment plant failure, line breaks in the distribution system, or other disruptions to water supply caused by disasters, when the rapid detection methods are needed.

15.2.1 7 Hour Fecal Coliform Test

Similar to the fecal coliform MF test, the 7 h fecal coliform test uses a separate medium and incubation temperature to yield results in 7 h that are generally comparable to those obtained by the standard fecal coliform method, see Chapter 18.

15.3 METHODS FOR MICROBIOLOGICAL EXAMINATION OF RECREATIONAL WATERS

Recreational waters include freshwater swimming pools, whirlpools, and naturally occurring fresh and marine surface waters. As the previous routine shows, these waters have been tested for coliform and heterotrophic bacteria. These results indicate that it may be safe or unsafe to drink. Other bacteria have been isolated from recreational water that may suggest public health risks through body contact, ingestion, or inhalation.

Possible indicators of unsafe recreational water quality include the coliform group, species of *Pseudomonas*, *Streptococcus*, *Staphylococcus*, and in rare cases, *Legionella.*

Organisms, as *Mycobacterium, Candida albicans, Naegleria,* and *Acanthamoeba* may produce spores and cysts that are more resistant than the indicator bacteria.

Viral diseases associated with untreated recreational water include those caused by *Coxsackie A* and *B*, *Adenovirus types 3* and *4*, *Hepatitis A*, and a variety of *gastroenteritis viruses.*

Routine examination for pathogenic microorganisms is not recommended except for special studies, or for investigations of water related illnesses. In these cases, microbiological assays should be focused on the known or suspected pathogen.

15.3.1 Swimming Pools

Microorganisms of concern typically are those from the bather's body, and its orifices, include those causing infections of the ear, upper respiratory tract, skin, and intestinal or genitourinary tracts. Water quality depends on the efficiency of disinfection, the number of bathers in the pool at any one time, and the total number of bathers per day.

15.3.1.1 Disinfected Indoor Pools

The heterotrophic plate count (HPC) is a primary indicator of disinfection efficiency. Check residual chlorine when turbidity value is above 1 NTU (nephelometric turbidity unit) and the HPC exceeds 500 CFUs (colony forming units). Other supporting indicators may include *Streptococcus, Staphylococcus*, and *Pseudomonas*. These organisms account for a large percentage of swimming pool associated illnesses and may be relatively resistant to the effect of chlorine. In special circumstances, *Mycobacterium, Legionella,* or *Candida albicans* may be significant.

15.3.1.2 Disinfected Outdoor Pools

Fecal coliform bacteria are the primary indicators of contamination. Supporting indicators are streptococcus, staphylococcus, and pseudomonas.

15.3.1.3 Untreated Pools

The primary indicator is fecal coliform and supporting organisms as listed in disinfected outdoor pools.

Methods for heterotrophic plate count are discussed in Chapter 16, methods for total coliform are discussed in Chapter 17, methods for fecal coliform, *Streptococcus*, and *Staphylococcus* are discussed in Chapter 18, and *Pseudomonas aeroginosa* is tested according to methods discussed in Chapter 19.

15.3.2 Whirlpools

A whirlpool is a shallow pool with a maximum water depth of 1.2 m and has a closed-cycle water system, a heated water supply, and usually a hydrojet recirculation system. Whirlpools accommodate one or more bathers. They are located in homes, apartments, hotels, athletic facilities, and hospitals (therapeutic use).

To ensure whirlpool safety, test water for proper germicide concentration, free and combined chlorine (or other halogen), pH, and temperature.

The standard indices of pollution, that is, coliform bacteria are insufficient to judge the microbiological quality of the pool water.

Analysis should include *Pseudomonas aeroginosa, Staphylococcus*, and *Streptococcus*.

15.3.3 Natural Bathing Beaches

A natural bathing beach is a shoreline area of a stream, lake, ocean impoundment, or hot spring that is used for recreation. A wide variety of pathogenic microorganisms can be transmitted to humans through the use of natural fresh and marine recreational waters that may be contaminated with wastewater.

Enteropathogenic agents, such as *Salmonella* and *Shigella, enteroviruses, protozoa*, and multicellular *parasites*, human pathogens, such as *Pseudomonas, Klebsiella, vibrio, Aeromonas hydrophyla,* and *Candida albicans* may multiply in recreational water in the presence of nutrients.

Other organisms may be carried into the water from the skin and upper orifices of the recreationists, such as staphylococcus, pathogenic mycobacteria, and pathogenic naegleria (causes amoebic meningoencephalitis). Methods suitable for routine examination are not available for these organisms.

Measure fecal contamination with the routine fecal coliform test, see Chapter 18.

An application of the fecal coliform and fecal streptococcus ratios of 4.0 or higher typically indicate domestic waste while ratios of 0.6 or lower are common in discharges from farm animals and stormwater runoff. Optimally measure this ratio near the point of waste discharge because the ratio changes with distance or time.

15.4 DETECTION OF PATHOGENIC BACTERIA

A wide variety of enteric pathogenic bacteria may occur in water supplies and in wastewater. With increasing demands on water resources, the potential for contamination of surface and groundwater by enteric microorganisms could be expected to increase. Water-borne outbreaks of disease continue to occur.

Routine examination of water and wastewater for pathogenic bacteria is not recommended. No single procedure is available that can be used to isolate and identify all pathogens. Therefore, negative findings for specific pathogens are provisional because state-of-the-art methodology may not be sufficiently sensitive to detect low levels of pathogens. Routine examination of water and wastewater for pathogens is limited by factors, such as the lack of facilities, untrained personnel, insufficient laboratory time, high costs, and inadequate methods. There is a strong need for intensive research in this area and such research should be encouraged at every opportunity.

The coliform test may not always be a suitable indicator of the microbiological safety of water. Pathogenic microorganisms can be isolated from water containing few, if any, coliform bacteria. For example, Legionellaceae, because they are not enteric, would not be expected to demonstrate a relationship to fecal bacteria.

Nevertheless, the coliform test has been, and continues to be, a useful tool for assessing the quality of water.

15.5 DETECTION OF SOIL MICROORGANISMS

Soil is a uniqe medium that contains a diverse community of organisms, representing many morphological and physiological types. Some organisms exist at relatively low levels numerically, but can exert profound influences on some soil biological processes. Organisms in the soil are never static in numbers or activity. Therefore, the enumeration of some population or microbial activity represents a point in time for that particular population that is in dynamic equilibrium with its physical, chemical, and biological environment.

Variation of microbial numbers and activities can occur with depth and soil type. Soil structure, texture, and moisture levels can drastically affect aeration within the soil environment from one characterized as aerobic to one of anaerobic properties, thus influencing the distribution of physiological types.

At least as important as soil physical properties on microbial diversity is the season of the year. Soil fertility also plays a role in governing microbial development. The diversity of organisms is dependent on the type of plant and the proximity of the organism to the plant root itself.

Within this highly complex environment, attempts to study soil organisms will be difficult. Therefore, the formation of study objectives and soil sampling and handling procedures are among the most critical steps in the process.

15.5.1 Sample Collection

Each sample collected should be a composite sample that adequately encompasses the variation within the sample unit. After the composite sample is thoroughly mixed, a portion may be placed directly in plastic bags for transport to the laboratory. During the mixing process, avoid sample drying or exposure to excessive high temperature, see Section 10.2.7.

15.5.2 Sample Handling and Storage

Sieve samples containing large quantities of extraneous material such as root debris or rock before use. It is important to prevent sample drying at this point because the alteration of sample conditions may drastically affect the number of microorganisms. Use the soil sample as soon as possible after sampling. Avoid long-term storage. For samples that cannot be used immediately, store under conditions cooler than ambient but avoid freezing temperatures.

15.5.3 Moisture Determination

Determination of the percent of moisture content of the sample is necessary for calculating the bacterial population expressed on the dry weight of the sample.

1. Fill a tared moisture tin about half full with a portion of the collected soil sample.
2. Weigh the moisture tin *and soil*, and place in a 105°C oven until a constant weight is obtained. For most soil samples, drying 24 h is usually satisfactory.

3. Determine the amount of water held in the sample and the oven-dry weight of the soil.
4. Express moisture content as percent of moisture.

$$\text{moisture } \% = \frac{\text{weight of water} \times 100}{\text{weight of original soil sample}} \tag{15.1}$$

$$\text{Moisture } \%_{\text{DW}} = \frac{\text{weight of water} \times 100}{\text{dry weight of the soil sample}} \tag{15.2}$$

For example:

Weight of the original soil sample		= 5.0000 g
Weight of the oven-dry soil		= 4.2836 g
Weight of the water		= 0.7164 g
Moisture%	= (0.7164 × 100)/5.0000	= 14.3280%
Dry weight%	= 100.00 − 14.33	= 85.67%
Moisture% DW	= (0.7164 × 100)/4.2836	= 16.72%

15.5.4 Plate Count

No microbiological technique is as widely accepted in enumerating soil organisms as the plate count. The underlying principles are almost simplistic: dispersing of the sample, distributing an aliquot to an appropriate medium, incubating under suitable conditions, and, finally, counting the developed colonies. Standardization of shaking for uniform dispersion is achievable through the use of mechanical shakers or through strict attention to the shaking motion and time of shaking.

Determine and report the number of colony-forming units per g on the dry base of the soil.

The heterotrophic plate count technique is discussed in Section 15.1.3 and Chapter 16.

15.5.5 Most Probable Number (MPN) Method

This method permits estimation of population density without an actual count of single cells or colonies as described in Chapter 18. It is sometimes called the method of ultimate or extinction dilution or, less descriptively, simply the dilution method.

16 Heterotrophic Plate Count

16.1 INTRODUCTION

Heterotrophic Plate Count (HPC), formerly known as the standard plate count (SPC) is a direct quantitative measurement of the viable aerobic and facultative anaerobic bacteria in water, measuring changes during water treatment, and distribution (see Sections 7.2.1 and 13.2). The test gives an indication of the effectiveness of disinfection as well as the existence of cross connection and other problems within the distribution line. Total bacterial densities greater than 500 to 1000 organisms per ml may indicate coliform suppression. It is also used to test swimming pool quality. Although no one set of plate count conditions can enumerate all organisms present, the method provides a uniform technique required for comparative testing and monitoring of water quality in selected situations.

Colonies may arise from pairs, chains, clusters, or single cells, all of which are included in the term colony forming units (CFU). Two techniques are available: the pour and spread technique.

The pour plate method is simple. Each colony that develops in the agar medium originates theoretically from one bacterial cell. Volumes of the sample or the diluted sample that were used ranged from 0.1 to 2.0 ml. The colonies produced are relatively small and compact. Submerged colonies are slower growing and difficult to transfer. The agar is transferred to the sample when the temperature is between 44 to 46°C and this heat transfer shock affects bacterial growth.

The spread plate method causes no heat shock, all colonies are on the surface and they can transfer quickly and easily. However, this method is limited by the small sample size, 0.1 to 0.5 ml, that can be absorbed by the agar.

16.2 POUR PLATE METHOD

An aliquot of the sample or its dilution is pipetted into a sterile petri dish and a liquefied agar medium is added, cooled, and solidified. Then it is incubated and counted.

16.2.1 Equipment and Material

The following is a list of the necessary equipment and material:

- Autoclave, see Sections 11.1.3 and 9.3.9.
- Incubator, maintain a stable 35° ± 0.5°C temperature. The temperature is checked against a NBS certified thermometer and recorded. See Sections 11.1 and 9.3.11.
- Water bath for tempering agar is set at 44 to 46°C (see Section 9.3.12).
- Sterile 100 mm × 15 mm petri dishes, made from glass or plastic.
- Sterile TD (to deliver) bacteriological or Mohr pipets made from glass or plastic with the appropriate volumes (see Section 11.1.2).
- Dilution bottles (see Section 11.2.6), marked at 99 ml volume, screw cap with neoprene rubber liner.
- Bunsen gas burner.
- Sterile buffered dilution water (working phosphate buffer solution); preparation and specifications are given in Section 13.1.4.

16.2.2 Media Preparation

The following procedure should be used to prepare the media.

1. Used the dehydrated form of the sterile plate count agar (tryptone glucose yeast agar). Prepare the media by following the manufacturer's description on the bottle.
2. Measure the prescribed quantity of the dehydrated media.
3. Add the required quantity of laboratory pure water.
4. Dissolve the media by heating and constantly stirring until the media is totally dissolved.
5. Dispense the media in tubes (15 to 20 ml per tube) or in bulk quantities in screw cap flasks.
6. Cool and save in the refrigerator. Label properly with the name and date prepared.

16.2.3 Preparation of Agar

The following procedures should be followed in the preparation of agar.

1. Melt the prepared plate count agar media (see Section 16.2.2) by heating in boiling a water bath. Do not allow the medium to remain at this high temperature beyond the time necessary to melt it. Prepared agar should be melted once only.
2. Place the melted agar in a tempering water bath maintained at a temperature of 44 to 46°C. Do not hold agar at this temperature longer than 30 min. Melting the media in melting tubes in a water bath is shown in Figure 16.1 and testing the agar temperature is shown in Figure 16.2.

16.2.4 Dilution Preparation

The dilution technique is discussed in Chapter 13. Selection of the proper dilution depends on the origin of the sample or previous analytical results of the sample. For most potable water samples, plates suitable for counting will be obtained by plating 1 ml and 0.1 ml undiluted samples and 1 ml and 0.1 ml from the 100× dilution as shown in Figure 16.3.

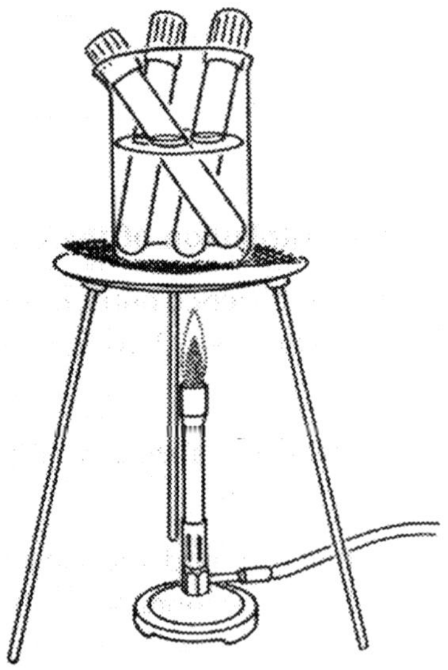

FIGURE 16.1 Melting media in tubes.

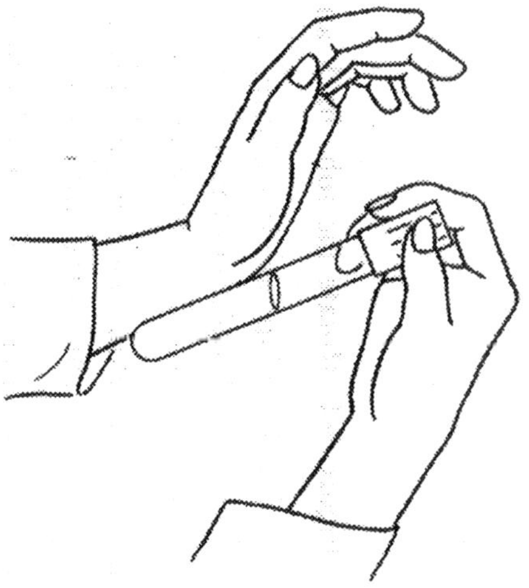

FIGURE 16.2 Testing agar temperature. It should be slightly warm to the touch but not hot.

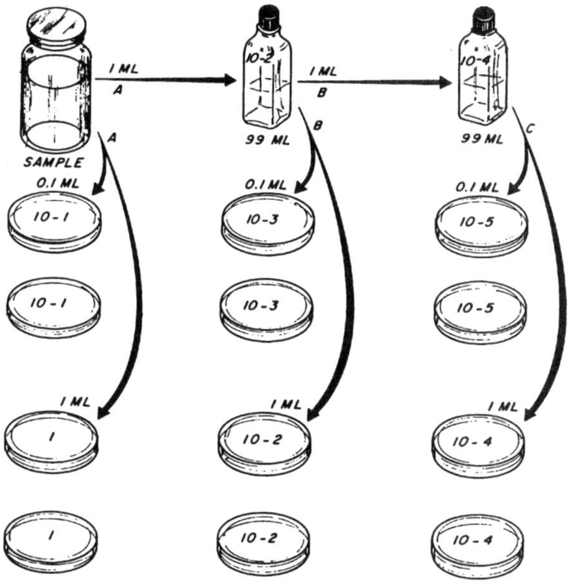

FIGURE 16.3 Typical dilution series for heterotrophic plate count.

Do not let more than 20 min elapse between starting pipetting and pouring the plates.

16.2.5 Plate Preparation

The following procedures should be followed in the preparation of plates.

1. Mark and arrange plates in a reasonable order for use. Prepare blank, duplicates, and positive control plate with one series of samples. For a sterility check, add 1 ml volume of sterile dilution water into the petri dish, add agar, and mix. This control plate will check the sterility of pipet, agar, and dilution water.

2. Arrange samples and dilutions according to the plates.
3. Vigorously shake each sample and each dilution before measuring the suitable volume into the empty sterile petri dishes.
4. Pipet the aliquot from the sample or dilution. Hold the pipet at an angle of about 45° with the tip touching the bottom of petri dish or inside neck of the dilution bottle. Lift the cover of the petri dish just high enough to insert pipet. Allow 2 to 4 s for the liquid to drain from the 1 ml graduation mark to the bottom of the petri dish. When the sample size is 0.1 ml, take 1 ml from the appropriate dilution the same way as previously described. The technique for handling the pipet and the petri dish cover is shown in Figure 16.4. Use a separate sterile pipet for each sample. Prepare two replicate plates for each sample and dilution.
5. Use a thermometer to check the temperature of the medium before pouring.
6. After the measurement of the samples for each series of plates, pour culture medium and mix carefully. Limit the number of samples to be plated in any one series so that no more than 20 min (preferably, 10 min) elapse between dilution of the first sample and pouring of the last plate in the series. Several precautions (i.e., aseptic techniques) are necessary to prevent contamination. When taking the melted agar from the holding water bath, the outside of the container should be wiped with a cloth or paper towel. Otherwise, the water will run into the plate and introduce contaminants. When removing the plug or cap to pour the agar, flame the mouth of the container to kill organisms on the outside lip. In pouring the agar from the container to the plate, raise the cover of the plate only on one side and just sufficiently to easily admit the mouth of the container. Add 12 to 15 ml of the agar medium to each petri dish and to the measured sample. The correct method to pour the melted media into the sterile petri plate with the sample by using the aseptic technique is shown in Figure 16.5.
7. Mix the inoculated medium carefully to prevent spilling. The recommended technique is to rotate plates 5× to the left, 5× to the right and 5× in a back and forth motion. Rotate the petri dish as shown is in Figure 16.6.
8. Allow plates to solidify (about 10 min), invert the plates, and than incubate at 35°C for 48 h. During incubation maintain humidity within the incubator so that the plates will have no moisture loss greater than 15 percent. A pan of water placed at the bottom of the incubator may be sufficient. To avoid rust on the inside walls and shelf in the incubator, the walls and shelf should be made of high quality stainless steel or anodized aluminum.

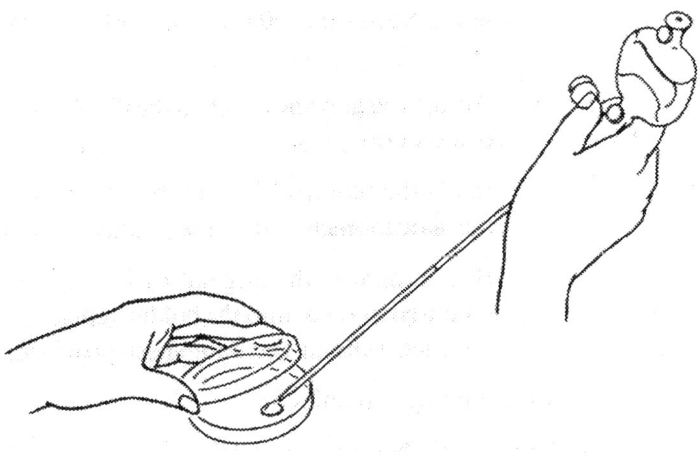

FIGURE 16.4 Pipetting sample into the petri dish. Technique for holding a pipet and a petri dish cover.

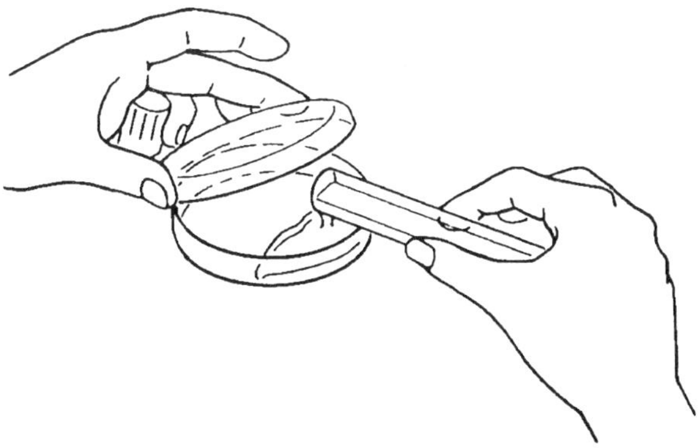

FIGURE 16.5 Using aseptic technique, lift the cover of the petri plate high enough to insert the mouth of the tube to pour the melted medium into it. Do not touch the plate with the tube.

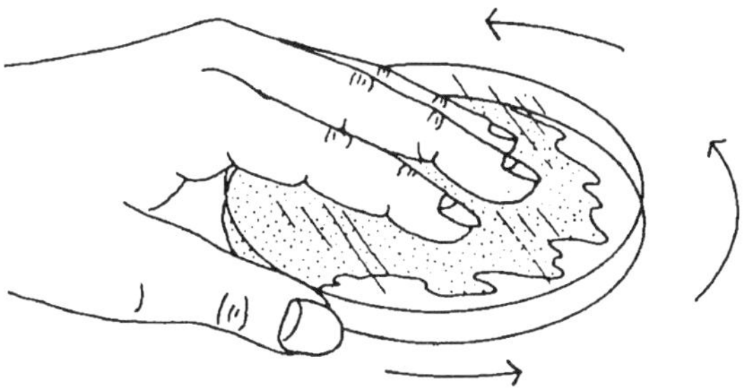

FIGURE 16.6 Rotate the petri plate so the medium covers the bottom and mix with the inoculum. Do not move the plate again until the medium has solidified.

16.2.6 Counting

Count colonies promptly after incubation. If counting must be delayed temporarily, store plates at 5 to 10°C for no more than 24 h. Use a counting aid, such as Quebec colony counter, or any other device that gives magnification and illumination.

16.2.6.1 Select Plates Having 30 to 300 Colonies

Count colonies and calculate as colonies per ml sample:

average number of colonies per plate × dilution
report as CFU per ml.

16.2.6.2 Plates Having More Than 300 Colonies

Use the plate(s) with counts nearest to 300 colonies. Compute colonies per ml:

average count × dilution
report as estimated CFU per ml.

16.2.6.3 Plates Have No Colonies

If plates of all dilutions have no colonies, calculate as colonies per ml:

greater than 1 × the corresponding lowest dilution
report as estimated CFU per ml.

For example, greater than 1 × 100 dilution is equal to or greater than 100 CFU per ml.

16.2.6.4 Number of Colonies per Plate Greatly Exceeding 300

When the number of colonies per plate greatly exceeds 300, the following procedures should be followed:

1. When there are fewer than 10 colonies per cm^2—select seven consecutive squares horizontally across the plate and six consecutive squares vertically. Multiply the counted colony number by 5 because the area of the plate is 65 cm^2 and the counted square number is 13 ($5 \times 13 = 65$).
2. When there are more than 10 colonies per cm^2—select four representative squares, take the average count per cm^2 and multiply by 57 for disposable plastic plates and 65 for glass plates.
3. When the bacterial count is greater than 100 per cm^2—for disposable plastic plates less than 100 × 57 × dilution and for glass plates less than 100 × 65 × dilution, report as estimated CFU per ml.

16.2.6.5 Spreading Colonies Are on the Plate(s)

Count colonies on the spreader-free area of the plate when colonies are well distributed in this area, and the spreader(s) do not exceed one half of the plate area. If plates have excessive spreader growth,

report as spreaders.

16.2.6.6 Plates Are Uncountable

Plates will be uncountable because of improper dilution, contamination, mistakes were made on control plates, plates were broken, etc.,

report as laboratory accident.

16.2.6.7 Examples for Counting and Reporting

Some examples for counting and reporting include:

1. Duplicate the plate count and average the two counts.
   280 and 34 colonies are counted in the 1 : 100 and 1 : 1000 dilution:
   280 equals 28,000 count per ml, using 1 ml sample from the 100 × dilution.
   34 equals 34,000 count per ml, using 1 ml from the 1000 × dilution.
   Report the average value: 31,000 counts per ml.
2. Using the original undiluted sample of 1 ml:
   If the count is less than 30 colonies, report the actual count, as count per ml.
3. Using 1 ml of undiluted sample, if there are no colonies, report as greater than 1 count per ml.
4. For plates with greater than 300 colonies, use the smallest count plate. For example, 500 and 340 count for 1 ml undiluted sample, the reported value is 340 count per ml.

16.3 SPREAD PLATE METHOD

A 0.1 ml or 0.5 ml of sample (inoculum) or 0.1 ml or 0.5 ml of appropriate dilution is added to the surface of a prepoured, solidified agar medium. The inoculum is then spread uniformly over the surface of the medium with a sterile glass rod. The inoculum is spread uniformly by holding the stick at a set angle on the agar and rotating the agar plate or rotating the stick until the inoculum is distributed. This method positions all the colonies on the surface and avoids contact of the cells with melted agar. After incubation at the proper temperature and time, count bacterial colonies on the surface of the agar (see Chapter 13). With the pour plate technique, colonies will grow within the nutrient agar (from cells suspended in the nutrient medium as the agar solidifies) as well as on the surface of the agar plate.

16 .3.1 EQUIPMENT AND MATERIAL

This is the same as the pour plate method, see Section 16.2.1. Add sterile glass rods (sterilize before using).

16.3.2 MEDIA AND AGAR PREPARATION, SAMPLE DILUTION

This is the same as the pour plate method, see Sections 16.2.2 and 16.2.4.

16.3.3 PREPARATION OF AGAR PLATES

The following procedure is followed for the preparation of agar plates.

1. Several precautions (i.e., aseptic technique) are necessary to prevent contamination. Follow the aseptic technique by pouring the melted agar into the plates as discussed in Section 16.2.5 and in Figure 16.5.
2. Pour 15 to 20 ml melted agar medium into the petri dish.
3. Rotate the poured media in the petri dish so that the medium covers the bottom as shown in Figure 16.6.
4. Do not move the plate before the medium has solidified.
5. Covers plates partially, leaving open slightly for excess moisture to evaporate, about 15 to 30 min.
6. When agar surfaces are dry, close the dishes.

16.3.4 PROCEDURE

The following procedures are followed for the preparation of plates for plate counts.

1. Mark and arrange plates in order. Prepare blank, duplicates, and positive control samples with one series of samples. For a sterility check, use 0.1 ml vol of dilution water to check the sterility of the pipet, agar, and dilution water.
2. Prepare sample dilutions as discussed in Section 16.2.4.
3. Arrange samples and dilutions according to the plates.
4. Shake samples and dilutions vigorously prior to pipetting onto the plates.
5. Pipet desired sample or dilution volume (0.1 ml or 0.5 ml) onto the surface of a prepoured, solidified agar medium while the dish is being rotated.

6. The inoculum is then spread uniformly over the surface of the medium with a sterile glass rod. The inoculum is spread uniformly by holding the glass rod at a set angle on the agar and rotating the agar plate or rotating the glass rod until the inoculum is distributed. This method positions all the colonies on the surface.
7. Let the inoculum absorb completely into the medium before incubating.
8. When agar surfaces are dry, close dishes, invert them, and incubate as required for the specified test. After incubation at the proper time and temperature, isolated surface colonies should develop.
9. Incubation is the same as in Section 16.2.5.(8.)

16.3.5 Counting and Reporting

Counting and reporting is the same as described in Section 16.2.6.

16.4 PLATE COUNT FROM SOILS AND SEDIMENTS

No microbiological technique is as widely accepted in enumerating soil organisms as the plate count using a variety of media as was discussed previously in Section 15.5. Sample collection, handling and preparation techniques were also mentioned in Section 15.5.

16.4.1 Moisture Determination

See Section 15.5.

16.4.2 Preparation of Dilutions

The following procedures are used in the preparation of dilutions.

1. For each sample, make a dilution blank containing 95 ml of diluent (see preparation of phosphate buffer diluent in Section 13.1.4.1) and use 15 to 20 2 mm glass beads. Also, make separately 7 dilution blanks with the diluent, each one containing 90 ml.
2. Cap the bottles and autoclave at 121°C for 20 min.
3. Transfer 10 g of moist soil sample to the bottle containing the 95 ml of diluent and the 2 mm glass beads.
4. Cap the bottle, place on a mechanical shaker, and shake for 10 min. Alternatively shake by hand, moving the bottle in a large arc at least 200 times. For samples difficult to disperse, add a wetting agent at a concentration of 0.1 percent (w/v) to the diluent in the initial dilution bottle before autoclaving.
5. After removing the sample from the shaker and just before using, shake the bottle vigorously.
6. Immediately thereafter, transfer 10 ml of the soil suspension taken from the center of the suspension to a fresh 90 ml blank. This establishes the 10^{-2} dilution.
7. Cap and vigorously shake this bottle, and remove 10 ml of the suspension as previously described.
8. Continue the sequence until the dilution of 10^{-7} is reached. On the basis of previous experience with the samples, it may be necessary to continue diluting until the 10^{-8} and 10^{-9} dilution is reached. Dilution techniques are described in Section 13.1.

16.4.3 Calculation and Reporting

Average the number of the colonies per plate for the dilutions, giving between 30 and 300 colonies, and determine the number of colony forming units (CFU) per gram of soil using the following:

$$\text{CFU/g} = \frac{\text{(mean plate count)(dilution factor)}}{\text{dry weight soil, initial dilution}} \tag{16.1}$$

$$\text{dry weight soil} = \text{(weight moist soil, initial dilution blank)} \times (1 - \%\ \text{moisture}/100) \tag{16.2}$$

For example, the average count of a plate is 46 colonies. The plate is prepared from a 10^{-2} dilution (10 g wet soil measured into 95 ml diluent, and after shaking, 10 ml of the soil suspension is pipetted into a 90 ml diluent), so the dilution factor is 100. The moisture content of the soil is 12 percent. What will be the CFU per g of the soil sample?

By using equations (16.1) and (16.2):

$$\text{dry weight soil} = 10 \times (1 - 12/100) = 10 \times 0.88 = 8.8$$

$$\text{CFU/g} = (46 \times 100)/8.8 = 522.7 = 523$$

Without using the formula, the original wet soil is 10 g; according to the 100× dilution, the 46 bacterial counts are in 0.1 g soil.

46 counts/0.1 g sample

460 counts/1.0 g sample

If 12 percent is moisture, 88 percent is the dry weight. Therefore, 1 g wet soil is equivalent to 0.88 g dry soil. If there are 460 counts in 1 g soil, how much in 0.88 g?

$$460/0.88 = 522.7 = 523$$

The reporting form is

CTU/g = 460 on the wet base

CTU/g = 523 on the dry base

16.4.4 Pour Plates

Follow Section 16.2 by using soil suspensions prepared and diluted as in Sections 16.4.1, 16.4.2, and the results are calculated and reported as in Section 16.4.3.

16.4.5 Spread Count

The method is described in Section 16.3. Inoculate 0.1 ml soil suspension prepared and diluted as in Sections 16.4.1, 16.4.2, and the results are calculated and reported in as in Section 16.4.3.

17 Determination of Total Coliform

The chapter describes the enumerative techniques for total coliform bacteria in water and wastewater. The method chosen depends upon the characteristics of the sample.

The coliform or total coliform group includes all of the aerobic and facultative anaerobic, Gram-negative, nonspore-forming rod shaped bacteria that ferment lactose in 24 to 48 h at 35°C. The definition includes the genera: *Escherichia, Citrobacter, Enterobacter,* and *Klebsiella.*

17.1 MEMBRANE FILTER TECHNIQUE

17.1.1 INTRODUCTION

The membrane filter (MF) technique is a quick, accurate, highly reproducible technique for the enumeration of bacterial density. General introduction, advantages, and limitations of the membrane filter technique have been discussed in Section 13.3.

The total coliform test can be used for any type of water or wastewater, but since the development of the fecal coliform procedure there has been increasing use of this more specific test as an indicator of fecal pollution. However, the total coliform test remains the primary indicator of bacteriological quality for potable water, distribution system waters, and public water supplies because a broader measure of pollution is desired for these waters. It is also a useful measure in shellfish-raising waters.

Determination of total coliform for water safety is in use today and is used to detect particular indicator organisms (see Section 7.1.3).

There are several criteria for an indicator organism. The most important criterion is that the organism be consistently present in human feces in substantial numbers so its detection will be a good indication that human wastes are entering the water. The indicator organisms should also survive in the water at least as well as the pathogenic organisms would. The usual indicator organisms are the coliform bacteria. Coliform are defined as aerobic or facultative anaerobic, Gram-negative (Section 7.2.2), nonendospore-forming, rod-shaped bacteria.

Usually bacteria are of medical, industrial, and environmental importance. They live in soil, water, and intestinal tracts, but there are many important pathogens, too.

Some important genera are *Pseudomonas, Neisseria, Brucella, Legionella, Agrobacterium, Salmonella, Shigella, Klebsiella, Vibrio, Escherichia,* and *Enterobacter.* They ferment lactose in 24 to 48 h at 35°C to produce acid and gas.

Because some coliforms are not solely enteric bacteria, but are more commonly found in plant and soil samples, many standards for food and water specify the determination of coliforms. The predominant fecal coliform is *Escherichia coli,* which constitutes a large portion of the human intestinal population. There are specialized tests to distinguish between fecal coliforms and nonfecal coliforms. It is important to note that coliforms are not themselves pathogenic under normal conditions, although they can cause diarrhea and opportunistic urinary tract infections.

17.1.2 APPLICATION

As applied to the membrane filter (MF) technique, the coliform group may be defined as comprising all aerobic and facultative anaerobic, Gram-negative, nonspore-forming, rod-shaped bacteria that develop a red colony with a metallic sheen within 24 h at 35°C on an Endo-type medium containing lactose. Generally, all red, pink, blue, white, or colorless colonies lacking sheen are considered noncoliforms by this method. Low coliform estimates may be caused by the presence of high numbers of noncoliforms or of toxic substances.

The MF technique is applicable to saline waters, but not wastewaters that have received only primary treatment followed by chlorination because of turbidity or the wastewaters containing highly toxic metals or toxic organics such as phenol.

The standard volume to be filtered for drinking water samples is 100 ml. Because treated drinking water should contain no coliforms per 100 ml, water plant laboratories should consider testing 1 liter (L) samples of finished water. Smaller or larger samples may be used for other samples or special analysis.

17.1.3 Equipment and Glassware

Specifications for equipment and glassware used in microbiological laboratories are discussed in Sections 11.1 and 11.2.

- Incubator that maintains 35°C ± 0.5°C temperature; daily temperature checks recorded; thermometer checked against a NBS certified thermometer; incubator must have humidity control if petri dishes with loose-fitted lids are used. The specification is given in Section 11.1.1.
- Graduated cylinders, sterile 50 to 100 ml, covered with aluminum foil, see Section 11.2.3.
- Sterile bacteriological pipets, TD, see Section 11.2.2.
- Membrane filtration unit (filter holding assembly, constructed from glass, autoclavable plastic, or stainless steel), sterile, unassembled; it consists of a seamless funnel fastened to base by a locking device or held in place by a magnetic force. The design should permit the membrane filter to be held securely on the porous plate of the receptacle without mechanical damage and allow all fluid to pass through the membrane during filtration. Separately wrap the two parts of the filtration unit in aluminum foil, sterilize by autoclaving, and store until used. Sterile, disposable field units may be used. Membrane filtration setup is shown in Figure 17.1 and different filtration units are seen in Figure 11.2.
- Vacuum source.
- Vacuum filter flask with appropriate tubing, or filter manifold which holds a number of filters.
- Safety trap flask between the filter manifold (or flask) and the vacuum source; see Figure 11.12.
- Ultraviolet sterilizer.
- Forceps with smooth tips.
- Methanol or ethanol 75 percent in small vial for flaming forceps.
- Bunsen burner.
- Sterile petri dishes with tight-fitting lids, 50 × 12 mm or loose-fitting lids 60 × 15 mm, glass or plastic, see Section 11.2.1.
- Dilution bottles (pyrex) marked at 99 ml volume, screw cap with neoprene rubber liner, see Section 11.2.6.
- Membrane filters, white, gridded with 0.45 μm pore size, presterilized, with the manufacturer's certificate, that the sterilization technique has neither induced toxicity, nor altered the chemical or physical properties of the membrane. Quality requirements and specifications are discussed in Section 9.4.7.

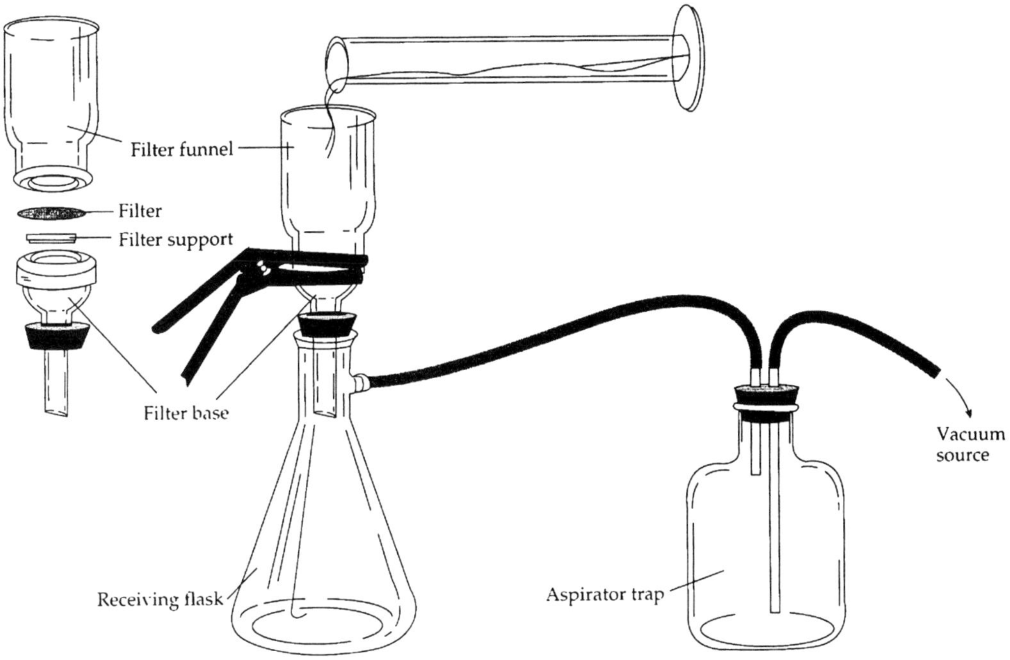

FIGURE 17.1 Membrane filtration setup.

- Absorbent pads, in case broth media is used, each lot should be certified by the manufacturer to be of high quality and free of sulfites or other substances that could inhibit bacterial growth. Quality requirements and specifications are discussed in Section 9.4.7.
- Fermentation tubes and vials, there should be sufficient media in the outer fermentation tube to fill the enclosed vial after sterilization, see Section 11.2.7.

17.1.4 Culture Media

The need for uniformity dictates the use of dehydrated media. Follow manufacturer's directions for rehydration and sterilization. The culture media may be in liquid form, broth, or solid form, agar.

Dehydrated media used for detection and enumeration of total coliform bacteria are DICFO *M-Endo broth MF* (No. 0742) and BBL *M-coliform (MC) broth* (No. 11119), or equivalent.

Confirmation of the coliform colonies is based on the lactose fermentation and gas production ability of the bacterial group.

The media used are

DIFCO *lauryl tryptose broth* (No. 0241) or BBL *lauryl sulfate broth* (No. 11338 or the equivalent) and DIFCO *brilliant green bile* 2 percent (No. 0007-02) or BBL brilliant green bile broth (No. 11079) or the equivalent.

17.1.4.1 Preparation of the M-Endo Broth Media

Add 48 g of broth to 1 l of laboratory pure water containing 20 ml of 95 percent ethanol (denatured alcohol should not be used). Heat in boiling water bath or heat on hot plate to near boiling to dissolve medium. Cool. Do not autoclave. Final pH should be between 7.1 and 7.3. Store prepared medium in the dark at 4°C. Discard unused medium after 96 h.

17.1.4.2 Preparation of the M-Endo Agar Medium

Prepare M-Endo agar by adding 15 g of agar per l of M-Endo Broth media (1.5 percent). Heat to boiling. Cool to about 45°C and dispense into petri dishes to provide a minimal agar depth of 2 to 3 mm and allow to solidify. Protect prepared medium from light. It can be stored at 4°C for up to 2 weeks, in a plastic bag with plates in an inverted position, see Section 12.1.8.

17.1.4.3 Preparation of Lauryl Tryptose Broth

Add 35.6 g of medium to 1 l of laboratory pure water and mix to dissolve. Dispense a 10 ml vol into a fermentation tube. Sterilize for 15 min at 121°C with 15 lb pressure by autoclaving.

17.1.4.4 Preparation of Brilliant Green Bile Broth

Add 40 g of media to 1 l of laboratory pure water. Mix until dissolved. Dispense 10 ml vol into fermentation tubes. Sterilize for 15 min at 121°C with 15 lb pressure by autoclaving.

17.1.4.5 Preparation of Dilution Water

Prepare from stock phosphate buffer solution, stored in the refrigerator (see preparation in Section 13.1.4). Let the stock solution warm up to room temperature before using. Add 1.25 ml stock phosphate buffer solution and mix with 5 ml of stock magnesium chloride solution, then fill up to 1000 ml with DI water. Final pH should be 7.2 ± 0.1. Sterilize before use by autoclaving for 15 min.

17.1.4.6 Preparation of Magnesium Chloride Solution

Dissolve 81 g of magnesium chloride hexahydrate ($MgCl_2 \cdot 6H_2O$) in laboratory pure water and dilute to 1 l.

17.1.5 Set Up Filtration Apparatus

The following procedure is used in setting up filtration apparatus.

1. The receiving flask, filter base, and filter support should be wrapped in aluminum foil and sterilized by autoclaving.
2. Set up filtration apparatus as shown in Figure 17.1.
   - Remove wrappers as each piece is fitted into place.
   - Attach the filter trap to the vacuum source.
   - Place the filter holder base (with stopper) on the filtering flask. Attach the flask to the filter trap.

17.1.6 Suggested Sample Volumes for Membrane Filter Total Coliform

The following table provides sample volumes.

Water Source	Volume (ml) to Be Filtered
Drinking water	100
Swimming pools	100
Wells, spring	100, 50, 10
Lakes, reservoirs	100, 50, 10
Water supply intake	10, 1, 0.1
Bathing beaches	10, 1, 0.1
River waters	1, 0.1, 0.01, 0.001
Chlorinated sewage	1, 0.1, 0.01
Raw sewage	0.1, 0.01, 0.001, 0.0001

17.2 MEMBRANE FILTRATION (MF) PROCEDURE

17.2.1 Procedure Using Broth Media

The membrane filtration procedure using a broth media includes:

1. Mark petri dishes with sample ID (identification number) and volumes filtered on the bottom of the petri dish.
2. Place sterile absorbent pad in the bottom of each dish.
3. Pipet 2 ml of M-Endo broth onto the pad to saturate it.
4. Disinfect the forceps by burning off the alcohol. Store the forceps in a beaker immersed in 70 percent alcohol. Keep the beaker of alcohol away from flame.
5. Using the sterile forceps, place the filter on the filter holder with gridded side up and attach the funnel to the base of the filter unit. The membrane filter is now held between the funnel and the base.
6. Shake the sample bottle vigorously about 10× and measure the desired volume. If the sample volume is less than 10 ml, add 10 ml of sterile dilution water to the filter before adding the sample.
7. If necessary (if the sample looks dirty or should the history of the sample indicate this) use a dilution of the sample (see Section 13.1) for the proper count.
8. Filter samples in order of increasing sample volume, filter potable waters first.
9. Turn on the vacuum and allow the sample to pass into the filtering flask. Leave the vacuum on.
10. Filter sample or proper dilution of the sample and rinse the sides of the funnel at least twice with 20 to 30 ml of sterile dilution water.
11. Turn off the vacuum and remove the funnel from the filter base.
12. Aseptically (flamed forceps) remove the membrane filter from the filter base. Holding the filter at its edge with the sterilized forceps gently lift and place the filter grid-side up in the culture dish. Slide the filter onto an absorbent pad (or agar medium) using a rolling action to avoid trapping air bubbles between the membrane filter and the underlying pad (or agar). Reseat the filter membrane if air bubbles are present; see Figure 17.2.
13. Start test with blank (empty filter), dilution water blank (filter 100 ml dilution water), and finish the set of samples with a positive control of *E. coli*.
14. Each set of samples should have a duplicate sample.
15. Invert the plate and incubate at 35°C for 24 h in an atmosphere with close to 100 percent relative humidity. If loose-lid dishes are used, place the inverted dishes into plastic boxes containing wet towels for humidity and cover. If tight-lid dishes are used there is no requirement for near saturated humidity.
16. After incubation, remove dishes from the incubator and examine for red metallic sheen colonies.

17.2.2 Procedure Using Agar Media

This procedure is the same as the preceding procedure for using broth media except absorbent pads are not used. After filtration, remove the filter from filter base and place grid-up position on the prepared agar. The inoculation process is shown in Figure 17.2. Reseat the membrane if bubbles occur. Invert the dish and incubate for 24 h at 35°C in an atmosphere with near saturated humidity.

17.2.3 Counting and Recording Colonies

After incubation, count colonies on those membrane filters containing 20 to 80 pink to dark-red color with a metallic surface sheen colonies and less than 200 total bacterial colonies.

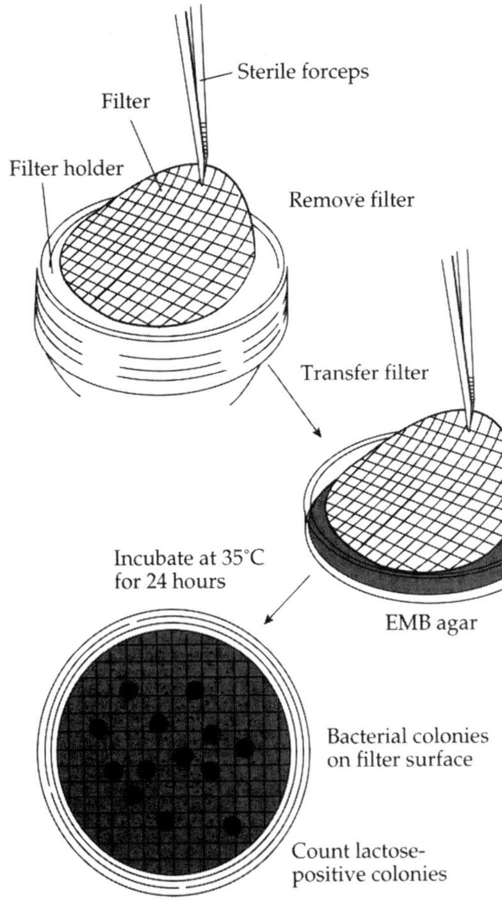

FIGURE 17.2 Inoculation in membrane filter technique. Using sterile forceps, remove the filter from the filter holder. Place the filter on the culture medium, gradually laying it down from one edge to the other.

17.2.3.1 General Rules for Counting and Reporting

The following is a list of the general rules for counting and reporting.

1. For countable membranes with 20 to 80 sheen colonies and less than 200 total bacterial colonies

$$\text{counts/100 ml} = \frac{\text{number of coliform colonies}}{\text{volume of sample filtered, ml}} \times 100 \tag{17.1}$$

2. When counts are greater than the upper limit of 80 colonies—for example, 1, 0.5, and 0.01 ml samples are filtered and produce total coliform colonies too numerous to count (TNC), greater than 200, 150, and 110 colonies. Use the count from the smallest filtered volume and report it as a greater than counts/ml:

$$\text{counts/100 ml} = \frac{110}{0.01} \times 100 = 1{,}100.000 \tag{17.2}$$

Report as greater than 1,100.000 coliforms per 100 ml

3. Membranes with more than 200 total colonies (coliforms and noncoliforms)—estimate sheen colonies. If possible, calculate the total coliform density as discussed in steps 1 and 2 and report as:

$$\text{estimated counts/100 ml}$$

If the estimation of coliform colonies is not possible, report as:

$$\text{Coliform count/100 ml} = \text{TNTC (too numerous to count)}$$

4. Membranes with confluent growth—report as confluent growth and specify the presence or absence of sheen.

17.2.3.2 Special Rules in Counting and Reporting for Potable Waters

The following is a list of special rules in counting and reporting for potable waters.

1. For countable membranes with 0 to 80 sheen colonies and less than 200 total colonies—count the sheen colonies and calculate as count per 100 ml coliform, as described in Section 17.2.3.1, equation (17.1).
2. Uncountable membranes for potable water samples—if a 100 ml sample cannot be tested because of the high background counts or confluency, less than 100 ml vol can be filtered. For example, two 50 ml potions of the samples filtered, one of the membrane has 60 colonies, and the other has 50 colonies, the report will be the average of the duplicate counts and calculated for 100 ml, so the report is

$$\text{total coliform count/100 ml} = 110$$

3. Membranes with confluent growth—potable waters with confluence will need resampling and retesting. If a filtration of less than 100 ml still causes confluency, samples should be analyzed by a MPN test. This MPN check should be made on at least one sample for each problem water once every three months.

17.2.4 Verification

The typical coliform colony has a pink to dark-red color with a metallic surface sheen. The sheen area may vary in size from a small pinhead to complete coverage of the colony surface. Colonies which lack sheen may be pink, red, white, or colorless and are considered to be noncoliforms. Typical sheen colonies may be produced occasionally by noncoliform organisms. If questionable sheen occurs, verification procedures are also necessary.

Verification is required for all positive samples from potable waters. Verification is also recommended for nontypical colonies to determine that false negative colonies do not occur, and for identifying unusual colony types and support for data used in legal actions. Verification is also used for establishing quality control with new test waters, new procedures, or new technicians.

17.2.4.1 Verification Procedures

The following is a list of verification procedures.

1. Using a sterile inoculating needle, pick growth from the centers of at least 10 well-isolated sheen colonies (5 sheen colonies per plate for potable waters). Inoculate each into

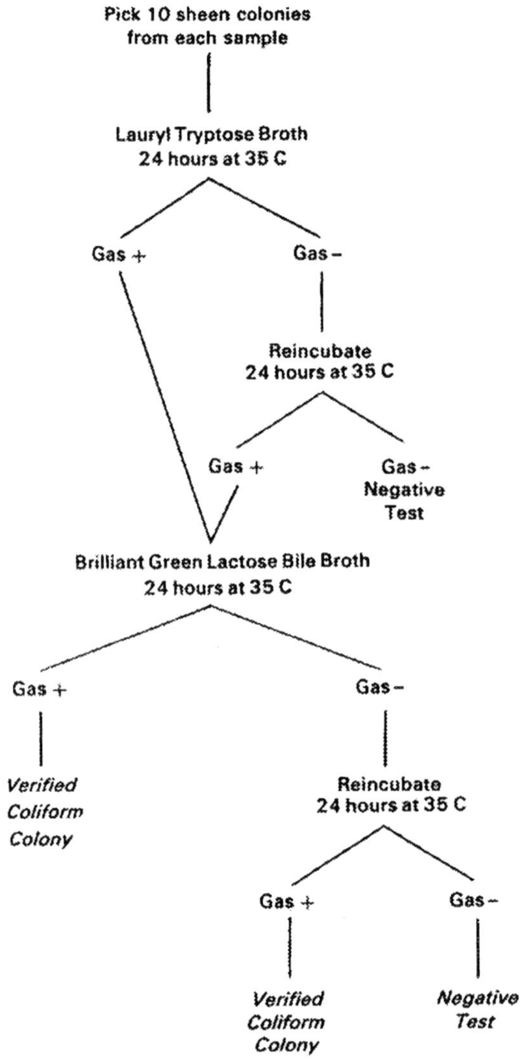

FIGURE 17.3 Verification outline for total coliform using the MF technique.

a tube of lauryl tryptose broth and incubate 24 to 48 h at 35°C. After the incubation time, check tubes for gas forming.

2. After the 24 and 48 h readings, confirm gas-positive lauryl tryptose broth tubes by inoculating a loopful of growth into brilliant green bile broth and incubate for 24 to 48 h at 35°C. After incubation check tubes for gas forming. Cultures that are positive in brilliant green bile broth are interpreted as verified coliform colonies. A verification outline is shown in Figure 17.3.

A verified membrane filter test establishes the validity of colony differentiation on a selected medium and provides support evidence of colony interpretation.

17.3 DELAYED INCUBATION BY THE MF METHOD

Modification of the standard membrane filter technique gives the ability to filter samples at the field immediately after collection, place filter in to a holding medium, and ship to the laboratory.

Complete the coliform determination in the laboratory by transferring the filter from the transport, also called the holding media, to the M-Endo media, incubate at 35°C for 24 h, and count typical metallic sheen red coliform colonies that develop. Transport media are designed to keep organisms viable and generally do not permit visible growth during transit time. The bacteriostatic agents in the holding media suppress growth of microorganisms on route but allow normal coliform growth after transfer to a fresh medium.

17.3.1 Application

Filter sample at the field immediately after collection through a 0.45 μm membrane filter by using a field filtration unit. Disinfect filtration apparatus by adding methyl alcohol to the filtering chamber, igniting the alcohol, and covering the unit to produce formaldehyde. A glass or metal filtration unit may be sterilized by immersing in boiling water for 5 min. Use a hand aspirator to obtain the necessary vacuum. Place filter on M-Endo preservative medium and send the filter to the laboratory where the filter will be transferred onto an M-Endo medium. Incubate at 35°C for 24 h, count, and report coliform colonies as described in Section 17.2.3.

17.3.2 Equipment and Glassware

Use the same equipment and glassware as discussed in Section 17.1.3. Instead of the laboratory filtration unit, use a field filtration unit and a hand aspirator to obtain the vacuum.

17.3.3 Culture Media

Use the same culture media as discussed in Section 17.1.4 and an M-Endo preservative medium.

17.3.3.1 Preparation of M-Endo Preservative Medium

The following is the list of steps in the preparation of the M-Endo preservative medium.

1. Prepare M-Endo agar medium as described in Section 17.1.4.
2. After cooling the medium to 45°C, aseptically add 3.84 g sodium benzoate per 1 of medium or 3.2 ml (12 g/100 ml) 12 percent sodium benzoate solution per 100 ml of medium.
3. Mix well, and dispense into petri dishes as discussed in the preparation of M-Endo agar medium, see Section 17.1.4.

A 12 percent sodium benzoate solution is prepared by dissolving 12 g of sodium benzoate in distilled water and filling to 100 ml. Sterilize the solution by autoclaving or filtering through a 0.22 μm membrane filter. Discard solution after 6 months.

17.3.4 Procedure, Counting, and Recording

The field filtered sample on the M-Endo preservative medium is shipped to the laboratory where the filter paper is transferred to an M-Endo medium, incubated at 35°C for 24 h and the developing colonies are counted and reported as in Sections 17.2.3 and 17.2.4.

17.4 MOST PROBABLE NUMBER (MPN) METHOD

17.4.1 Introduction

This method detects and estimates the total coliforms in water samples by the multiple fermentation tube technique. The method has three stages: the presumptive, the confirmed, and the completed tests.

In the *presumptive test*, a series of lauryl tryptose broth fermentation tubes are inoculated with decimal dilution of the sample. The formation of gas at 35°C within 48 h constitutes a positive presumptive test for coliform groups.

In the *confirmed test*, inocula from positive presumptive tubes are transferred to tubes of brilliant green lactose bile (BGLB) broth. The BGLB medium contains selective and inhibitive agents to suppress the growth of all noncoliform organisms. Gas production at 35°C within 24 or 48 h gives a positive confirmed test. It is the point at which most MPN tests are terminated.

In the *completed test*, place inoculum from the positive confirmed tubes onto eosine methylene blue (EMB) agar plates for 24 h at 35°C. Make the differentiation and isolation of lactose fermenting and nonlactose fermenting entero-bacilli (rod-shaped bacteria). Typical and atypical colonies are transferred into lauryl tryptose broth fermentation tubes and onto nutrient agar slant. Gas formation in the fermentation tubes and presence of Gram-negative rods constitute a positive completed test for total coliforms.

The MPN per 100 ml is calculated from the MPN table (see Table 17.1).

17.4.2 Application

The MPN procedure is a tube-dilution method using a nutrient rich medium, which is less sensitive to toxicity and supports the growth of stressed organisms. This method applies to chlorinated effluents, for turbid samples, soils, sediments, and sludges instead of the membrane filter technique.

The MPN test also has some limitations. Certain noncoliform bacteria may suppress coliforms or act synergistically to ferment lauryl tryptose broth and yield false positive results. Brilliant green lactose bile broth gives false positive results when chlorinated primary effluents are tested, specifically when stormwater is mixed with sewage. False negatives may also occur with waters containing nitrates. False positives are more common in sediments.

17.4.3 Equipment and Glassware

The following equipment and glassware are used in the MPN method.

- Incubator set at 35 ± 0.5°C.
- Inoculation loops.
- Disposable sterile applicator sticks or plastic loops as an alternative to the inoculating loops.
- Bunsen burner.
- Sterile TD Mohr or bacteriological pipets, glass or plastic of appropriate size.
- Pyrex culture tubes, containing inverted fermentation vials.
- Culture tube racks to hold 50 and 25 mm diameter tubes.
- Dilution bottles, 99 ml mark, screw cap with neoprene rubber liner.

17.4.4 Culture Media

17.4.4.1 Lauryl Tryptose Broth

Lauryl tryptos broth is prepared as described in Section 17.1.4.

17.4.4.2 Brilliant Green Lactose Bile Broth (BGLB)

Brilliant green lactose bile broth is prepared as described in Section 17.1.4.

17.4.4.3 Eosine Methylene Blue Agar (EMB Agar)

The procedure for making eosine methylene blue agar includes:

TABLE 17.1
MPN TABLE WITH 95 PERCENT CONFIDENCE LIMITS

Most Probable Number Index and 95 Percent Confidence Limits for Five Tubes and Three Dilution Series

No. of Tubes Giving Positive Reaction out of			MPN Index	95% Confidence Limits		No. of Tubes Giving Positive Reaction out of			MPN Index	95% Confidence Limits	
5 of 10 ml Each	5 of 1 ml Each	5 of 0.1 ml Each	per 100 ml	Lower	Upper	5 of 10 ml Each	5 of 1 ml Each	5 of 0.1 ml Each	per 100 ml	Lower	Upper
0	0	0	< 2								
0	0	1	2	< 0.5	7	4	2	1	26	9	78
0	1	0	2	< 0.5	7	4	3	0	27	9	80
0	2	0	4	< 0.5	21	4	3	1	33	11	93
						4	4	0	34	12	93
1	0	0	2	< 0.5	7						
1	0	1	4	< 0.5	11	5	0	0	23	7	70
1	1	0	4	< 0.5	11	5	0	1	31	11	89
1	1	1	6	< 0.5	15	5	0	2	43	15	110
1	2	0	6	< 0.5	15	5	1	0	33	11	93
						5	1	1	46	16	120
2	0	0	5	< 0.5	13	5	1	2	63	21	150
2	0	1	7	1	17						
2	1	0	7	1	17	5	2	0	49	17	130
2	1	1	9	2	21	5	2	1	70	23	170
2	2	0	9	2	21	5	2	2	94	28	220
2	3	0	12	3	28	5	3	0	79	25	190
						5	3	1	110	31	250
3	0	0	8	1	19	5	3	2	140	37	340
3	0	1	11	2	25	5	3	3	180	44	500
3	1	0	11	2	25	5	4	0	130	35	300
3	1	1	14	4	34	5	4	1	170	43	490
3	2	0	14	4	34	5	4	2	220	57	700
3	2	1	17	5	46	5	4	3	280	90	850
3	3	0	17	5	46	5	4	4	350	120	1000
4	0	0	13	3	31	5	5	0	240	68	750
4	0	1	17	5	46	5	5	1	350	120	1000
4	1	0	17	5	46	5	5	2	540	180	1400
4	1	1	21	7	63	5	5	3	920	300	3200
4	1	2	26	9	78	5	5	4	1600	640	5300
4	2	0	22	7	67	5	5	5	≥ 2400		

1. Add 37.5 grams dehydrated media to 1 l of laboratory pure water and heat in a boiling water bath until completely dissolved. Final pH is 7.2 ± 0.2.
2. Dispense into tubes or flasks and sterilize for 15 min at 121°C (15 lb pressure).
3. A flocculent precipitate may form after autoclaving. Resuspend the precipitate by gently shaking the flask prior to pouring the medium into sterile petri dishes.

17.4.5 Dilution Water

Prepare as in Section 17.1.4. Dispensed in 99 ± 2 ml amounts in dilution bottles.

17.4.6 PROCEDURE

17.4.6.1 Presumptive Test

The following procedures are used in a presumptive test.

1. Arrange a series of culture tubes containing lauryl tryptose broth in rows in a tube rack, providing for five replicates in each row. Use five rows for samples.
2. Select sample volumes and label tubes with sample identification number (ID) and sample volume.
   - Potable water—5 tubes of 10 ml each or 100 ml each.
   - Relatively unpolluted waters—the 5 tube rows might be 100, 10, 1.0, 0.1, and 0.01 ml. The last two volumes: 0.1 and 0.01 ml are derived as dilutions of the original sample.
   - Polluted waters—0.1, 0.01, 0.001, 0.0001, and 0.00001 ml of the original sample (see dilution techniques in Section 13.1).
3. Shake the sample and dilutions vigorously. Inoculate each five tube row with predetermined sample volumes in increasing decimal dilutions. The series of sample volumes listed in step 2 would yield determinate results from test waters containing up to 16 million organisms per 100 ml sample by use of the MPN table.
4. Incubate tubes for 24 h at 35°C. A positive presumptive test is gas production from the coliforms. If there is no gas or growth, reincubate these negative tubes for an additional 24 h.
5. If the presumptive tubes are negative after 48 h incubation, discard them. If the presumptive tests are positive, the cultures are verified by the confirmed test.

17.4.6.2 Confirmed Test

The following procedures are used in the confirmed test.

1. Carefully shake each positive presumptive tube. With a sterile loop or sterile applicator stick, transfer growth from each tube to brilliant green lactose bile broth. Gently agitate the tubes to mix the inoculum and incubate at 35°C.
2. After 24 $\pm$ 2 h incubation, examine the tubes for gas. Any amount of gas in BGLB constitutes a positive confirmed test. Reincubate negative tubes for an additional 24 h.
3. Count the positive and negative tubes, and hold the positive tubes for the completion test if required, and incubate the negative tubes for an additional 24 h at 35°C.

Most sample analysis is terminated after the completion of the confirmed test.

Usually, the confirmed test should be verified by analyzing 5 percent of the confirmed tests with a minimum of one sample per test run with the completion test. For the certification of water supply laboratories, they should verify 10 percent of the positive confirmed samples and they should verify at least one sample quarterly.

17.4.6.3 Completion Test

The following procedures are used in the completion test.

1. Using the aseptic technique, streak one or more EMB (Eosine Methylene Blue) agar plates from each positive test. Incubate plates in inverted position for 24 h at 35°C.
2. The colonies developing are typical (pink to dark red with a green metallic surface sheen) or atypical (pink, red, white, or colorless without sheen). From each plate pick up typical colonies and transfer to nutrient agar slant.

3. Microscopically examine Gram-stained preparations from those 24 h nutrient agar slant cultures. Presence of Gram-negative rod form bacteria is the positive completed test for total coliforms. The Gram-stain procedure is discussed in Section 13.5.

The flow chart for the total coliform MPN test is shown in Figure 17.4.

17.4.7 CALCULATION

The results of the confirmed or completed test may be obtained from the MPN table based on the number of positive tubes in each dilution. Table 17.1 illustrates the MPN indices and the 95 percent confidence limits for general use. Table 17.2 lists the MPN indices and limits for potable water testing.

17.4.7.1 Calculation of Reported Value When 10, 1.0, and 0.1 ml Portions Are Used

The following procedure is used to calculate reported values of 10, 1.0, and 0.1 ml portions.

1. Three dilutions are necessary to formulate the MPN code. If 5 portions each of 10 1.0 ml, and 0.1 ml are used as inocula and positive results are observed in 5 of the 10 ml inocula, 3 of the 1.0 ml, and 0 of the 0.1 ml inocula. The coded results of the test is 5-3-0.
2. The code is located in the MPN table. The corresponding MPN index is 79/100 ml (because the largest sample size was 10 ml).

17.4.7.2 Calculation of Reported Values When the Series of Decimal Dilutions Is Other Than 10, 1.0, and 0.1 ml

The following procedure is used to calculate reported values when the dilutions are not 10, 1.0, or 0.1 ml.

1. Determine the coded result from the MNP table.
2. Calculate MPN per 100 ml according to the following formula:

$$\text{MPN/100 ml} = \text{MPN (from table)} \times \frac{10}{\text{largest quantity tested}} \tag{17.3}$$

For example,

Sample sizes:	0.01,	0.001,	and	0.0001 ml
Positive:	5	2		0
Code:	5 –	2	–	0
MPN index:	49			
Result:	calculated by using equation (17.3)			

$$\text{MPN/100 ml} = 49 \times \frac{10}{0.01} = 49{,}000$$

17.4.7.3 More Than Three Sample Volumes Are Inoculated

If more than three sample volumes are inoculated, the three significant dilutions must be determined. The significant dilutions are selected using the following rules:

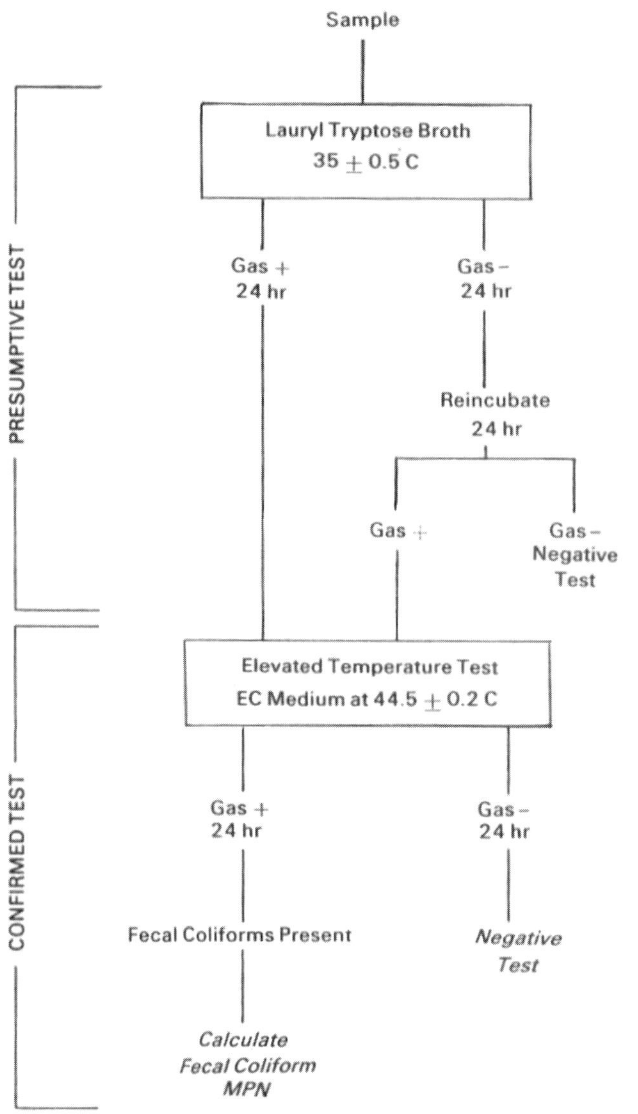

FIGURE 17.4 Flow chart for total coliform determination by MPN method.

TABLE 17.2
MPN Indices and 95 Percent Confidence Limits

Number of Positive Tubes from Five Portions of size 10 ml	MPN Index/100 ml	95 Percent Confidence Limits Lower	Upper
0	< 2.2	0	6.0
1	2.2	0.1	12.6
2	5.1	0.5	19.2
3	9.2	1.6	29.4
4	16	3.3	52.9
5	> 16	8.0	Infinite

- Only three dilutions are used in the MPN test.
- To obtain the proper three dilutions, select the smallest sample volume giving all positive results and the following two smaller volumes. For example,

Sample sizes:	10,	1.0,	0.1,	0.01,	0.001, and	0.0001 ml
Positive:	5	5*	4*	0*	0	0
Code:	5 –	4 –	0.			

- If less than three dilutions show positive tubes, select the three highest sample volumes which will include the dilutions with the positive tubes. For example,

Sample sizes:	1.0,	0.1,	0.01,	0.001, and	0.0001 ml
Positive:	4*	1*	0*	0	0
Code:	4 –	1 –	0.		

- If there are positive tubes in the dilutions higher than the dilutions selected, positive results are moved up from the dilutions sample volume to increase the positive tubes in the highest dilution selected. For example,

Sample sizes:	10,	1.0,	0.1,	0.01,	0.001, and	0.0001 ml
Positive:	5*	5*	4*	1*	1	0
Code:	5 –	4 –	1.			

- There should be no negative results in higher sample volumes than those chosen. However, if negative tubes are present, for example, 4/5, 5/5, 3/5 and 0/5 the highest sample volume with all positive tubes must be used along with the next two lower sample volumes.

Sample sizes:	10,	1.0,	0.1,	0.01,	0.001, and	0.0001 ml
Positive:	4	5*	3*	0*	0	0
Code:	5 –	3 –	0.			

- If all tubes are positive, choose the three highest dilutions. For example,

Sample sizes:	1.0,	0.1,	0.01,	0.001, and	0.0001 ml
Positive:	5	5	5*	5*	5*
Code:	5 –	5 –	5.		

- If all tubes are negative, choose the three lowest dilutions. For example,

Sample sizes:	1.0,	0.1,	0.01,	0.001, and	0.0001 ml
Positive:	0*	0*	0*	0	0
Code:	0 –	0 –	0.		

- If positive tubes skip a dilution, select the highest dilution with positive tubes and the two lower dilutions.

Sample sizes:	1.0,	0.1,	0.01,	0.001, and	0.0001 ml
Positive:	4*	0*	2*	0	0
Code:	4 –	0 –	2.		

- If only the middle dilution is positive select this dilution and one higher and one lower dilution.

Sample sizes:	1.0,	0.1,	0.01,	0.001, and	0.0001 ml
Positive:	0*	1*	0*	0	0
Code:	0 –	1 –	0.		

The numbers marked* are the selected results to establish the code and calculate the MPN index.

17.5 DETERMINATION OF *KLEBSIELLA*

17.5.1 Introduction

Klebsiella, a genus included in the coliform group, may be associated with coliform regrowth in water supply distribution systems and is often a major component of the coliform population on vegetation and in paper mill, textile, and other industrial wastes. The normal coliform population in human and other warm-blooded animal feces may contain 30 to 40 percent *Klebsiella* strains. Approximately 4 percent of bacterial pneumonia cases and 18 percent of urinary tract infections are caused by pathogenic strains of *Klebsiella.* Of these clinical strains, 85 percent are fecal coliforms.

Environmental sources, such as vegetation, paper mills, textile production, sugar cane, and farm produce contribute 71 to 88 percent of all *Klebsiella* in the total coliform population. *Klebsiella* occasionally may become established in the sediments of water supply distribution networks as a result of inadequate source water protection, unsatisfactory treatment protocols, or changes in the integrity of the pipe environment.

17.5.2 Membrane Filter Procedure

Rapid quantitation may be achieved in a MF procedure by modifying the M-FC agar base through substitution of inositol for lactose and adding carbenicillin or by using *M-Kleb agar*.

17.5.2.1 Apparatus

See Section 17.4.3.

17.5.2.2 Culture Medium

17.5.2.2.1 Modified M-FC agar (M-FCIC agar)

This medium may not be available in dehydrated form and may require preparation from basic ingredients.

Tryptose 3or biosate	10.0 g
Proteus peptone No.3 or polypeptone	5.0 g
Yeast extract	3.0 g
Sodium chloride, NaCl	5.0 g
Inositol	10.0 g
Bile salt No.3 or bile salt mixture	1.5 g
Aniline blue	0.1 g
Agar	15.0 g
Reagent-grade water	1.0 l

The following procedure is used to prepare modified M-FC agar.

1. Heat medium to boiling and add 10 ml of 1 percent rosolic acid solution. The 1 percent rosolic acid is prepared by dissolving 1 g of rosolic acid in 0.2 *N* NaOH (8 g NaOH dissolved in reagent grade water and diluted to 1 l).
2. Cool to below 45°C and add 50 mg carbonicillin (available from Geopen, Roerig-Pfizer, Inc., New York, NY).
3. Dispense aseptically into petri dishes. Store in refrigerator until needed. Discard unused medium after two weeks. Do not sterilize by autoclaving. Final pH should be 7.4 ± 0.1.

17.5.2.2.2 M-Kleb agar

This medium may require preparation from basic ingredients.

Phenol red agar	31.0 g
Adonitol	5.0 g

Aniline blue	0.1 g
Sodium lauryl sulfate	0.1 g
Reagent grade water	1.0 l

The following procedure is used to prepare M-Kleb agar.

1. Sterilize by autoclaving for 15 min at 121°C.
2. After autoclaving, cool to 50°C in a water bath.
3. Add 20 ml of 95 percent ethyl alcohol (not denatured) and 0.05 g filter sterilized carbonicillin per l.
4. Shake thoroughly and dispense aseptically into petri dishes. The final pH should be 7.4 ± 0.2.
5. Prepared medium can be held at 4°C for 20 days.

17.5.2.3 Procedure

This procedure is the same as the one described in Section 17.2. Select a sample volume that will yield 20 to 60 colonies per membrane.

17.5.2.4 Counting Colonies

After incubation at 35°C for 24 h, Klebsiella colonies on *M-FCIC agar* are blue or bluish-gray. Most atypical colonies are brown or brownish. Klebsiella colonies on *K-agar* are deep blue or deep gray, whereas other colonies most often are pink or pale-yellow. Count the colonies as discussed in Section 17.2.3.

17.5.2.5 Verification

Verify Klebsiella colonies from samples from ambient waters and effluents and when Klebsiella is suspect in the water supply distribution systems.

Verify a minimum of five typical colonies by transferring growth from a colony or pure culture to a commercial multitest-system for Gram-negative speciation, see Chapter 20.

1. Key tests for Klebsiella are:

Citrate	positive
Indole	negative
Motility	negative
Lysine decarboxylase	positive
Ornithine decarboxylase	negative
Urease	positive

2. Klebsiella of nonfecal origin can be identified by correlation of indole production and liquefaction of pectin with the ability to grow at 10°C and with a negative fecal coliform response.

17.6 APPLICATION FOR SOIL, SEDIMENT, AND SLUDGE SAMPLES

Collecting, handling, analyzing, and reporting microbiological densities of solid type samples are discussed in Section 16.4. Calculation of these microbiological values on the basis of 1 g of dry weight sample is also discussed in Chapter 16.

17.7 ANALYTICAL QUALITY CONTROL

Each analytical batch requires quality control (QC) checks such as blank, negative, and positive controls (approve accuracy), and duplicate samples (approve precision). These QC steps are detailed in the procedure for the total coliform MF test in Section 17.2.1. These steps differ with specified incubation temperature and time, as well as with the representative color of the counted colonies in each parameter.

Analytical QC requirements are discussed in Section 9.6.

18 Determination of Fecal Coliform

18.1 DEFINITION OF THE FECAL COLIFORM GROUP

The fecal coliforms are part of the total coliform group (Chapter 17). They are defined as Gram-negative nonspore-forming rods that ferment lactose in 24 ± 2 hours at 44.5°C with the production of gas in a multiple-tube fermentation procedure and produce acidity with blue colonies in a membrane filter procedure. The major species of the group is *Escherichia coli (E. coli),* a species indicative of fecal pollution and the possible presence of enteric pathogens.

The direct membrane filter (MF), the delayed-incubation MF, the multiple-tube, and the most probable number (MPN) methods can be used to enumerate fecal coliforms in water, wastewater, soil, sediment, and sludge. The method chosen depends upon the characteristics of the sample.

The fecal coliforms are part of the total coliform group. The major species in the fecal coliform group is *E. coli,* a species indicative of fecal pollution and the possible presence of enteric pathogens.

18.2 MEMBRANE FILTER METHOD

18.2.1 APPLICATION

An appropriate water sample or its dilution is passed through a membrane filter that retains the bacteria present in the sample. The filter containing the bacteria is placed on an absorbent pad and saturated with *M-FC broth* or on *M-FC agar* in a petri dish. After being incubated at 44.5°C for 24 h, the typical blue colonies are counted and reported as fecal coliform count per 100 ml sample.

The advantage of the MF test is that the results are obtained in 24 h. The test is applicable to the examination of lakes and reservoirs, wells and springs, public water supplies, natural bathing waters, secondary nonchlorinated effluents from sewage treatment plants, farm ponds, stormwater runoff, raw municipal sewage, and feedlot runoff. The membrane filter test is also used successfully in marine water testing.

The single step MF fecal coliform procedure may produce lower results than those obtained with the MPN procedure, particularly for chlorinated samples. Disinfection and toxic materials such as metals, phenols, acids, or caustics also affect recovery of fecal coliforms on the membrane filter.

18.2.2 EQUIPMENT AND GLASSWARE

The same equipment and glassware are used as in the total coliform determination discussed in Section 17.1.3, except that the incubator should be maintained at a stable 44.5 ± 0.2°C.

18.2.3 CULTURE MEDIA

The proper rehydration of the dehydrated media is supplied by the manufacturers. The needed media are DIFCO M-FC broth (No. 0883-02) or BBL M-FC broth (No. 11364), DIFCO lauryl tryptose broth (No. 0241-02) or BBL lauryl sulfate broth (No. 11338), and DIFCO EC-medium (No.0314-02) or BBL EC broth (No. 11187).

18.2.3.1 Preparation of M-FC Broth

Add 37 g of M-FC medium to 1 l of laboratory pure water containing 10 ml of the rosolic acid solution. Heat to near boiling to dissolve and cool. Do not sterilize by autoclaving. Store at 4°C in refrigerator. Discard unused medium after 96 h.

18.2.3.1.1 Rosolic acid solution

Dissolve 1 g of rosolic acid in 100 ml of 0.2 *N* NaOH. Keep in the refrigerator. Discard after 2 weeks or if the color changes from dark red to muddy brown.

18.2.3.1.2 0.2 **N** *NaOH*

Dissolve 8 g of NaOH in DI water and fill to 1 l.

18.2.3.2 Preparation of M-FC Agar

The preparation of M-FC agar is the same as for M-FC broth. Add 15 g of agar per 1 l of medium (1.5 percent). Heat to boiling, then cool to about 44 to 46°C and add to petri dishes to a minimum agar-depth of 2 to 3 mm. Allow to solidify. Protect from light. Store in the refrigerator, place dishes in an inverted position in a plastic bag. Discard unused medium after 2 weeks.

18.2.3.3 Preparation of Lauryl Tryptose Broth

Add 35.6 g of the medium to 1 l of laboratory pure water and mix to dissolve. Dispense 10 ml vol into fermentation tubes. Sterilize for 15 min at 121°C (15 lb pressure).

18.2.3.4 Preparation of EC Medium

Add 37 g of medium to 1 l of laboratory pure water. Mix for dissolution. Dispense 10 ml vol into fermentation tubes. Sterilize by autoclaving for 15 min at 121°C (15 lb pressure).

18.2.3.5 Preparation of Sterile Dilution Water

The preparation of sterile dilution water is the same as in the total coliform membrane filter method, see Section 17.1.4.

18.2.4 PROCEDURE

18.2.4.1 Set Up Filtration Apparatus

Filtration apparatus is set up as described in Section 17.1.5.

18.2.4.2 Suggested Sample Volumes for the Membrane Filter Fecal Coliform Test

Suggested sample volumes for the MF fecal coliform test include:

Water Source	Volume To Be Filtered, ml
Lakes, reservoirs	100, 50
Wells, springs	100, 50
Water supply intake	50, 10, 1
Natural bathing waters	50, 10, 1

Water Source	Volume To Be Filtered, ml
Sewage secondary effluent	10, 1, 0.1
Farm ponds, rivers	1, 0.1, 0.01
Stormwater runoff	1, 0.1, 0.01
Raw municipal sewage	0.1, 0.01, 0.001
Feedlot runoff	0.1, 0.01, 0.001

18.2.4.3 Procedure Using Broth Media

The procedure using broth media is the same as the procedure is in the total coliform membrane filter method, see Section 17.2.2 with the exception that incubation temperature for fecal coliform bacteria is 44.5°C and, after incubation, count the blue colonies.

18.2.4.4 Procedure Using Agar Media

The procedure using agar media is the same as the procedure is in the total coliform membrane filter method detailed in Section 17.2.1. The incubation temperature for fecal coliform bacteria is 44.4°C. After incubation, count blue colonies.

18.2.5 Counting and Recording Colonies

Colonies produced by fecal coliform bacteria on the M-FC medium are various shades of blue. Pale yellow colonies are atypical for *E. coli,* and need verification. Nonfecal colonies are gray to cream colored and are not counted. Pinpoint blue should be counted and confirmed. Count filters with 20 to 60 colonies and report as count per 100 ml. Fecal coliform levels are generally lower than total coliform densities in the same sample.

18.2.5.1 Countable Membranes with 20 to 60 Colonies

Count all blue colonies and calculate by using the formula:

$$\text{counts/100 ml} = \frac{\text{Number of colonies counted}}{\text{sample filtered, ml}} \times 100 \qquad (18.1)$$

18.2.5.2 Countable Membrane Filters with Less Than 20 Blue Colonies

Report countable membrane filters with less than 20 blue colonies as estimated count per 100 ml and specify the reason.

18.2.5.3 Membranes with No Colonies

Counts are based upon the largest volume filtered. For example, if 10, 5, and 1 ml are filtered and all plates show 0 counts, select the largest volume (10 ml), apply the general formula and report the count as a less than value.

$$\frac{1}{10} \times 100 = 10$$

Report this count as fecal coliform count per 100 ml is less than or equal to 10.

18.2.5.4 Countable Membranes with More Than 60 Colonies

Calculate the count from the highest dilution and

Report it as a greater than value.

18.2.5.5 Uncountable Membranes with More Than 60 Colonies

Use 60 colonies as the basis of calculation with the smallest filtration volume, for example, 0.01 ml,

$$\frac{60}{0.01} \times 100 = 600{,}000$$

Report as greater than 600,000 fecal coliforms per 100 ml.

18.2.6 Verification

Verification of the membrane filter test for fecal coliforms establishes the validity of colony differentiation by blue color and provides supporting evidence of colony interpretation.

18.2.6.1 Procedure

The procedure for verifying the membrane filter test for fecal coliforms includes:

1. Pick from the centers of at least 10 well-isolated blue colonies. Inoculate into lauryl tryptose broth and incubate 24 to 48 hours at 35°C.
2. Confirm gas-positive lauryl tryptose broth tubes at 24 and 48 h by inoculating a loopful of growth into EC-tubes and incubating for 24 h at 44.5°C. Cultures that produce gas in EC-tubes are interpreted as verified fecal coliform colonies. Verification of Fecal coliform colonies is shown in Figure 18.1.

18.3 DELAYED INCUBATION MF METHOD

18.3.1 Application

To conduct the delayed incubation test filter sample in the field immediately after collection, the filters are placed on the *M-VFC broth* (a minimum growth medium) and transported from the field sites to the laboratory where the filters are transferred to the *M-FC medium* and are incubated at 44.5°C for 24 h. Blue colonies are counted as fecal coliforms.

Transport the culture dish containing the membrane filter in an appropriate shipping container. Fecal coliform bacteria can be held on the M-VFC medium for up to 72 h with little effect on the final count. Of course, the holding time should be kept to a minimum. The M-VFC medium keeps fecal coliform organisms viable but prevents visible growth during transportation. The delayed incubation MF method is useful in survey monitoring or emergency situations when the standard fecal coliform test cannot be performed at the sample site, or when time and temperature limits for sample storage cannot be met. Consistent results have been obtained with this method using water samples from a variety of sources, but it is not as effective in saline waters.

18.3.2 Equipment and Glassware

The equipment and glassware is the same as listed in Section 17.2.2. A field filtration unit is available, with the same specification as described in Section 17.2.2.

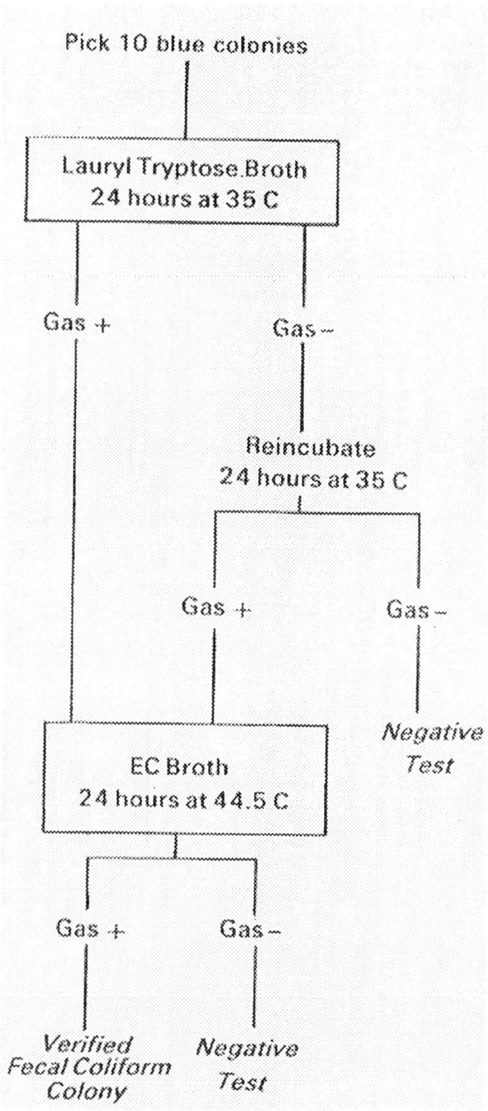

FIGURE 18.1 Verification of fecal coliform colonies on the membrane filter.

18.3.3 Culture Media

The culture media is the same as listed in Section 17.2.3 and includes the M-VFC holding medium.

18.3.3.1 Preparation of the M-VFC Medium

This medium may not be available in dehydrated form and may require preparation from the basic ingredients:

Casitone, vitamin free	0.2 g
Sodium benzoate	4.0 g
Sulfanilamide	0.5 g
Ethanol (95 percent)	10.0 ml
Distilled water	1.0 l

Heat medium to dissolve and sterilize by filtration through 0.22 μm membrane filter. If only 100 ml medium are to be prepared, preferably add 2 ml of a 1:100 aqueous solution (1 g dissolved in 100 ml distilled water) of casitone rather than 0.02 g dry reagent. Do not use denatured alcohol. Final pH should be 6.7.

Store finished medium in refrigerator at 2° to 10°C and discard unused medium after 2 weeks.

18.3.4 Counting and Recording Colonies

Counting and recording colonies were discussed in Section 17.2.3. The report should include the elapsed time.

18.4 THE MOST PROBABLE NUMBER (MPN) METHOD

18.4.1 Application

Culture from positive tubes in the *lauryl tryptose broth* (same as presumptive test in total coliform MPN determination) inoculated into *EC broth* and incubated at 44.5°C for 24 h. Formation of gas in any quantity in the inverted vial is a positive reaction confirming fecal coliforms. Fecal coliform densities are calculated from the MPN table on the basis of the positive EC tubes.

18.4.2 Equipment and Glassware

Incubator or a water bath should maintain a 44.5 ± 0.2°C temperature, otherwise the equipment and glassware is same as listed in Section 17.4.3.

18.4.3 Media

18.4.3.1 Lauryl Tryptose Broth

The lauryl tryptose broth is the same broth as the broth used in the total coliform presumptive test, see Section 17.1.4.

18.4.3.2 EC Medium

The EC medium is the same medium as the medium used in the fecal coliform membrane filter method, see Section 18.2.3.

18.4.3.3 Dilution Water

Sterile buffered dilution water is dispensed in 99 ml vol in dilution bottles. See Section 17.1.4 for the preparation of dilution water.

18.4.4 Procedure

Determination of total coliform by MPN test, Section 17.4 describes the general MPN procedure in detail.

1. Prepare lauryl tryptose broth and EC medium. Clearly mark each bank of tubes identifying the sample and the sample volume inoculated.

2. Inoculate the lauryl tryptose broth (presumptive test) medium with appropriate quantities of sample following the presumptive test total coliform procedure in Section 17.4.6.
3. Gently shake the presumptive tubes. Using a sterile inoculating loop transfer inocula from positive presumptive test tubes to EC confirmatory media. Gently shake the rack of inoculated EC tubes to ensure mixing the inoculum with medium.
4. Incubate inoculated EC tubes at 44.5°C for 24 ± 2 h. Tubes must be placed in the incubator within 30 min after inoculation. If incubate in a water bath, the water depth must come to the top level of the culture medium in the tubes.
5. The presence of gas in any quantity in the EC confirmatory tubes after 24 h constitutes a positive test for fecal coliforms.

The flow chart for the fecal coliform MPN test is shown in Figure 18.2.

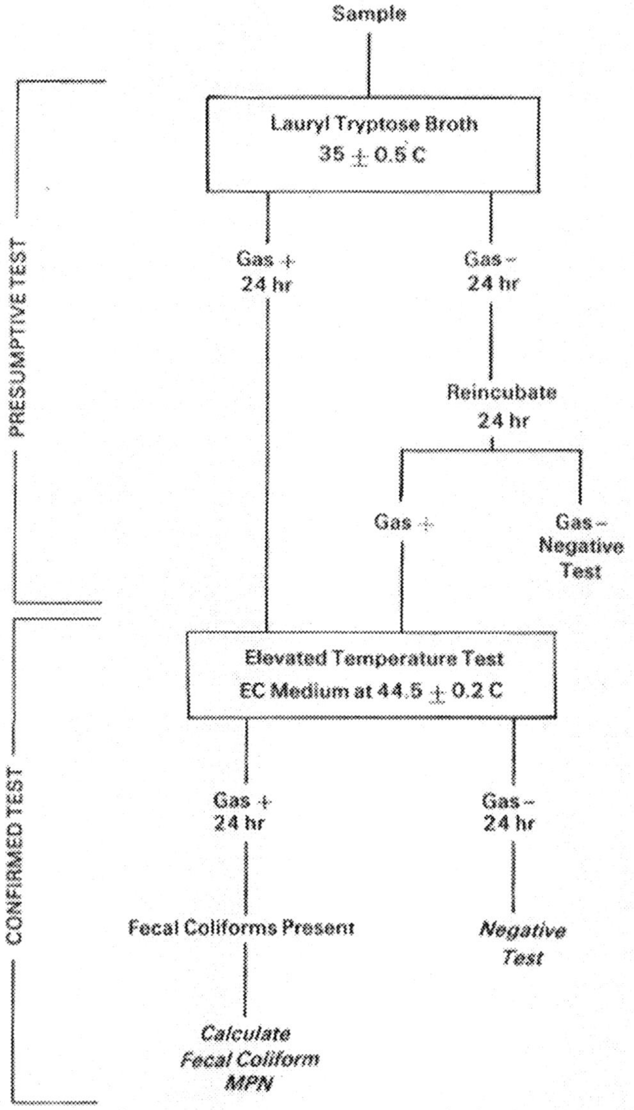

FIGURE 18.2 Flow Chart for the fecal coliform MPN tests.

18.4.5 Calculation and Reporting

Calculate fecal coliform densities on the number of positive EC fermentation tubes using the MPN table, Table 17.1. The MPN results are computed from three dilutions that include the highest dilution with all positive tubes and the next two higher dilutions. For example, if 5 samples of 10 ml, 5 samples of 1.0 ml, and 5 samples of 0.1 ml are inoculated initially into presumptive test medium, and positive EC confirmatory results are obtained from 5 of the 10 ml portions, 3 of the 1.0 ml portions, and 0 of the 0.1 ml portions, the coded result of the test is 5-3-0. The code is located in the MPN table and the MPN per 100 ml is recorded. Detailed calculation and rules of MPN reporting is given in the total coliform MPN test, see Section 17.4.7.

18.5 THE MUG TEST

A *fluorophore* is a compound that absorbs the light of a short, excitatory wavelength and emits the light of a longer wavelength. The emitted, fluorescent light, can be seen visually and can be measured quantitatively by special photometers, called fluorometers.

Fluorophores are used in biological reactions by binding them to a substrate that effectively inhibits their fluorescence. When the substrate is acted on by the metabolic enzymes of microorganisms, the fluorophore is released and fluoresces under ultraviolet (UV) or other short wave light. Substrate-fluorophore combinations are stable and allow the detection of tiny amounts of reactant.

The first practical test using this technology was the MUG-test (*4-methylum-belliferyl- -D-glucuronide*) for rapid identification of *E. coli.* The enzyme -glucuronidase produced by *E. coli* and a few species of salmonella and shigella, breaks the bond holding MUG together and releases the potent *fluorophore 4-methylum-belliferone*. By observing the fluorescence, one can identify the organisms, often within 30 min. Commercially available MUG-tests have been found to be specific and convenient.

18.6 APPLICATION FOR SOIL, SEDIMENT, AND SLUDGE SAMPLES

Collection, preparation, analysis, and reporting microbiological density of solid type samples is detailed in Section 16.4.

18.7 ANALYTICAL QUALITY CONTROL (QC) PROCEDURE

See Section 17.6.

19 Determination of Fecal Streptococcus

The membrane filter (MF), most probable number (MPN), and direct pour plate procedures can be used to enumerate and identify *Fecal streptococci* in water and wastewater. The method selected depends on the characteristics of the sample. The multiple-tube technique is applicable primarily to raw and chlorinated wastewater and sediments, and can be used for marine and fresh waters. The membrane filter technique also may be used for fresh and marine water samples but is unsuitable for highly turbid waters.

19.1 INTRODUCTION

The terms *Fecal streptococcus* and *Lancefield's group D streptococcus* have been used synonymously. When used as indicators of fecal pollution, the following species and varieties are implied: *S. faecalis, S. faecalis liquefaciens, S. faecalis zymogenes, S. faecium, S. bovis,* and *S. equinus.* For sanitary analyses, media and methodology for quantification are selective for these organisms. The *Lancefield's group Q* occur in the feces of humans and other warm-blooded animals, especially chickens, called *S. avium.* The *Fecal streptococci* group includes the serological groups D and Q.

Fecal streptococci data verify fecal pollution and may provide additional information concerning the recency and probable origin of pollution. In combination with data on coliform bacteria, Fecal streptococci are used in the sanitary evaluation as a supplement to fecal coliforms when a more precise determination of the sources of contamination is necessary.

The occurrence of *Fecal streptococci* in water indicates fecal contamination by warm-blooded animals. Further identification of streptococci types present in the sample may be obtained by biochemical characterizations. For example, *S. bovis* (cow) and *S. equinus* (horse) are found in the fecal excrement of nonhuman warm-blooded animals. High numbers of these organisms are associated with pollution from meat processing plants, dairy wastes, and run-off from feed lots and farmlands. Because of limited survival time outside the animal intestinal tract, their presence indicates very recent contamination from farm animals.

19.1.1 THE FECAL COLIFORM AND FECAL STREPTOCOCCI, FC/FS RATIOS

The relationship of the fecal coliform and fecal streptococci density may provide information on the potential source of contamination.

Origin of Pollution	FC/FS ratio
Man	4.4
Duck	0.6
Sheep	0.4
Chicken	0.4
Pig	0.4
Cow	0.2
Turkey	0.1

A ratio greater than 4.1 indicates domestic wastes; a ratio less than 0.7 indicates livestock and poultry wastes.

The value of this ratio has been questioned because of the variable survival rates of fecal streptococcus group species. *S. bovis* and *S. equinus* die rapidly once exposed to aquatic environments, whereas *S. faecalis* and *S. faecium* tend to survive longer.

Furthermore, disinfection of wastewaters appears to have a significant effect on the ratio of these indicators, which may result in misleading conclusions regarding the source of contaminants. The ratio is also affected by the methods for enumerating fecal streptococci.

The KF membrane filter procedure has a false-positive rate ranging from 10 to 90 percent in marine and fresh waters. For this reason, the FC/FS ratio cannot be recommended and should not be used as a means of differentiating human and animal sources of pollution.

19.2 ENTEROCOCCI PORTION OF THE FECAL STREPTOCOCCUS GROUP

The enterococcus group is a subgroup of the fecal streptococci that includes *S. faecalis, S. faecium, S. gallinarum,* and *S. avium.* The enterococci portion of the fecal streptococcus group is a valuable bacterial indicator for determining the extent of fecal contamination of recreational surface waters. Enterococci are the most efficient bacterial indicators of water quality. Studies at marine and fresh water bathing beaches indicated that swimming associated gastroenteritis is related directly to the quality of the bathing water.

Water quality guidelines based on enterococcal density have been proposed for recreational waters. For recreational fresh waters, the guideline is 33/100 ml and for marine waters, it is 35/100 ml. Each guideline is based on the geometric mean of at least 5 samples per 30 day period during the swimming season.

The enterococci are differentiated from other streptococci by their ability to grow in a 6.5 percent sodium chloride solution at a pH of 9.6, and at 10°C and 45°C.

19.3 MEMBRANE FILTER (MF) TECHNIQUE

The membrane filter technique is recommended as the standard method for assaying fecal streptococci in fresh and marine waters and in nonchlorinated sewage. Wastewaters from food processing plants, slaughter houses, canneries, sugar processing plants, dairy plants, feed-lots and farmland runoff may be analyzed by this method.

19.3.1 APPLICATION

A suitable volume of a sample is passed through the 0.45 μm membrane filter which retains the bacteria. The filter is placed on *KF streptococcus agar* and incubated at 35°C for 48 h. Red and pink colonies are counted as streptococci. The general advantages and limitations of the MF method are given in Chapters 17 and 18.

19.3.2 EQUIPMENT AND GLASSWARE

The equipment and glassware are the same as that listed in total coliform membrane filter method determination, Section 17.1.3.

19.3.3 Culture Media

19.3.3.1 Preparation of KF Streptococcus Agar

DIFCO 0496-02 or BBL 11313 :

1. Add 76.4 grams of the medium per 1 l laboratory pure water. Dissolve by heating in a boiling water bath with agitation. After solution is ready, heat an additional five minutes in the water bath. Do not autoclave.
2. Cool to 60°C and add 1 ml of a filter sterilized 1 percent aqueous solution of 2,3,5-triphenyl-tetrazolium-chloride (TTC) per 100 ml of agar. (More information about preparation will be provided later on.)
3. If necessary, adjust pH to 7.2 with 10 percent sodium carbonate (Na_2CO_3) solution. (More information about preparation will be provided later on.)
4. Cool to 44 to 46°C and add 0.015 g Brom Cresol Purple indicator. Do not hold the completed medium (with indicator) at 44 to 46°C for more than 4 h before using.
5. If the medium is not used immediately after preparation, do not add the indicator until just prior to use. Store the prepared medium (without indicator) in the dark for as long as 30 days at 4°C (the TTC solution is light sensitive).

19.3.3.1.1 1 percent solution of 2,3,5-triphenyl-tetrazolium-chloride (TTC)

Use 1 g of 2,3,5-triphenyl-tetrazolium-chloride per 100 ml laboratory pure water. After dissolution and mixing, filter and sterilize the solution.

19.3.3.1.2 10 percent sodium carbonate, Na_2CO_3

Use 10 grams of sodium carbonate in 100 ml of laboratory pure water. After dissolution and mixing, filter and sterilize the solution.

19.3.3.2 Preparation of Brain Heart Infusion Broth (BHI)

DIFCO 0037-02 or BBL 11058:

1. Dissolve 37 grams of medium in 1 l of laboratory pure water.
2. After complete dissolution, dispense in 8 to 10 ml vol in screw-cap tubes and sterilize for 15 min at 121°C (15 lb pressure).
3. If the medium is not used the same day as prepared and sterilized, heat at 100°C for several minutes to remove absorbed oxygen and cool quickly without agitation, just prior the inoculation.

19.3.3.3 Brain Heart Infusion Agar (BHI Agar)

DIFCO 0418-02 or BBL 11064: It is the same composition as BHI with the addition of 15 g agar per 1 l of the medium.

1. Heat in boiling water bath until dissolved.
2. Dispense 10 to 12 ml in screw-capped fermentation tubes.

3. Sterilize for 15 min at 121°C (15 lb pressure).
4. Slant after sterilization.

19.3.3.4 Brain Heart Infusion Broth with 40 Percent Bile

The preparation for brain heart infusion broth with 40 percent bile is the same as BHI broth preparation with the addition of 40 ml of sterile 10 percent oxgall to 60 ml of medium or 688 ml of oxgall to each liter of medium, as follows:

1. Prepare the broth medium as described previously.
2. Cool after sterilization.
3. Add the filter sterilized 10 percent oxgall solution in the 40 to 60 ml ratio and mix.
4. Dispense aseptically in 10 ml vol into sterile culture tubes.

19.3.3.4.1 Preparation of 10 percent oxgall solution

Use 10 g oxgall per 100 ml laboratory pure water. After dissolving and mixing, filter and sterilize the solution.

19.3.3.5 Dilution Water

The procedures are the same as in the total coliform test, See Section 17.1.4.

19.3.4 Procedures

The procedures are the same as in total coliform MF test, see Section 17.1.

19.3.4.1 Sample Volume

Filter appropriate volumes of water sample to obtain 20 to 100 colonies on the membrane surface. At least three sample increments should be filtered in order of increasing volume. Where no background information is available, more research may be necessary.

19.3.5 Counting and Recording Colonies

Select plates with 20 to 100 pink to dark red colonies. These may range in size from barely visible to about 2 mm in diameter. Colonies of other colors are not counted.

Report fecal streptococcal density as organisms per 100 ml using the general formula:

$$\text{counts/100 ml} = \frac{\text{colonies counted}}{\text{volume of sample filtered, ml}} \times 100 \tag{19.1}$$

See Section 17.2.3 for calculation and reporting the results.

19.3.6 Verification

Colonies growing on the membrane filter should be verified.

1. Plates for verification should contain 20 to 100 colonies. At least 10 typical colonies from the membrane should be picked and transferred into the BHI broth or onto BHI agar slants.
2. After a 24 to 48 h incubation at 35°C, transfer a loopful of growth from the BHI slant to a clean glass slide and add a few drops of 3 percent hydrogen peroxide (H_2O_2) to the smear. If the catalase enzyme is present, it cleaves the H_2O_2 to water and visible oxygen gas. Bubbles constitute a positive catalase test and indicates nonstreptococcal species (see the catalase test, Section 14.2.1). Confirmation need not continue.
3. Final confirmation of fecal streptococci is achieved by determining growth of the catalase negative isolates in BHI broth incubated at 45°C and BHI broth plus 40 percent bile incubated at 35°C. Growth within two days indicates fecal streptococcal species.
4. Further identification of streptococcal types present in the sample may be obtained by biochemical characterization. Such information is useful for investigating sources of pollution.

Verification procedure for fecal streptococci is shown in Figure 19.1.

19.4 DELAYED MF PROCEDURE

Because of the stability of the KF agar (see Section 19.3.3) and its extreme selectivity for fecal streptococci, it is possible to filter water samples at a field site, place membranes on the KF agar medium in tight-lidded petri dishes (see Section 11.2.1), and hold these plates for up to 3 days. After the holding period, plates are incubated for 48 h at 35°C and are counted in the normal manner.

Verification, counting, and reporting are the same as discussed in Sections 19.3.5 and 19.3.6.

19.5 MOST PROBABLE NUMBER (MPN) METHOD

The multiple-tube procedure estimates the number of fecal streptococci by inoculating decimal dilutions of the sample into broth tube media. Positive tubes in the presumptive test are indicated by growth (turbidity) in *azide-dextrose broth* after incubation at 35°C for 24 to 48 h. To confirm the presence of fecal streptococci, a portion of the growth from each positive azide-dextrose broth tube is streaked onto *Pfizer selective enterococcus (PSE) agar* and incubated at 35°C for 24 h. The presence of brownish-black colonies with brown halos confirms fecal streptococci.

This method can be used for detection of fecal streptococci in water and sewage, but it is more time consuming and less direct than the other procedures.

The MPN must be used for samples which cannot be examined by the MF or direct plating techniques because of turbidity, high numbers of background bacteria, metallic compounds, the presence of coagulants, the chlorination of sewage effluents, or sample volume limitations of the plating technique.

19.5.1 Culture Media

19.5.1.1 Azide Dextrose Broth

DIFCO 0837-02, BBL 11000:

1. Add 34.7 g of azide dextrose broth to 1 l of laboratory pure water.
2. Dissolve and dispense into tubes. Sterilize at 118°C for 15 min (15 lb pressure).

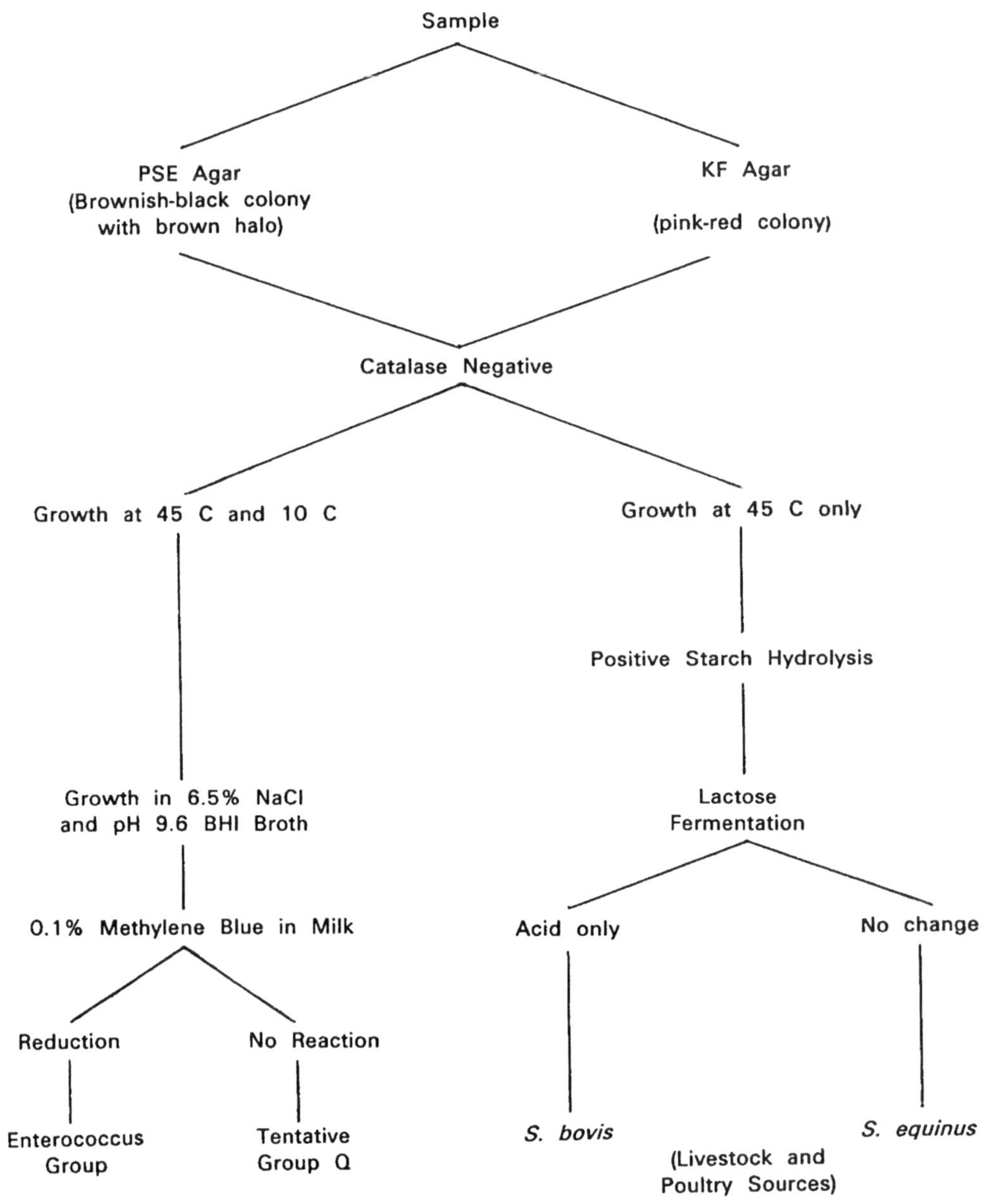

FIGURE 19.1 Isolation and identification of fecal streptococci, general scheme.

Prepare medium in multiple strength for larger inocula to preserve the correct concentration of ingredients. For example, if 10 ml of inoculum is to be added to 10 ml of medium, the medium should be prepared double strength.

19.5.1.2 Pfizer Selective Enterococcus, (PSE) Agar

Grand Island Biological Company, GIBCO

1. Add 58 g of PSE agar to 1 l of laboratory pure water. Heat in a boiling water bath to complete solution.

2. Dispense into flasks and sterilize for 15 min at 121°C (15 lb pressure). pH should be 7.1 ± 0.2 after sterilization.
3. Hold medium for not more than 4 h at 45 to 50°C before plates are poured.

19.5.2 Procedure

19.5.2.1 Presumptive Test

The following procedure is used for the presumptive test.

1. Mark culture tubes and the azide dextrose tubes to identify samples and sample volumes.
2. Shake sample vigorously before inoculating the *azide dextrose broth* with the appropriate sample volumes. The number of fecal streptococci in a water polluted with municipal wastes is generally lower than the number of coliforms. Therefore, larger sample volumes must be used to inoculate MPN tubes for fecal streptococci than for coliforms. For example, if the sample volumes of 1.0, 0.1, 0.01, and 0.001 ml are used for the coliform test, a series of 10, 1.0, 0.1, and 0.01 ml volumes are inoculated for the fecal streptococci test. Use double strength broth when inoculating 10 ml samples. Sample volumes from feedlots, meat packing plants, and stormwater run-off with more streptococci than coliforms must be adjusted accordingly.
3. Shake the rack of inoculated culture tubes to mix well and incubate at 35°C for 24 h.
4. Examine incubated tubes for turbidity or a button of sediment at the bottom of the culture tube or both. If no definite turbidity or sediment is present, reincubate and read again at the end of 48 h.

19.5.2.2 Confirmed Test

The following procedure is used for the confirmed test.

1. Numbered *PSE agar* plates correspond to the positive azide dextrose tube used.
2. Streak a portion of growth from each positive azide dextrose broth tube onto PSE agar plates.
3. Incubate the inverted dishes at 35°C for 24 h.
4. Brownish-black colonies with brown halos confirm the presence of fecal streptococci.
5. Read and record the results of each plate corresponding to the positive azide dextrose tube.

19.5.3 Calculation

Using Table 17.1 in Chapter 17, calculate fecal streptococci densities on the basis of the number of positive confirmed tests from the PSE agar plates. To calculate and record the results, see Section 17.4.7.

19.6 POUR PLATE METHOD

Aliquots of the water sample are delivered to the bottom of a petri dish and liquified PSE agar, or KF agar, is added and thoroughly mixed with the water sample. Fecal streptococci on PSE agar are 1 mm in diameter and brownish-black with brown halos after 24 h incubated at 35°C. On KF agar, fecal streptococci are red or pink after 48 h incubated at 35°C.

The pour plate method is recommended as an alternate procedure to the MF technique when chlorinated sewage effluent and water samples with high turbidity are encountered.

PSE agar, the medium of choice, has several advantages: it requires only 24 h incubation compared to 48 h for other media, and it exhibits consistent recovery, regardless of sources.

With the pour plate technique, only small volumes of the sample may be analyzed. This is the disadvantage when the fecal density is low and a large volume of sample would be required for an accurate density determination. Consequently, the MF technique should be used unless the water is so turbid that filtration is impossible.

19.6.1 MEDIA

Use sterile *PSE agar* (see Section 19.5.1) or *KF streptococcus agar* (see Section 19.3.3).

19.6.2 PROCEDURE

The following procedure is used to test for fecal streptococci.

1. Prepare duplicate petri dishes for each sample increment. Mark each petri dish with the number of the sample, the dilution, the date, and other necessary information.
2. Shake the sample bottle vigorously to disperse the bacteria. Take care that the sample bottle is tightly closed to prevent sample leakage during shaking.
3. Dilute the sample to obtain final plate counts between 30 and 300 colonies. The number of colonies within this range gives the most accurate estimation of the microbial population. Because the magnitude of the microbial population in the original sample is not known beforehand, a range of dilutions must be prepared and plated to obtain a plate within this range of colony count. The dilution technique is explained in Section 13.1.
4. Transfer 0.1 and 1.0 ml from the undiluted sample to each of two separate petri dishes. Deliver the liquid into the dish, and touch the tip once against a dry area in the petri dish bottom while holding the pipet vertically.
5. Prepare the initial 1:100 or 10^{-2} dilution by pipeting 1 ml of the sample into a 99 ml dilution water blank.
6. Vigorously shake the 1:100 dilution and pipet 0.1 and 0.01 ml into each of two petri dishes with the technique described in Section 16.2.5.
7. Pour 12 to 15 ml of liquified cooled agar medium into each petri dish containing the sample or its dilution. Mix the medium and the sample thoroughly by gently rotating and tilting the petri dish. Not more than 20 min should elapse between dilution, plating, and addition of medium.
8. Allow agar to solidify as rapidly as possible after pouring, and place the inverted PSE plates at 35°C for 24 h and KF plates for 48 h.

19.6.3 COUNTING AND RECORDING

After the specified incubation period, select those plates with 30 to 300 fecal streptococcal colonies. Fecal streptococci on PSE agar are brownish-black colonies, about 1 mm in diameter with brown halos. On KF agar, fecal streptococci are pink or red and of varying sizes. Counting and reporting of the fecal streptococci are similar to that discussed in Section 16.2.6.

19.6.3.1 Plates with 30 to 300 Colonies

Select plates with 30 to 300 colonies. Calculate the average count for these plates, correcting for the dilution.

$$\text{FS count/ml} = \frac{\text{sum of colonies}}{\text{volume tested (ml)}} \tag{19.2}$$

19.6.3.2 All Plates Greater than 300 Colonies

When counts for all dilutions contain more than 300 colonies; for example, the count is greater than 500 for 1.0 ml, greater than 500 for 0.1 ml, and 340 for 0.01 ml, count the plate having nearest to 300 colonies.

$$\text{FS count/ml} = \frac{340}{0.01} = 34{,}000$$

Report as estimated FS count is greater than or equal to 34,000 per ml.

19.6.3.3 All Plates with Fewer than 30 Colonies

If all plates have less than 30 colonies, record the actual number of colonies on the lowest dilution plate.

Report as estimated FS count per ml.

19.6.3.4 Plate with No Colonies

If plates from all dilutions show no colonies, assume a count of one colony; then divide one by the largest volume filtered and report the value as a less than count.

For example, if 0.1, 0.01 and 0.001 ml were filtered with no reported colonies, the count would be:

$$\text{FS count/ml} = 1/0.1 = 10$$

Report as *FS count is less than or equal to 10 per ml.*

19.6.3.5 All Plates Are Crowded

It is possible to use the grid on the Quebec or similar colony counter.

- With less than 10 colonies per cm^2—Count the colonies in 13 squares with representative distribution of colonies. Select seven consecutive horizontal squares and six consecutive vertical squares for counting. Sum the colonies in these 13 cm^2 and multiply by 5 to estimate the colonies per plate for glass plates (area of 65 cm^2), or multiply by 4.32 for plastic plates (area of 57 cm^2).

Report FS value as estimated colonies/plate.

- More than 10 colonies per cm^2—Count 4 representative squares, average the count per cm^2, multiply by the number of cm^2 per plate, usually 65 for glass plates and 57 for plastic plates, to estimate the colonies per plate. Then multiply by the reciprocal of the dilution to determine the count per ml.

- Bacterial counts are greater than 100 colonies per cm^2—The smallest sample size is 0.1 ml. Report as:

$$\text{estimated FS count/ml} \geq 100 \times 65 = > 6{,}500/0.1 \geq 65{,}000$$

19.7 PROCEDURES FOR SOILS, SEDIMENTS, AND SLUDGES

Collecting, handling, analyzing and reporting microbiological densities of solid-type samples are discussed in Section 16.4. Calculation of these microbiological values on the basis of 1 g of dry weight sample is also detailed.

19.8 ANALYTICAL QUALITY CONTROL (QC) PROCEDURES

See Section 17.6.

20 Enterobacteriaceae

20.1 INTRODUCTION

One very important group of bacteria is the Gram-negative, facultatively anaerobic rods. Many of them cause diseases of the intestinal tract, as well as other organs. One important family of this group is called *Enterobacteriaceae* or enterics, as they are commonly called, includes bacteria that inhabit the gastrointestinal tracts of humans and other animals. Some species are permanent common residents, others are found in only a fraction of the population, and still others are present only under disease conditions. Most enterics are active fermenters of glucose and other carbohydrates.

Enterobacteriaceae are divided into two groups: a group of bacteria which ferments lactose with gas formation (coliforms) and a group of bacteria that does not ferment lactose with gas formation. Enteric, Gram-negative bacteria are a large heterogeneous group of microbes whose natural habitat is the intestinal tract of human and animals. Among the enteric bacteria are members of the genera *Escherichia, Enterobacter, Shigella* and *Salmonella.*

Escherichia and *Enterobacter* can be distinguished from *Salmonella* and *Shigella* according to their ability to ferment lactose: the former ferment lactose to produce acid and gas whereas the latter do not.

20.1.1 Coliform Organisms

Coliform bacteria are Gram-negative, nonspore forming rods that ferment lactose with gas formation within 48 h at 35°C, and as applied to the membrane filter method, produce a dark red colony with a metallic sheen within 24 h on an Endo-type medium containing lactose. However, anaerogenic (non-gas producing) lactose fermenting strains of *Escherichia coli* and coliforms that do not produce metallic sheen on an Endo medium may be encountered. These organisms, as well as typical coliforms, can be considered indicator organisms, but they are excluded from the current definition of coliforms.

Collectively, the coliforms are referred to as indicator organisms because they indicate the presence of human or animal feces. All types of coliform organisms may occur in feces. Although *Escherichia coli (E. coli)* is nearly always found in fresh pollution from warm-blooded animals, other coliform organisms may be found in fresh pollution in the absence of *E. coli.*

The genera *Enterobacter, Klebsiella, Citrobacter,* and *Escherichia* usually represents the majority of the bacteria isolated from raw and treated municipal water supplies with *Enterobacter* being isolated most frequently, but all coliforms do not necessarily originate from sewage. *E. coli* is most readily affected by conventional water treatment than other coliforms. Coliform organisms can multiply on wood, swimming pool ropes, and may produce slime inside pipes. Differentiation of coliform types is valuable in determining the source and nature of the pollution.

Industrial wastes containing high concentrations of bacterial nutrients are capable of promoting large after-growths of coliforms in effluents and receiving waters. After-growths also may appear in treated water distribution systems. *Aeromonas* and other oxidase-positive, Gram-negative bacteria, as well as *Erwinia,* can be expected in raw and treated waters.

20.1.2 Differentiation of Enterobacteriaceae

Differentiation of potential indicator organisms of the Enterobacteriaceae is based on specified biochemical tests. The traditional IMViC (indol, methyl red, Voges-Proskauer, and citrate utilization) test is useful for differentiation, but does not provide complete identification, and additional biochemical tests are necessary. Detailed discussion of these biochemical tests are discussed in Section 14.2.

Enterics can also be distinguished from each other according to antigens present in their substrate. Preparing differential media and reagents may not be as economical for laboratories as using commercially prepared and prepackaged multiple test kits. These kits are simple to store, easy to use, and give reproducible and accurate results.

20.2 ENTEROTUBE II

ENTEROTUBE II is a prepared, sterile multimedia tube for rapid differential identification of Gram-negative bacteria (Enterobacteriaceae). The system is a self-contained, sterile, compartmented plastic tube containing twelve conventional media and an enclosed inoculating wire. It permits the inoculation of all media and the subsequent performance of 15 standard biochemical tests from a single bacterial colony. The resulting combination of reactions together with the computer coding system allow identification of Enterobacteriaceae. The computer coding and identification system for ENTEROTUBE II was constructed using the percentage data contained in the recent publication of Ewing entitled "Differentiation of Enterobacteriaceae by biochemical reactions." A simple pad permits a rapid check of the positive reactions obtained with ENTEROTUBE II. The checked positive numbers are totaled, and the composite number is then located in the coding manual to identify the organism. Where two or more organisms are listed, the confirmatory tests required to further identification.

The tube should be stored in refrigerator at 2° to 8°C. Do not freeze. Any of the following conditions may interfere with the accuracy of ENTEROTUBE II: dehydration or liquefication of media, lifting of wax from media surfaces, any change of media colors from those shown on uninoculated tube, and any indication of growth on media surfaces.

20.2.1 Operation of ENTEROTUBE II

The following procedure for the operation of ENTEROTUBE II includes:

1. Remove both caps from the ENTEROTUBE II. One end of the wire is straight and is used to pick up the inoculum; the bent end of the wire is the handle. Do not heat the wire.
2. Pick: Holding the ENTEROTUBE II, pick a well-isolated colony from the plate with the inoculating (straight) end of the wire. Avoid touching the agar with the needle.
3. Inoculate the ENTEROTUBE II by holding the bent end of the wire and twisting; then draw the wire through all twelve compartments using a turning motion
4. Reinsert the wire into the ENTEROTUBE II, and using a turning motion, draw the wire through all twelve compartments; draw wire until the tip is in H_2S per indole compartment. Break wire at the notch by bending. Discard handle and loosely replace caps on the tube. The portion of the wire remaining in the tube maintains anaerobic conditions necessary for fermentation of glucose, production of gas, and decarboxylation of lysine and ornithine.
5. Strip off blue tape after inoculation, but before incubation, to provide aerobic conditions in adonitol, lactose, arabinose, sorbitol, V-P, dulcitol/phenylalanine, urea and citrate compartments. Slide clear band over glucose compartment to contain any small amount of sterile wax that may escape owing to excessive gas produced by some bacteria.
6. Incubate the tube lying on its flat surface at 35 to 37°C for 24 h. Separate each tube slightly to allow for sufficient air circulation.
7. After incubation, compare the tube to an uninoculated one as in Interpret and record all reactions with the exception of indole and Voges-Proskauer, as described in Section 20.2.2. All other tests must be read before the indole and V-P tests are performed because these tests may alter the remainder of the ENTEROTUBE II reactions.

8. To perform the indole test, place the ENTEROTUBE II in a rack with the glucose compartment pointing downward and add 1 or 2 drops of Kovacs reagent through a plastic film of H_2S per indole compartment using either a needle and syringe or by melting a small hole using a warm inoculating needle or disposable pipet and dropping reagent into the compartment. Allow the reagent to contact the surface of the medium or the inner surface of the plastic film. A positive test is indicated by the development of a red color in the added reagent on surface of the media within 10 s.
9. To perform the Voges-Proskauer test, place the ENTEROTUBE II in a rack with the glucose compartment pointing downward or place it horizontally and add two drops of 20 percent potassium hydroxide (KOH) containing 0.3 percent creatine and three drops of 5 percent alpha-naphthol in absolute ethyl alcohol into the Voges-Proskauer compartment. A positive test is indicated by the development of a red color within 20 min.

The operation outline of the ENTEROTUBE II is shown in Figure 20.1.

20.2.2 To Read ENTEROTUBE II

The following procedures are used to read ENTEROTUBE II.

1. If no color change or an orange color is observed in the glucose compartments, the organism is not a member of the family Enterobacteriaceae. Reactions in ENTEROTUBE II is shown in Figure 20.2.
2. To read ENTEROTUBE II, interpret all media reactions in a sequential fashion by comparing reactions in the ENTEROTUBE II with color reactions as indicated in Table 20.4.

20.2.3 To Indicate Positive Reactions

The procedure for indicating positive reactions includes:

1. Indicate each positive reaction by circling the number appearing below the appropriate compartment of the ENTEROTUBE II outline on the pad (see Figure 20.3).
2. Add circled numbers within each bracketed section and enter this sum in the space provided below the arrow.
3. Read the five numbers in these spaces across as a five-digit number.
4. Find this five-digit number in the computer coding manual in the column titled "ID Value." This number will identify the genus and/or species of the organism.
5. Record the name of the organism and the date on the lines provided on pad, see Figure 20.4.

20.2.4 Ordering Information

ENTEROTUBE II is available in boxes of 25 tubes. The complete computer coding and identification system for ENTEROTUBE II may be obtained free of charge from:

Roche Diagnostic Systems
Technical Consultation Services Department
11 Franklin Avenue
Bellville, NJ 07109
(800) 631-0160

or

Becton Dickinson Microbiology Systems
Cockeysville, MD 21030

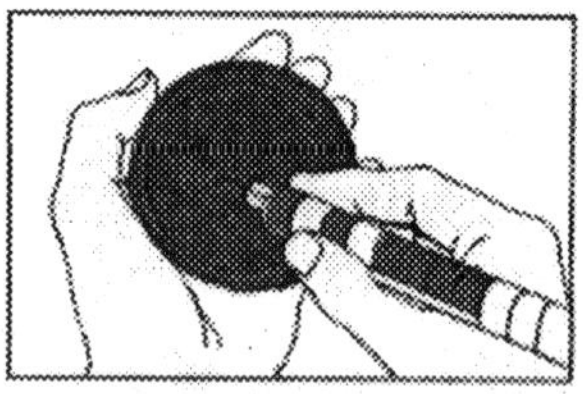

Remove both caps. Do not flame wire.

Pick a well isolated colony directly with the tip of the ENTERO-TUBE II inoculating wire.

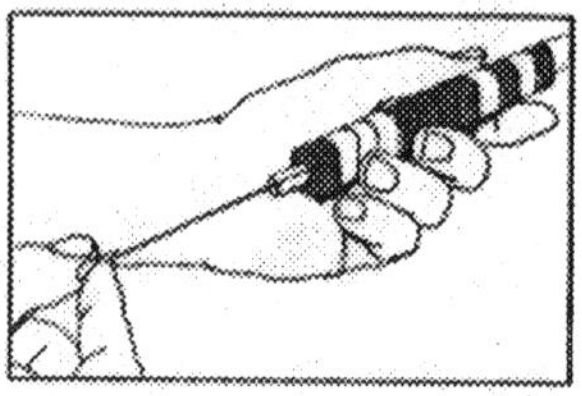

Inoculate ENTEROTUBE II by first twisting wire, then drawing wire through all twelve compartments using a turning motion.

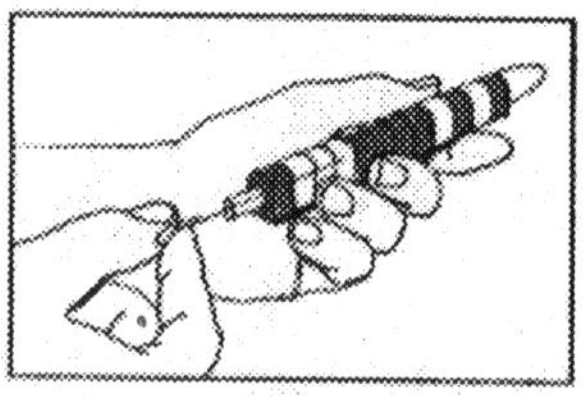

Reinsert wire (without sterilizing) into ENTEROTUBE II using a turning motion through all 12 compartment. Withdraw wire until the tip is in the H_2S/indole compartment. Break wire at notch by bending, discard handle and replace cap on tube loosely.

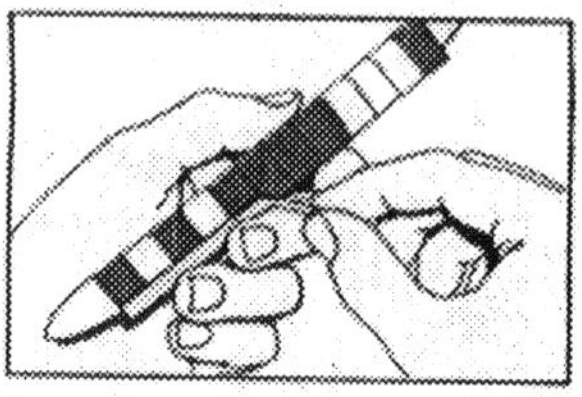

Strip off blue tape after inoculation—but before incubation—to provide aerobic conditions in lactose, arabinose, sorbitol, Voges-Proskauer, dulcitol, phenylalanine, urea, and citrate compartments. Slide clear band over glucose compartment to contain any small amount of sterile wax that may escape owing to excessive gas production. Incubate ENTERO-TUBE II at 35 to 37°C for 18 to 24 h by lying on the flat surface.

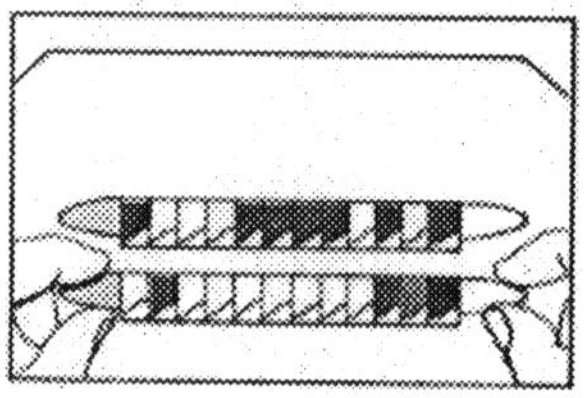

After incubation, compare the inoculated ENTEROTUBE II reacted colors with the colors of an uninoculated tube. Interpret and record all reactions except for the indole and Voges-Proskauer.

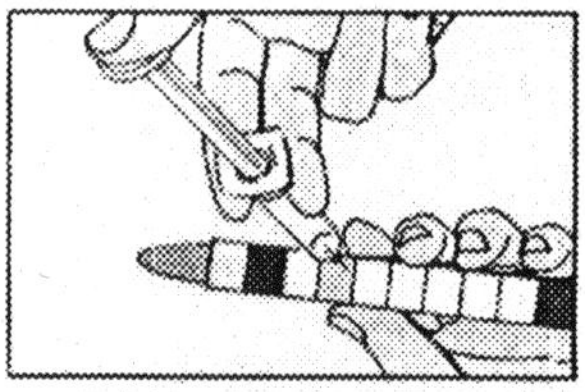

Perform indole test by injecting 1 to 2 drops of indole reagent into the H_2S/indole compartment. Positive test is indicated by a red color development after 10 s of addition of the reagent.

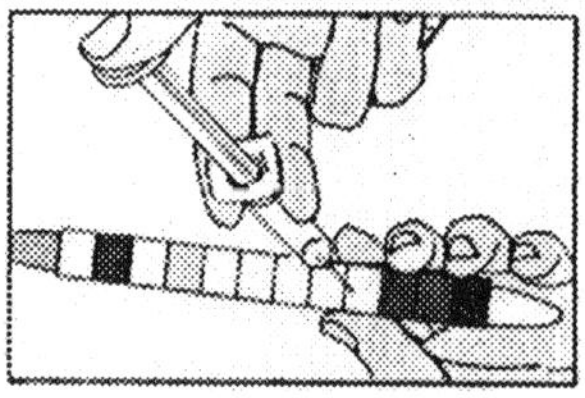

Perform the Voges-Proskauer test by injection of two drops of 20 percent KOH containing 0.3 percent creatine and 3 drops of 5 percent alpha-naphthol in ethanol. A positive test is indicated by the development of a red color within 20 min.

(From ENTEROTUBE II Manufacturer's Manual, Roche Diagnostic Systems, a Division of Hoffmann-La Roche Inc., Nutley, NJ 07110-1199).

FIGURE 20.1 To use the ENTEROTUBE II for the identification of Gram-negative bacteria (Enterobacteriaceae).

GLUCOSE (GLU): The end products of bacterial fermentation of glucose are either acid or acid and gas. The shift in pH owing to the production of acid is indicated by a change in the color of the indicator in the medium from red (alkaline) to yellow (acidic). Orange should be considered negative.

GAS PRODUCTION (GAS): This is the evidence by a definite and complete separation of the wax overlay from the surface of the glucose medium but not by bubbles in the medium.

LYSINE DECARBOXYLASE (LYS): Bacterial decarboxylation of lysine, which results in the formation of the alkaline end product cadaverine, is indicated by a change in the color of the indicator in the medium from pale yellow (acidic) to purple (alkaline).

ORNITHINE DECARBOXYLASE (ORN): Bacterial decarboxylation of ornithine, which results in the formation of alkaline end product putrescine, is indicated by a change in the color of the indicator from pale yellow (acidic) to purple (alkaline).

H_2S PRODUCTION (H_2S): Hydrogen sulfide is produced by bacteria capable of reducing sulfur-containing compounds, such as peptones and sodium thiosulfate, present in the medium. The hydrogen sulfide reacts with the iron salts, also present in the medium, to form a black precipitate of ferric sulfide usually along the lines of inoculation.

INDOLE FORMATION (IND): The production of indole from the metabolism of tryptophan by the bacterial enzyme tryptophanase is detected by the development of a pink to red color after the addition of Kovacs indole reagent.

ADONITOL (ADON): Bacterial fermentation of adonitol, that results in the formation of acidic end products, is indicated by a change in color of the indicator present in the medium from red (alkaline) to yellow (acidic).

LACTOSE (LAC): Bacterial fermentation of lactose, that results in the formation of acidic end products, is indicated by a change of color of the indicator present in the medium from red (alkaline) to yellow (acidic).

ARABINOSE (ARAB): Bacterial fermentation of arabinose, that results in the formation of acidic end products, is indicated by a change of color of the indicator present in the medium from red (alkaline) to yellow (acidic).

SORBITOL (SORB): Bacterial fermentation of sorbitol, that results in the formation of acidic end products, is indicated by the color change of the indicator in the medium from red (alkaline) to yellow (acidic).

VOGES-PROSKAUER (V-P): Acetylmethylcarbinol (acetoin) is an intermediate in the production of butylene glycol from glucose fermentation. The development of a red color after 20 min after the addition of the Vogel-Proskauer reagent indicates the presence of acetoin.

DULCITOL (DUL): Bacterial fermentation of dulcitol, which results in the formation of acidic end products, is indicated by the color change of the indicator in the medium from green (alkaline) to yellow or pale yellow (acidic).

PHENYLALANINE DEAMINASE (PA): This test detects the formation of pyruvic acid from the deamination of phenylalanine. The pyruvic acid formed reacts with a ferric salt present in the medium to produce a characteristic black to smoky gray color.

UREA (UREA): Urease, an enzyme possessed by various microorganisms, hydrolyzes urea to ammonia causing the color change of an indicator in the medium from yellow (acidic) to red-purple (alkaline).

CITRATE (CIT): This test detects those organisms that are capable of utilizing citrate in the form its sodium salt as the sole source of carbon. Organisms capable of utilizing citrate produce alkaline metabolites which change the color of the indicator from green (acidic) to deep blue (alkaline).

FIGURE 20.2 Biochemical reaction of the compartments of ENTEROTUBE II. Each ENTEROTUBE II provides 15 biochemical reactions in 12 separate compartments. When interpreting reacted ENTEROTUBE II, comparison to an uninoculated control is recommended.

TABLE 20.1
ENTEROTUBE II Color Reactions

Compartment No.	Uninoculated Color	Reacted Color	Reagents
1	red	yellow	glucose (GLU)
	red-yellow wax separation		gas production (GAS)
2	yellow	purple	lysine decarboxylase (LYS)
3	yellow	purple	ornithine decarboxylase (ORN)
4	beige	black	H_2S production precipitate
	beige	red	indole (IND)
5	red	yellow	adonitol (ADON)
6	red	yellow	lactose (LAC)
7	red	yellow	arabinose (ARAB)
8	red	yellow	sorbitol (SORB)
9	beige	red	Voges-Proskauer (VP)
10	green	yellow	dulcitol (DUL)
	green	black to smoky gray	phenylalanine deaminase (PA)
11	yellow	red–purple	urea (UREA)
12	green	deep blue	citrate (CIT)

Note: Each ENTEROTUBE II provides the preceding 15 biochemical reactions in 12 separate compartments.

20.3 BBL OXI/FERM TUBE II

BBL OXI/FERM Tube II is a system similar to ENTEROTUBE II. It is a ready-to-use in vitro diagnostic system for the identification of the oxidative-fermentative Gram-negative bacteria.

The oxidative-fermentative group of Gram-negative rods resembles the Enterobacteriaceae in its colonial and microscopic morphology. However, it differs from them biochemically in that carbohydrates are utilized principally by oxidation and, only occasionally, by fermentation.

The BBL OXI/FERM Tube II System is available from Becton Dickinson Microbiology Systems, Cockeysville, MD 21030.

20.4 API 20E SYSTEM

The API 20E System is a standardized, miniaturized version of conventional procedures for the identification of Enterobacteriaceae and other Gram-negative bacteria. It is a ready-to-use, microtube system designed for the performance of 23 standard biochemical tests from isolated colonies of bacteria on plating medium.

The API 20E System consists of a microtube containing dehydrated substrates. These substrates are reconstituted by adding a bacterial suspension, incubated so the organisms react with the contents of the tubes and read when the various indicator systems are affected by the metabolites or added reagents, generally after 18 to 24 h incubation at 35 to 37°C. See Table 20.2.

The API 20E strips should be stored at 2 to 8°C (refrigerator) in the dark until used. The incubation trays and lids do not require refrigeration. Figure 20.5 shows that the API 20E System Microtube consists of a tube and a cupula section.

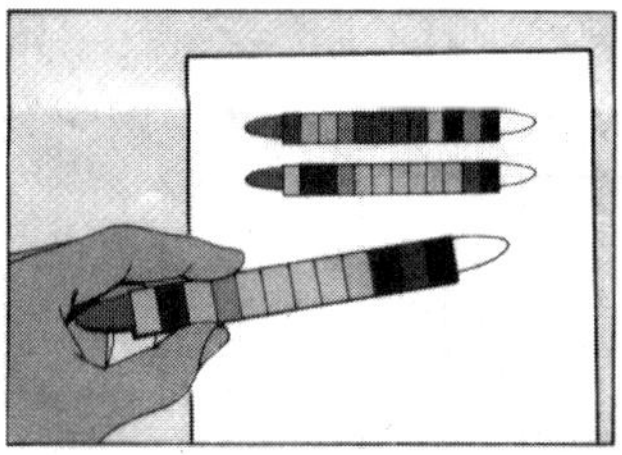

Interpret all media reaction in a sequential fashion by comparing reactions in the ENTEROTUBE II with color reactions.

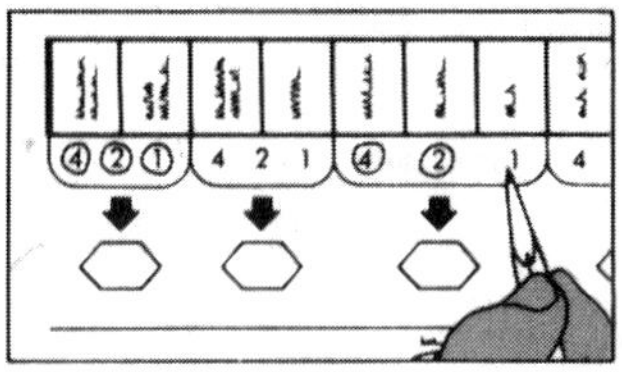

Indicate each positive reaction by circling the number appearing below the appropriate compartment of the ENTEROTUBE II outline on the pad.

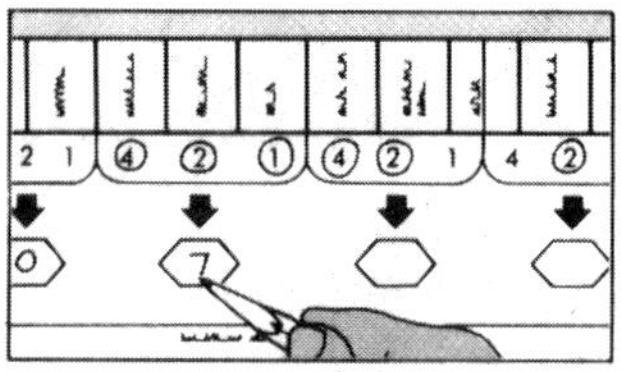

Add circled numbers only within each bracketed section and enter this sum in the space provided below the arrow.

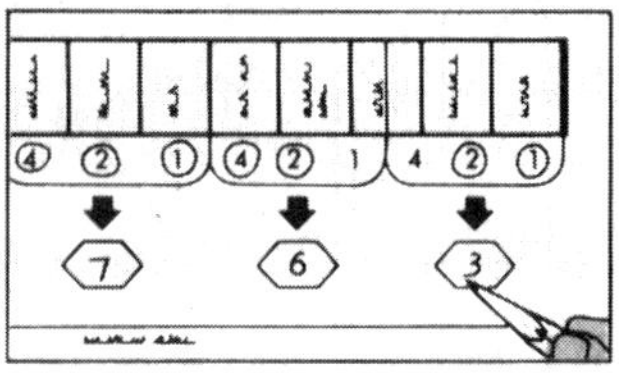

Read the five numbers in these spaces across as a five-digit number (ex. 70763).

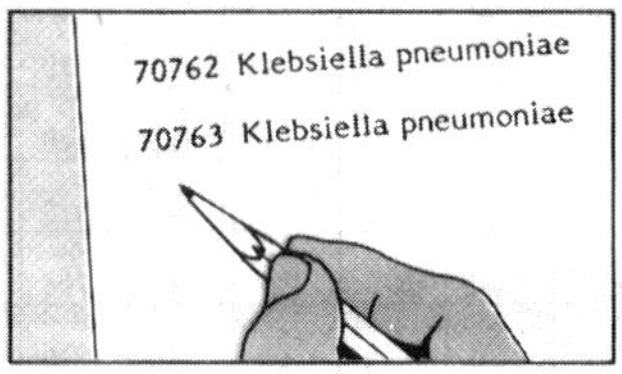

Find this five digit number in the Computer Coding Manual in the column entitled "ID value." This number will identify the genus and/or species of the organism.

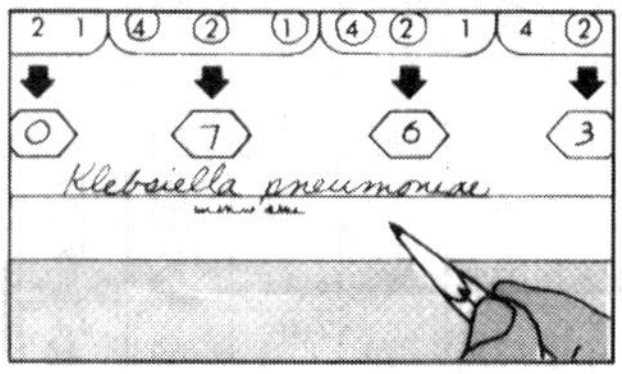

Record culture number or sample number, name of the organism, and the date on lines provided on pad.

(From ENTEROTUBE II manufacturer's Manual, Roche Diagnostic Systems, a Division of Hoffmann-La Roche Inc., Nutley, NJ 07110-1199).

FIGURE 20.3 ENTEROTUBE II pad for interpretation, indication and reporting result.

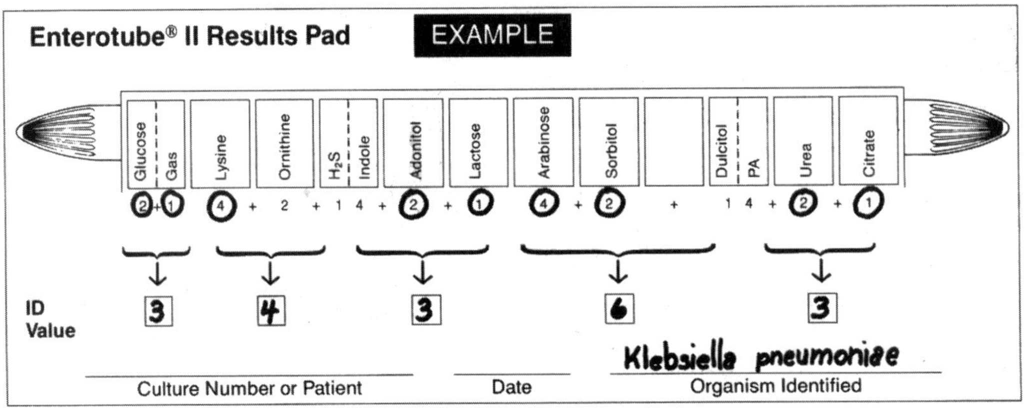

FIGURE 20.4 To read the ENTEROTUBE II System

TABLE 20.2
Summary of Tests Results, API 20E System

Tube	Positive	Negative
ONPG	yellow (any shade)	colorless
ADH	red or orange	yellow
LDC	red or orange	yellow
ODC	red or orange	yellow
CIT	turquoise or dark blue	light green or yellow
H_2S	black precipitate	no black precipitate
URE	red or orange	yellow
TDA	brown red	yellow
IND	red ring	yellow
VP	red	colorless
GEL	diffusion of pigment	no diffusion of pigment
GLU	yellow or gray	blue or blue green
MAN	yellow	blue or blue green
GLU (nitrate)	red	yellow
(Zn red)	yellow	red
MAN	bubbles	no bubbles

Note: ONPG (orthonitrophenyl-galactopyranoside), ADH (arginine dehydrolase), LDC (lysine decarboxylase), ODC (ornithine decarboxylase), CIT (citrate), H_2S (hydrogen sulfide), URE (Urea), TDA [triptophane deaminase (10 percent $FeCl_3$)], IND [indol (Kovacs reagent)], VP [Voges-Proskauer (40 percent KOH, alpha-naphtol)], GEL (gelatin), GLU [glucose (fermentation occurs in the bottom, oxidation primarily in the top)], GLU [nitrate (nitrite reagent) reduction (Zn)], MAN [mannitol (fermentation occurs in the bottom, oxidation primarily in the top)], add H_2O_2 and observe for bubbles.

20.4.1 Operation of API 20E

The following list explains the operation of the API 20E System.

1. Preparation of bacterial suspension: add 5 ml of buffer solution or 0.85 percent saline solution, pH 5.5 to 7.0, to a sterile test tube.
2. Gently touch the center of a well-isolated colony with the tip of a wooden applicator stick. Insert the applicator stick into the tube of buffer or saline solution and, with the tip of the stick at the base of the tube, rotate the stick in a vortex-like action. Recap the tube.
3. Set up an incubation tray and lid. Record the sample identification on the elongated flap of the tray.
4. Dispense 5 ml of tapwater into the incubation tray to provide a humid atmosphere during incubation.
5. Remove the API strips from the sealed envelope and place one strip in each incubation tray. The API 20E contains 20 microtubes each of which consists of a tube and a cupule section as shown in Figure 20.4. The API 20E microtube system is shown in Figure 20.5.
6. Using a sterile *Pasteur* pipet, tilt the API 20E tray and fill the tube section of the microtubes with bacterial suspension. Also, fill the cupule section of the CIT (citrate), VP (Voges-Prokauer), and GEL (gelatin) tubes.
7. After inoculation, completely fill the cupule section of the ADH (arginine dihydrolase), LDC (lysine decarboxylase), ODC (ornithine decarboxylase), and URE (urea) tubes with mineral oil.
8. Place the plastic lid on the tray and incubate the strip for 18 to 24 h at 35 to 37°C. If the strip cannot be read after 24 h, the strips should be removed from the incubator and refrigerated until the reaction can be read.
9. Read and record all reactions and follow the tests for TDA (tryptophane deaminase), VP, and indole reactions.
10. Read the number according to the accompanied interpretation guide.
11. The entire incubation unit must be autoclaved, incinerated, or immersed in a germicide prior to disposal.

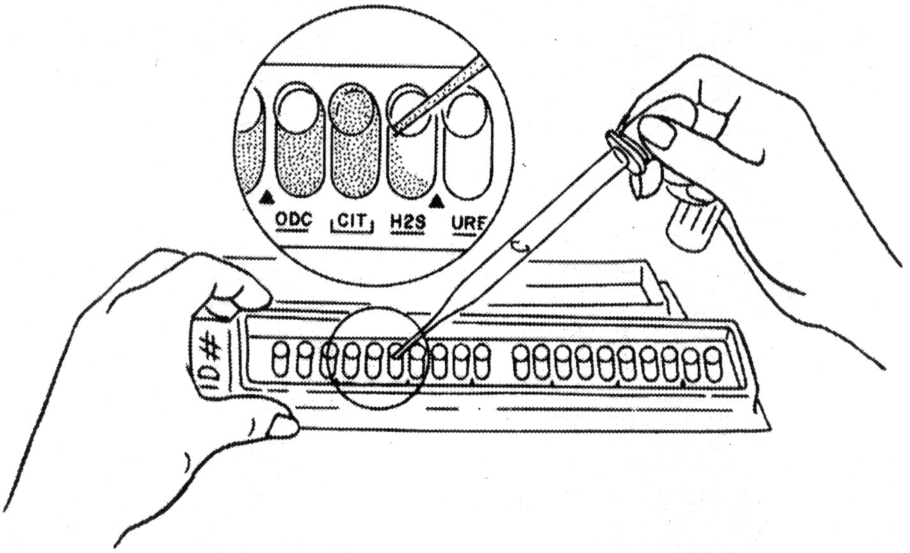

FIGURE 20.5 API 20E System, microtube consists of a tube and a cupule section.

21 Iron and Sulfur Bacteria

21.1 INTRODUCTION TO IRON AND SULFUR BACTERIA

The group of nuisance organisms collectively designated iron and sulfur bacteria is morphologically and physiologically heterogenous, having in common the ability to transform or deposit significant amounts of iron or sulfur, usually in the form of objectionable slime. However, iron and sulfur bacteria are not the sole producers of bacterial slimes and in some cases may be associated with slimes of other bacteria. General characteristics of iron and sulfur bacteria are discussed in Section 7.4.

The organisms in this group may be filamentous or single celled, autotrophic or heterotrophic, aerobic or anaerobic. According to conventional bacterial classification, these organisms are assigned to a variety of orders, families, and genera.

They are studied as iron and sulfur bacteria, because these elements and their transformations may be important in water treatment and distribution systems and may be especially bothersome in waters for industrial use, such as cooling and boiler waters.

Iron bacteria may cause fouling and plugging of wells and distribution systems and sulfate-reducing bacteria may cause rusty water and tuberculation of pipes. These organisms also may cause odor, taste, frothing, color, and increases in turbidity in waters.

The nutrient supply of iron and sulfur bacteria may be wholly or partly inorganic and they may extract it, if attached in a gelatinous substrate, from a low concentration in flowing water. This seems quite important in the case of certain sulfur bacteria utilizing a small amount of H_2S or in the case of organisms such as *Gallionella,* which obtain their energy from the oxidation of ferrous iron. *Thiobacillus ferrooxidans* and *Ferrobacillus ferrooxidans,* that contribute to the problem of acid mine drainage, can be identified by tests for transformation of ferrus to ferric iron or oxidation of reduced sulfur compounds. Temperature, light, pH, and oxygen supply also affect the growth of these organisms.

21.2 IRON BACTERIA

The specific form iron takes in water depends on the amount of oxygen in the water and the pH. In natural groundwater systems where oxygen concentrations are low or absent and the pH is from 6.5 to 7.5, the iron occurs primarily as a dissolved ferrous ion (Fe^{2+}). Ferrous ions are unstable when they come into contact with oxygen. Ferrous ions change to ferric iron (Fe^{3+}) and precipitate as ferric hydroxide $Fe(OH)_3$. Ferric oxides or oxyhydroxides come out of the solution and coat the surrounding surfaces. Iron bacteria widely occur in wells open to the atmosphere when sufficient iron and/or manganese are present. The principal forms of iron bacteria plug wells by enzymatically catalyzing the oxidation of iron, using the energy to promote the growth of treadlike slime, and large amounts of ferric hydroxide in the slime. Precipitation of the iron and rapid growth of the bacteria create a voluminous material that quickly plugs pipes or screen pores of the sediment surrounding the well bore. Sometimes the quick growth of iron bacteria can render a well useless within a few months. The large amount of brown slime thus produced will impart a reddish tinge and an unpleasant odor to drinking water and may render the supply unsuitable for domestic or industrial purposes.

Some bacteria that do not oxidize ferrous iron, nevertheless, may cause ferrous oxide to be dissolved or deposited indirectly. In their growth, they either liberate iron by utilizing organic radicals to which the iron is attached or they alter environmental conditions to permit the solution or deposition of iron. Consequently, less ferric hydroxide may be produced, but taste, odor, and fouling may be engendered.

Gallionella is a common enzymatic iron bacteria composed of twisted bands resembling a ribbon or chain. It can be recognized by the twisted stalks and the bean shaped bacterial cell at the end of the twisted stalks. Figure 21.1 shows the fragment of stalks of *Gallionella ferruginea.* Figure 21.2 gives the pictures of the filamentous iron bacteria *Spherotylus natans* and *Crenothrix polispora,* respectively.

21.2.1 Identification of Iron Bacteria

Identification of iron bacteria usually has been made by microscopic examination of the suspected material. Directly examine bulked activated sludge, masses of microbial growth in lakes, rivers, and streams, and slime growths in cooling tower waters.

Water samples should be filtered, settled, or centrifuged, and the sediment examined directly under the microscope.

1. Place a portion of the sediment on the microscope slide, cover with a cover slip.
2. Examine under a low-power microscope for filaments and iron encrusted filaments.
3. Water pumped from wells may be passed directly through a 0.45 *m*m membrane filter and the filter examined directly under microscope after drying and clearing with immersion oil applied directly to the membrane filter.
4. Phase contrast microscopes are excellent tools for examination of unstained culture material.
5. When a conventional light-microscope is used, the material should be stained with india ink or lactophenol blue.

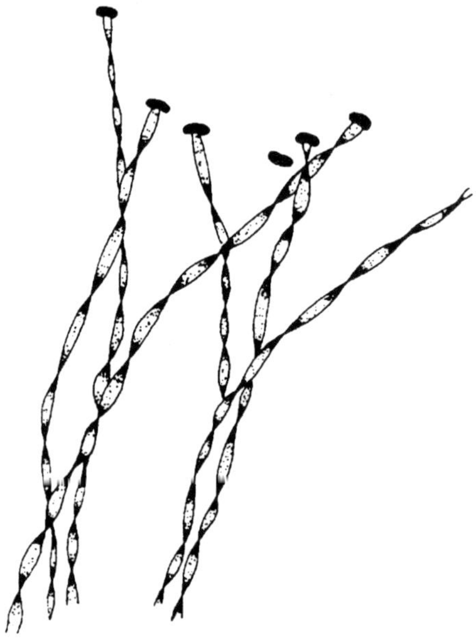

FIGURE 21.1 Iron bacteria, *Gallionella ferruginea,* with twisted bands. A precipitate of inorganic iron on and around the stalks often blurs the outlines. The average cells are 0.4 and 0.6 *m*m in width and 0.7 to 1.1 *m* in length.

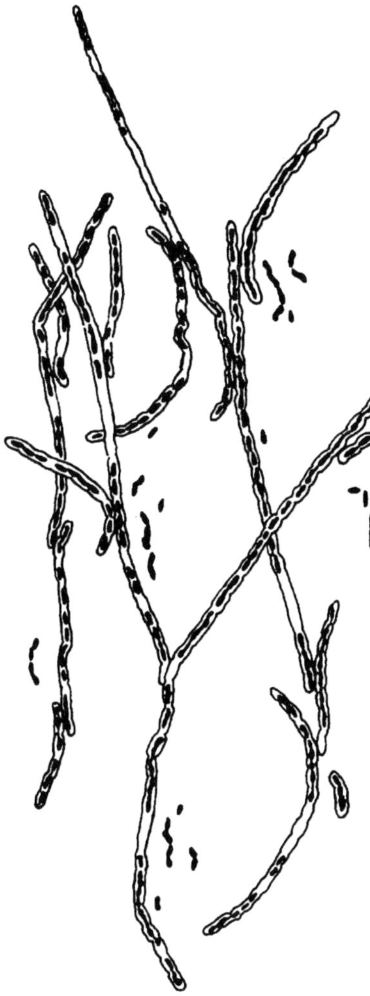

FIGURE 21.2 Filamentous iron bacteria, *Spherotylus natans.* Cells are within the filaments. Some of them free swarmer cells are shown. Filaments also show areas devoid of cells. Individual cells within the sheet may vary in size from 0.6 to 2.4 *mm* in width by 2.0 to 4.0 *mm* in length.

6. To dissolve iron deposits, place several drops of 1 *N* HCl at one edge of cover slip and draw it under the cover slip by applying filter paper to the opposite edge.
7. To verify that the material is iron, add a solution of potassium ferrocyanide to a sample on a slide and cover. Draw 1 *N* HCl under the cover slip. A blue precipitate of Prussian blue will form as iron around the cells and filaments is dissolved.
8. Identify organisms by comparing them with available drawing or photographs of iron bacteria, see Figures 21.1, 21.2, and 21.3.

21.3 SULFUR BACTERIA

21.3.1 Common Forms of Sulfur Bacteria

The bacteria that oxidize or reduce significant amounts of sulfur compounds exhibit a wide diversity of morphological and biochemical characteristics.

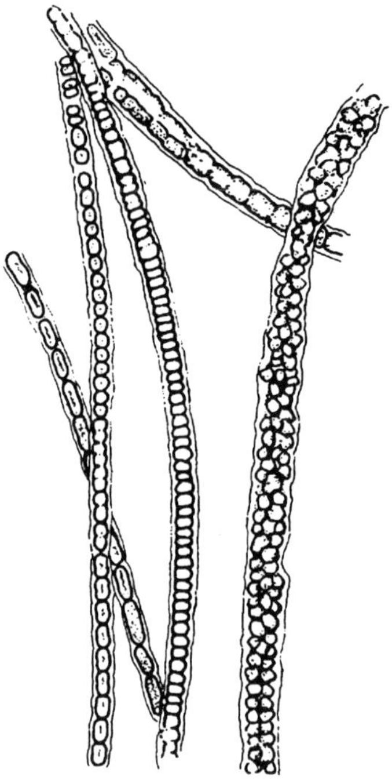

FIGURE 21.3 Filamentous iron bacteria, *Crenothrix polyspora.* Cells within the filaments show variations of size and shape, giving the name polyspora. Young growing colonies are usually not encrusted with iron or manganese. Older colonies often exhibit empty sheets that are heavily encrusted. Cells may vary considerably in size. Rod-shaped cells average 1.2 to 2.0 *mm* in width by 2.4 to 5.6 *mm* in length. Cocoid cells of conidia average 0.6 *mm* in diameter. (Cocoid means spherical or ovoid shaped; conidia means dust; spores freely detach from a chain and float in the air like dust).

21.3.1.1 Sulfate-Reducing Bacteria

Sulfate-reducing bacteria reduce sulfate, SO_4^{2-} to hydrogen sulfide, H_2S. This group consists mainly of single cells and grows anaerobically.

21.3.1.2 Photosynthetic Green and Purple Sulfur Bacteria

Photosynthetic sulfur bacteria grow anaerobically in the light and use H_2S as a source of hydrogen for photosynthesis. The sulfide is oxidized to sulfur or sulfate.

21.3.1.3 Colorless Filamentous Sulfur Bacteria

Members of this group are myxotropic and utilize organic sources of carbon, but may get their energy from the oxidation of reduced sulfur compounds.

21.3.1.4 Aerobic Sulfur Oxidizers

Bacteria in this group are chemoautotrophic. They aerobically oxidize reduced sulfur compounds to obtain energy for their growth.

Sulfur bacteria in water and wastewater are sulfur-reducing bacteria, such as *Desulfovibrio* and the single-celled aerobic sulfur oxidizer, *Thiobacillus.* Sulfur-reducing bacteria contribute greatly to the tuberculation and galvanic corrosion of water mains and to the taste and odor problem of water. *Thiobacillus* genus produce sulfuric acid (H_2SO_4), which destroy concrete sewers and causes corrosion problems of metals.

21.3.2 Identification of Sulfur Bacteria

Identification will mostly be made by microscopic examination. Directly examine samples of slimes, scrapings from surfaces, or sediments.

Any one of three groups of sulfur bacteria may be recognized microscopically: green and purple sulfur bacteria, large, colorless filamentous sulfur bacteria, and large, colorless nonfilamentous sulfur bacteria. Members of the genus *Thiobacillus* cannot be identified by appearance alone.

21.3.2.1 Green and Purple Sulfur Bacteria

Green and purple sulfur bacteria live in waters containing H_2S. Green sulfur bacteria are small, ovoid to rod shaped, nonmotile, less than 1 *m*m with a yellowish green color. Purple sulfur bacteria are large, contain large sulfur globules, and purple pigments and are large, dense, highly colored masses, easily detected by the naked eye.

21.3.2.2 Colorless Filamentous Sulfur Bacteria

Colorless filamentous sulfur bacteria live in waters where both oxygen and H_2S are present. These bacteria are large, yellowish-white in appearance, and contain yellow sulfur globules, see Figure 21.4.

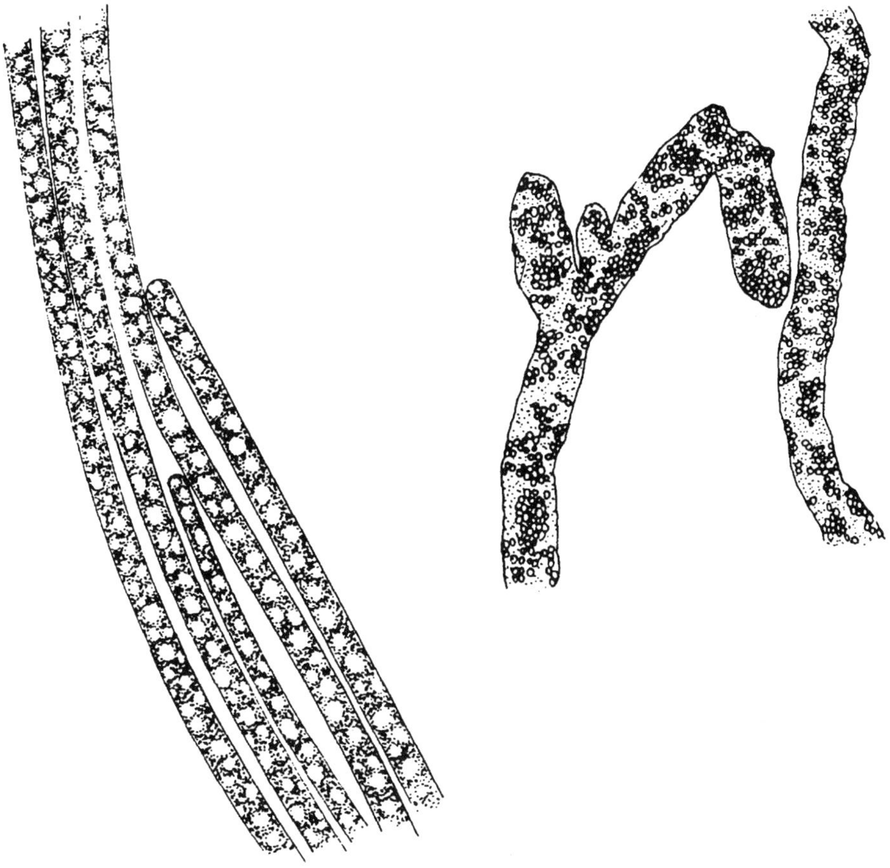

FIGURE 21.4 Filamentous sulfur bacteria, *Thiodendron mucosum* (left) and *Beggiatoa alba* (right). A portion of a colony, with branching mucoid filaments is shown. Individual cells are within the jelly-like material of the filaments. The filaments of *Beggiatoa alba* are composed of a linear series of individual rod-shaped cells that may be visible when not obscured by light reflecting from sulfur granules.

21.3.2.3 Colorless Nonfilamentous Sulfur Bacteria

Colorless nonfilamentous sulfur bacteria live together with algae. These bacteria are very large and contain sulfur globules and calcium deposits. They are ovoid to rod shaped and extremely motile. Colorless nonfilamentous and photosynthetic purple sulfur bacteria are shown in Figure 21.5.

21.3.2.4 Colorless Small Sulfur Bacteria and Sulfate Reducing Bacteria

The small single-celled bacteria, *Thiobacillus* and the sulfate-reducing bacteria, such as *Desulfovibrio,* cannot be identified by direct microscopic examination. They can only be identified physiologically. Thiobacillus types are small, colorless, motile, and rod shaped and are found in an environment containing H_2S. Sulfur globules are absent.

21.4 ENUMERATION, ENRICHMENT, AND ISOLATION OF IRON AND SULFUR BACTERIA

There are no good means of enumerating iron and sulfur bacteria other than the sulfur-reducing bacteria and the thiobacilli. Laboratory cultivation and isolation of pure cultures is difficult and successful isolation is uncertain. This is especially true of attempts to isolate filamentous bacteria from activated sludge or other sources where many different bacterial types are present.

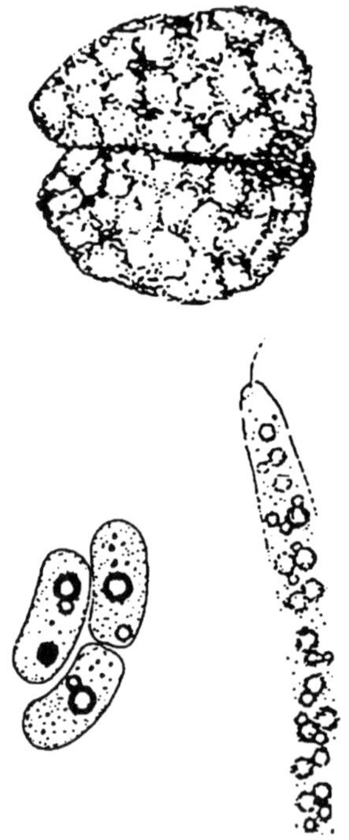

FIGURE 21.5 Nonfilamentous and photosynthetic sulfur bacteria. Dividing cell of colorless, nonfilamentous sulfur bacteria, *Thiovolum majus,* containing sulfur granules is shown (top). They are generally found in nature in a marine littoral zone rich in organic matter and hydrogen sulfide (H_2S). *Chromatum okenii,* photosynthetic purple sulfur bacteria is shown (bottom left). *Thiospirillum jenense,* photosynthetic sulfur bacteria (bottom right) uses hydrogen sulfide (H_2S) and produces granules of sulfur (S). These purple and green sulfur bacteria most frequently occur in waters containing hydrogen sulfide. Green sulfur bacteria are small, ovoid or rod shaped. Purple sulfur bacteria are large, generally stuffed with sulfur globes.

22 Detection of Actinomycetes

22.1 GENERAL INTRODUCTION

22.1.1 ACTYNOMYCETES

Actinomycetes are long, branched filamentous bacteria. The most striking property of Actinomycetes is fungal-type morphology. Although Actinomycetes were looked upon initially as fungi, later research revealed that Actinomycetes were filamentous, branching bacteria. Masses of these filaments are called mycelia, and a singular mass is called mycelium. They may be defined as Gram-positive bacteria that form branching hyphae usually 0.5 to 1 μm diam) that may develop into a mycelium. They are placed in the order of *Actinomycetales.*

Reproduction is by almost total fragmentation of the hyphae or by production of spores in specialized areas of the mycelium.

Most species are chemo-organotrophic, aerobic, mesophilic, and grow optimally at a pH near neutrality. The Actinomycetes are represented most commonly by saprophytic forms that have an extensive impact on the environment by decomposing and transforming a wide variety of complex organic residues. Actinomycetes are widely distributed in nature, in freshwater, seawater, cold- and warm-blooded animals, and compost. The soil, however, is their most important habitat. Viable counts of several millions per/g are common, and over 20 genera have been isolated from soil. Their numbers in waterlogged, anaerobic, and acidic soils are often found to be low (10^2 to 10^3 per g of dry weight soil).

One genus, *Frankia,* causes nitrogen-fixing nodules to form on older tree roots. *Rhizobium* bacteria causes nodules to form on the roots of legumes.

Actinomycetes are responsible for the earthy musty odors that affect the quality and public acceptance of municipal water supplies in many parts of the world. They are among the naturally occurring odors that plant operators find the most difficult to remove by conventional treatment. It was assumed that these odors could be attributed to volatile metabolites formed during normal Actinomycete development. Two such compounds, *geosmin* and *2-methylisoborneol,* have been isolated and identified as the agents responsible for earthy-musty odor problems. Geosmin and 2-methylisoborneol have threshold odor concentrations well below the μg per l concentration. Thus, traces of these products are sufficient to impart a disagreeable odor to water, soil, sediment, and give a muddy flavor to the fish.

Actinomycetes also have been recognized as a cause of the disruptions of wastewater treatment. Their massive growth rate is capable of producing thick foam in the activated sludge process.

21.1.2 STREPTOMYCES

The best known genus of Actinomycetes is Streptomyces. Streptomyces are one of the bacteria that are most commonly isolated from soil and are also considered to be significant in water supply problems. These organisms are strict aerobes. They often produce extracellular enzymes that enable them to utilize proteins, polysaccharides such as starch and cellulose, and many organic materials found in the soil. The most important function of various species of streptomyces is the production of most of our commercial antibiotics.

Aminoglucosides are a group of antibiotics in which amino sugars are linked by glycosidic bonds, hence the name. Probably the best known aminoglucoside is Streptomycin. Streptomycin was discovered in 1946.

Streptomycine was the first antibiotic effective in the treatment of tuberculosis and Gram-negative bacteria. Streptomycin is still used as an alternative drug in the treatment of tuberculosis, but rapid development of resistance and the appearance of serious toxic effects have diminished its usefulness.

Tetracyclines are a group of closely related, broad spectrum antibiotics, produced by streptomyces, that inhibit protein synthesis. Three of the more commonly encountered tetracyclines are *oxytetracyclines (Terramycin)*, *chlortetracycline (Aureomycin)* and *tetracycline* itself.

Chloramphenicol is a broad spectrum bacteriostatic antibiotic. Because of its relatively simple structure, it is cheaper for the pharmaceutical industry to synthesize it chemically rather than to isolate it from streptomycin.

22.2 DETERMINATION OF ACTINOMYCETES DENSITY

22.2.1 Plating Method

In determining the density of Actinomycetes, a plating method using a double-layer agar technique has been adapted. Because only the thin top layer of the medium is inoculated with the sample, surface colonies predominate and the identification and the counting of colonies is facilitated. Use of selected antifungal and antibacterial antibiotics to obtain cleaner plates are recommended.

No microbiological technique is as widely accepted in enumerating soil organisms as the plate count using a variety of media.

22.2.2 Preparation and Dilution of Samples

Prepare and dilute water samples as directed in the sample preparation for the heterotrophic plate count methods discussed in Chapter 16. Suitable dilutions are:

Raw water samples	1 : 1,000 dilution
Treated waters	examine directly
Soil samples	1 : 1,000 to 1 : 10,000 dilution

22.2.3 Preparation of Soil Samples

Soil sample preparation consists of the determination of moisture in the soil sample, followed by the preparation of the soil suspension, followed by the appropriate dilutions, as discussed in Section 16.4.

1. Mix soil sample thoroughly and weigh out a 10 to 20 g sample and dry at 105 to 110°C to constant weight and calculate the moisture content of the soil sample in percents.

$$\text{percent moisture} = \frac{\text{weight of water (weight loss), g} \times 100}{\text{weight of original sample, g}} \tag{22.1}$$

2. For each sample, make a dilution blank containing 95 ml of dilution water and 15 to 20 mm glass beads. Also make 7 dilution blanks with 90 ml of dilution water.
3. Cap the bottles and autoclave at 121°C for 20 min. Dilution blanks should be cooled to at least 42°C before using.
4. Transfer 10 g of the moist soil sample to the bottle containing the 95 ml of dilution water and the glass beads. Cap the bottle in a horizontal position on a mechanical shaker, and shake for 10 min. Alternatively shake by hand, moving the bottle in a large arc at least 200×.

5. After removing the sample from the shaker, and just before using, shake the bottle vigorously. Immediately thereafter, transfer 10 ml of the soil suspension, taken from the center of the suspension, to a fresh 90 ml blank. This establishes the 10^{-2} (1 : 100) dilution.
6. Cap and vigorously shake this bottle, and remove 10 ml of the suspension as previously described. This is the 10^{-3} (1 : 1000) dilution. Continue the sequence until a dilution of 10^{-7} (1 : 10,000,000) is reached. On the basis of previous experience with the samples, it may be necessary to continue diluting until the 10^{-8} and 10^{-9} dilutions are reached.

22.2.4 Medium

Starch-casein agar contains the following:

Soluble starch	10.0 g
Casein	0.3 g
Potassium nitrate, KNO_3	2.0 g
Sodium chloride, NaCl	2.0 g
Dipotassium hydrogen phosphate, K_2HPO_4	2.0 g
Magnesium sulfate, hydrate, $MgSO_4 \cdot 7\ H_2O$	0.05 g
Calcium carbonate, $CaCO_3$	0.02 g
Ferrous sulfate, hydrate, $FeSO_4 \cdot 7\ H_2O$	0.01 g
Agar	15.0 g
Reagent-grade water	1 l

No pH adjustment is required. The medium is used to prepare double-layer plates. Store medium for bottom layer in bulk or in tubes in 15 ml amounts. Store medium for surface layer in tubes in 17.0 ml amounts.

22.2.5 Procedure

The following procedure should be followed.

1. Prepare three plates for each dilution to be examined.
2. Aseptically, transfer 15 ml of the sterile *starch-casein agar* into each petri dish and let the agar solidify to form the bottom layer.
3. To each test tube containing the 17 ml medium (liquified at 45 to 48°C), add 2 ml of the appropriate diluted sample and 1 ml of the antifungal antibiotic, *cycloheximide* (Actidione, Upjohn and Co., Kalamazoo, MI, or the equivalent). The antibiotic is prepared in reagent grade water, 1 mg per 1 ml and sterilized by autoclaving for 15 min at 121°C.
4. Pipet 5 ml of inoculated agar over the hardened bottom layer with gentle swirling to obtain even distribution of the surface layer.
5. After plates are solidified, invert, and incubate at 28°C until no new colonies appear. This usually requires 6 to 7 days.

22.2.6 Counting

The following procedure is used in counting.

1. Acceptable plates are suitable for counting 30 to 300 colonies.
2. Identify Actinomycetes by gross colony appearance.
3. If necessary, verify by microscopic examination at a magnification of 50 to 100×. Actinomycetes colonies, because of filamentous growth, typically have a fuzzy colonial

border. Cycloheximide usually suppresses fungal growth, however, fungal colonies, if present, can be recognized by a wooly appearance.

22.2.7 Calculation and Reporting

The following procedure should be used in counting and reporting.

1. Report Actinomycetes per ml of water or per g of soil.
2. If three plates are used per sample, give the average colony count divided by 2 (because a 2 ml sample was used) and multiply by the dilution factor to obtain the actual colony count per ml of the sample.
3. For solid or semisolid samples determine the moisture content of the sample (see Section 22.2.3) and calculate the colony numbers per g of the sample on the dry base.
4. If the sample size on pour plates was 1 ml, the colony count will not be divided by 2. If the sample size on spread plates was 0.1 ml, the calculations have to adjusted according to:

$$\frac{\text{colony count}}{\text{ml}} = \frac{(\text{av colony count}/2) \times \text{dil factor}}{\text{ml sample}} \tag{22.2}$$

$$\frac{\text{colony count}}{\text{g}} = \frac{(\text{av colony count}/2) \times \text{dil factor}}{\text{g soil sample in suspension}} \tag{22.3}$$

$$\frac{\text{colony count}}{\text{g dry weight soil}} = \frac{\text{colony count/g soil}}{\text{decimal fraction of dry weight of soil}} \tag{22.4}$$

For example,

1. The colony counts of a water sample are calculated according to Equation (22.2). If the reading was made on the 10,000 dilution plates:

Colony counts on plate 1	64
Colony counts on plate 2	60
Colony counts on plate 3	56

$$\text{average counts} = 180 / 3 = 60$$

$$\frac{\text{colony count}}{\text{ml}} = (60 \times 10{,}000)/2 = 300{,}000$$

The water sample has 300,000 Actinomycetes colonies in a 1 ml sample.

2. The colony counts per g of soil sample are calculated according to Equation (22.3). If the average counts on 100,000 dilution plates are 88 colonies and the original sample size is 10 g soil,

$$\frac{\text{colony count}}{\text{g on wet soil}} = \frac{(88/2) \times 100{,}000}{10} = 440{,}000$$

If the measured moisture content of the sample was found to be 18 percent, according to the Equation (22.5), the dry weight percent is:

$$\begin{aligned}\text{dry weight percent} &= 100 - \text{moisture percent} \\ &= 100 - 18 = 82 \text{ percent}\end{aligned} \tag{22.5}$$

Calculate the colony count per g dry soil sample using Equation (22.6):

$$\frac{\text{count}}{\text{g of the dry weight}} = \frac{\text{count / g wet soil}}{\text{decimal fraction of dry weight percent}}.$$

$$\frac{\text{colony count}}{\text{soil on dry weight}} = \frac{440{,}000}{0.82} = 536{,}585. \quad (22.6)$$

22.2.7.1 Alternative Calculations

The reported value may calculated according to Equations (22.7) and (22.8), see Section 16.4.3, Equations (16.1) and (16.2):

$$\text{dry weight soil} = (\text{weight of moist soil, initial dil}) \times (1 - \text{percent moisture}/100) \quad (22.7)$$

$$\frac{\text{colonies}}{\text{soil}} = \frac{(\text{mean plate count})(\text{dilution factor})}{\text{dry weight soil, initial dilution}}. \quad (22.8)$$

$$\text{dry weight soil} = 10 \times (1 - 18/100) = 8.2$$

$$(44 \times 100{,}000)/8.2 = 536{,}585.$$

The reported value is 536,585 Actinomycetes colonies per 1 g of dry soil sample, the same as previously calculated.

A Exponential Notation

A.1 GENERAL DISCUSSION

When numbers are very large or very small, it is often more convenient to express them in exponential notation or scientific notation.

The Exponential Notation System Is Based on the Power of 10.

For example, if we multiply $10 \times 10 \times 10 = 1000$, we express this as 10^3. The 3 in this expression is called the exponent or the power, and it indicates how many times we multiplied 10 by itself and how many 0 follow the 1.

There are also negative powers of 10. For example, 10^{-3} means 1 divided by 10^3:

$$10^{-3} = 1/10^3 = 1/1000 = 0.001$$

Numbers are frequently expressed like this: 6.4×10^3. In a number of this type, 6.4 is the coefficient and 3 is the exponent or power of 10. This number means exactly what it says:

$$6.4 \times 10^3 = 6.4 \times 1000 = 6400.$$

Similarly, we can have coefficients with negative exponents:

$$27 \times 10^{-5} = 2.7 \times 1/10^5 = 2.7 \times 0.00001 = 0.000027.$$

A.1.1 NUMBERS GREATER THAN 10 IN EXPONENTIAL NOTATION

Move the decimal point to the left, to just after the first digit. The positive exponent is equal to the number of places we moved the decimal point. For example,

$$37{,}500 = 3.75 \times 10^4$$

For the number 37,500, move the decimal point 4 places to the left to equal 3.75 .In the 10^4, the 4 indicates the 4 places moved to the left. The coefficient equals 3.75 and 4 equals the exponent or the power of 10.

$$628 = 6.28 \times 10^2$$

For the number 628, move the decimal point 2 places to the left to equal 6.28. The coefficient equals 6.28 and 2 equals the exponent or the power of 10.

$$859{,}600{,}000{,}000 = 8.596 \times 10^{11}$$

The decimal point is moved 11 places to the left and equals 8.596×10^{11}

A.1.2 Numbers Less Than 1 in Exponential Notation

For small numbers (less than 1), we move the decimal point to the right, to just after the first nonzero digit and use a negative exponent. For example,

$$0.00446 = 4.46 \times 10^{-3}$$

For the number 0.00446, move the decimal point 3 places to the right to equal 4.46. In the 10^{-3}, the -3 indicates that the decimal moved three places to the right. The coefficient equals 4.46 and -3 equals the exponent or the power of 10.

$$0.000004213 = 4.213 \times 10^{-6}$$

The decimal point is moved 6 places to the right and equals 4.213×10^{-6}.

A.1.3 Adding and Subtracting Numbers in Exponential Notation

We are allowed to add or subtract numbers expressed in exponential notation only if they have the same exponent. All we do is add or subtract the coefficients and leave the exponents as they are. For example,

$$3.6 \times 10^{-3} + 9.1 \times 10^{-3} = 12.7 \times 10^{-3}$$
$$1.95 \times 10^{-2} - 2.8 \times 10^{-2} = 2.23 \times 10^{-2}$$

A.1.4 Multiplying and Dividing Numbers in Exponential Notation

To multiply numbers in exponential notation, we first multiply the coefficients in the usual way and then algebraically add the exponents. For example,

$$[7.40 \times 10^{5}] \times [3.12 \times 10^{9}] = 23.1 \times 10^{14} = 2.13 \times 10^{15}$$
$$[4.6 \times 10^{-7}] \times [9.2 \times 10^{4}] = 42 \times 10^{-3} = 4.2 \times 10^{-2}$$

To divide numbers expressed in exponential notation, the process is reversed. We first divide the coefficients and then algebraically subtract the exponents. For example,

$$[6.4 \times 10^{8}] \div [2.57 \times 10^{10}] = 2.5 \times 10^{-2}$$
$$[1.62 \times 10^{-4}] \div [7.94 \times 10^{7}] = 0.204 \times 10^{-11} = 2.04 \times 10^{-12}$$

B International System of Units (Metric System)

The metric system is the measurement system utilized by scientists and virtually all countries in the world. It was first developed by the French Academy of Sciences in 1790 in response to a request from the French National Assembly for a simple and organized system of weights and measures.

The metric system has evolved since 1790, but its basic structure has remained the same. The metric system is a decimal system; one that requires only the movement of the decimal point to change larger to smaller units or vice versa.

A conference held in 1960 made significant modifications in the metric system; the changes were significant enough that the name metric system was dropped. The revised system is called Le System International d'Unites or International System of Units (SI). Even though we commonly speak of the metric system; in actuality, we are usually refering to the International System.

The seven base units of the International System are:

1. Meter (m)—unit of length.
2. Kilogram (kg)—unit of mass.
3. Second (s)—unit of time.
4. Kelvin (K)—unit of temperature.
5. Mole (mol)—unit of amount of substance.
6. Ampere (A)—unit of electric current.
7. Candela (cd)—unit of luminous intensity.

B.1 DERIVED SI UNITS

The other SI units are created from a combination of these basic units.

Square meter, m^2 (area).
Cubic meter, m^3 or cc (volume).
Kilograms per cubic meter, kg per m^3 (density).
Meters per second, m per s (velocity).

B.2 DERIVED SI UNITS WITH SPECIAL NAMES

Physical Quantity	Name of SI Unit	Symbol of SI Unit	Physical Quantity	Name of SI Unit	Symbol of SI Unit
Frequency	hertz	Hz	Electric capacitance	Farad	F
Energy	Joule	J	Magnetic flux	Weber	Wb
Force	Newton	N	Inductance	Henry	H
Power	watt	W	Magnetic flux density (magnetic induction)	tesla	T
Pressure	Pascal	Pa	Luminous flux	lumen	lm
Electric charge	coulomb	C	Illuminance	lux	lx
Electric potential difference	volt	V	Absorbed dose	gray	Gy
Electric resistance	ohm	Ω	Activity	Becquerel	Bq
Electric conductance	Siemens	S	Dose equivalent	sievert	Sv

B.3 PREFIXES

Prefixes are placed in front of the SI unit to change the size of the unit.

To magnify the meter 1000 times, it is only necessary to add the prefix, kilo meaning 1000 times, in front of meter to produce the unit kilometer (1 km equals 1000 m).

For measuring small distances, the meter is an awkward unit. The two most commonly used prefixes are centi, meaning one hundred, and milli, meaning one thousandth (1 m equals 100 cm, 1 m = 1000 mm).

Prefix	Symbol	Meaning	Prefix	Symbol	Meaning
exa	E	10^{18}	deci	d	10^{-1}
peta	P	10^{15}	centi	c	10^{-2}
tera	T	10^{12}	milli	m	10^{-3}
giga	G	10^{9}	micro	μ	10^{-6}
mega	M	10^{6}	nano	n	10^{-9}
kilo	k	10^{3}	pico	p	10^{-12}
hecto	h	10^{2}	femto	f	10^{-15}
deca	da	10^{1}	atto	a	10^{-18}

B.4 USEFUL CONVERSION FACTORS

Mass	Length	Time	Volume
1 kg = 1000 g	1 km = 1000 m	1 day = 8.64×10^{4} s	1 l = 1000 ml
1 g = 1000 mg	1 m = 100 cm	1 hr = 3.60×10^{3} s	1 ml = 1000 μl
1 mg = 1000 μg	1 mg = 1000 mm	1 hr = 60 min	1 μl = 1000 nl
1 μg = 1000 ng	1 mm = 1000 μm	1 min = 60 s	1 nl = 1000 pl
1 ng = 1000 pg	1 μm = 1000 nm		1 l = 10^{-3} cm^3
1 g = 10^{-3} kg	1 nm = 1000 pm		1 ml = 1 cm^3
1 mg = 10^{-6} kg	1 cm = 10^{-2} m		1 cm^3 = 10^{-6} m^3
1 μg = 10^{-6} g	1 mm = 10^{-3} m		
1 mg = 10^{-3} g	1 nm = 10^{-9} m		
	1 Å = 10^{-10} m		
	= 0.10 nm		

Temperature	Energy	Pressure	
0°C = 273.15 K	1 cal = 4.185 J (Joule)	1 atm = 101,325 kPa	atm = atmosphere
0°K = 255.37 K	1 J = 0.23901 cal	1 atm = 760 mmHg or 760 torr	Pa = Pascal
		1 torr = 760 mmHg or 1 atm	kPa = kiloPascal
		1 torr = 133.322 Pa	
		1 bar = 1×10^{5} Pa	

C Units and Conversion Factors

C.1 CONVERSION TO METRIC MEASURES

	Symbol	When you know	Multiply by	To find	Symbol
Length	in.	inches	2.54	centimeters	cm
	ft	feet	30.48	centimeters	cm
	yd	yards	0.9	meters	m
	mi	miles	1.6	kilometers	km
Area	$in.^2$	square inches	6.5	square centimeters	cm^2
	ft^2	square feet	0.09	square meters	m^2
	yd^2	square yards	0.8	square meters	m^2
	mi^2	square miles	2.6	square kilometers	km^2
		acres	0.4	hectares	ha
Mass	oz	ounces	28	grams	g
	lb	pounds	0.45	kilograms	kg
Volume	tsp	teaspoon	5	milliliters	ml
	Tbsp	tablespoon	15	milliliters	ml
	fl oz	fluid ounces	30	milliliters	ml
	c	cups	0.24	liters	l
	pt	pints	0.47	liters	l
	qt	quarts	0.95	liters	l
	gal	gallons	3.8	liters	l
	ft^3	cubic feet	0.03	cubic meters	m^3
	yd^3	cubic yards	0.76	cubic meters	m^3
Temperature	°F	Fahrenheit		Celsius	°C
			°C = 5/9°(F − 32)		

C.2 CONVERSION FROM METRIC MEASURES

	Symbol	Known	Multiply	To find	Symbol
Length	mm	millimeters	0.04	inches	in.
	cm	centimeters	0.4	inches	in.
	m	meters	3.3	feet	ft
	m	meters	1.1	yards	yd
	km	kilometers	0.6	miles	mi
Area	cm^2	square centimeters	0.16	square inches	$in.^2$
	m^2	square meters	1.2	square yards	yd^2
	km^2	square kilometers	0.4	square miles	mi^2
	ha	hectares	2.5	acres	
Mass	g	grams	0.035	ounces	oz
	kg	kilograms	2.2	pounds	lb
	t	tonnes	1.1	short tonnes	
Volume	ml	milliliters	0.03	fluid ounces	fl oz
	l	liters	2.1	pints	pt
	l	liters	1.06	quarts	qt
	l	liters	0.26	gallons	gal
	m^3	cubic meters	35	cubic feet	ft^3
Temperature	°C	Celsius		Fahrenheit	°F
			°F = 9/5°C + 32		

D Biochemical Oxygen Demand (BOD)

D.1 SCOPE AND APPLICATION

Biochemical oxygen demand (BOD) is an empirical test in which standardized laboratory procedures are used to determine the relative oxygen requirements of wastewaters, effluents, and polluted waters. The test has its widest application in measuring waste loadings to treatment plants and in evaluating the BOD-removal efficiency of such treatment systems.

The test measures the oxygen utilized during a specified incubation period for the biochemical degradation of organic material (carbonaceous demand) and the oxygen used to oxidize inorganic material such as sulfides and ferrous iron. It also measures the oxygen used to oxidize reduced forms of nitrogen (nitrogenous demand) unless the oxidation is prevented by an inhibitor. The seeding and dilution procedures provide an estimate of the BOD at pH of 6.5 to 7.5.

Although the 5-day BOD (BOD_5) is described here, many variations of BOD measurements exist and use shorter or longer incubation periods.

Standard Method 5210 (AWWA, 18th ed.) is the definitive test method for BOD. EPA Method 405.1 (Methods for chemical analysis of water and wastes, EPA 600/4-79-020, 1983) refers to Method 5210. The Corps of Engineers Method 3-373 (CE-81-1, 1981) is used for soil or sludge samples because it is a modification of the basic method written for these matrices.

D.1.1 Carbonaceous vs. Nitrogenous BOD

Oxidation of the reduced form of nitrogen, mediated by microorganisms, exerts nitrogenous demand. The interference from nitrogenous demand can be prevented by an inhibitory chemical.

Chemical inhibition of nitrogenous demand provides a more direct and more reliable measure of carbonaceous demand.

D.1.2 Dilution Requirements

The BOD concentration in most wastewaters exceeds the concentration of dissolved oxygen (DO) available in an air-saturated sample. Therefore, it is necessary to dilute the sample before incubation. Because bacterial growth requires nutrients such as nitrogen, phosphorus, and trace metals, these nutrients are added to the dilution water. The dilution water is buffered to ensure that the pH of the incubated sample remains in the range suitable for bacterial growth. Complete stabilization of a sample may require a period of incubation too long for practical purposes. Therefore, five days has been accepted as the standard incubation period.

The source of dilution water is not restricted and may be distilled, tap, or receiving-stream water free of biodegradable organics and bioinhibitory substances such as chlorine or heavy metals. Distilled waters may contain ammonia or volatile organics, deionized waters are often contaminated with soluble organics; leached from the resin bed. Use of copper-lined stills or copper fittings attached to distilled water lines may produce water containing excessive amount of copper.

D.1.3 SUMMARY OF THE METHOD

A sample is poured into incubation bottles (BOD or Winkler bottles) in duplicate, at three predetermined dilutions. The sample is seeded and an initial dissolved oxygen level (DO) is measured with a DO meter. The bottles are then sealed and stored in the dark for five days at a controlled temperature. After incubation, the dissolved oxygen level (DO) is measured. The reduction in the amount of DO during incubation gives a measure of the biochemical oxygen demand, BOD.

Because the initial DO is determined immediately after the dilution is made, all oxygen uptake, including that occurring during the first 15 min, is included in the BOD measurement.

D.1.4 SAMPLING AND STORAGE

Samples for BOD analysis may degrade significantly during storage (between collection and analysis), resulting in low BOD values. Minimize reduction of BOD by analyzing the sample promptly or by cooling it to near freezing temperature during storage. However, even at a low temperature, keep the holding time to a minimum. Warm chilled samples to room temperature before analysis. State length and temperature for storage together with the reported BOD values.

The measurement of DO is possible by membrane electrode method, and the iodometric titration called Winkler method. The membrane electrode method is suited for analysis in the field. However, the Winkler method is not convenient for field work.

Collect samples very carefully. Methods for sampling are highly dependent on the source to be sampled, and to a certain extent, on the method of analysis.

D.1.5 SAMPLE COLLECTION FOR THE WINKLER METHOD

Collect samples in narrow-mouth glass stopper "Winkler bottles" (commonly called BOD bottles) with a 300 ml capacity and with tapered and pointed ground-glass stoppers and flared mouths.

1. Rinse the bottle with the sample.
2. Pour the sample into the bottle and let the sample overflow the top of the bottle (at least 1/3 of the volume of the bottle should be allowed to overflow).
3. The bottle is then stoppered after all the air bubbles have left the bottle.
4. The temperature of the sample should be recorded.
5. Remove the stopper and introduce 2 ml manganese sulfate reagent followed by 2 ml of alkali-iodide azide reagent beneath the surface of the sample.

   Manganese sulfate reagent—480 g manganese sulfate tetrahydrate, $MnSO_4 \cdot 4\,H_2O$, dissolve and dilute to 1 l with DI water.

   Alkali-iodide-azide reagent—add 10 g sodium azide, NaN_3; 480 g sodium hydroxide, NaOH; 750 g sodium iodide, NaI; dissolve and dilute to 1 l).
6. Replace the stopper, being careful not to trap the air inside.
7. Mix by inverting the bottle at least 15×. Allow the floc to settle and invert again.
8. Allow the floc to settle again, remove the stopper. Immediately add 2 ml of concentrated sulfuric acid, H_2SO_4, allow the acid to run down the neck of the bottle, restop, and mix until the precipitate dissolves, leaving a clear yellow-orange colored solution. Dissolution should be complete.
9. Transfer samples to the laboratory to complete the test. A sample stored at this point should be protected from strong sunlight and analyzed as soon as possible.

D.1.6 Interferences

The following may cause interference problems.

1. Extremes in pH can have an effect on the growth of the microbes and so can the presence of residual chlorine. These are checked before the test begins and adjusted, if necessary, with dilute acid or base.
2. Samples that contain toxic metals, such as waste from plating processes, cannot be analyzed within the confines of this method.
3. The sample must not be supersaturated with dissolved oxygen. The level of DO can be a function of sample temperature. Therefore, all samples must be at ambient temperature before beginning the analysis.
4. If the initial reading is above 9.17 mg per l, the method directs you to aerate or shake the sample vigorously until the reading decreases. This procedure is not practical when the sample is already in the BOD bottle (with the seed added). Instead, the sample must be poured out to allow some headspace in the collection bottle and placed in a mechanical shaker for about 20 to 30 min. Allow the sample to rest for 15 min before pouring. DO readings usually go down to the 7 to 9 mg per l range.

D.2 APPARATUS AND MATERIAL

The following apparatus and materials are needed.

1. DO meter with probe.
2. BOD bottles, 300 ml, borosilicate glass with airtight ground-glass stoppers and plastic caps.
3. Glassware:
   Erlenmeyer flasks, 250 or 500 ml.
   Graduated cylinders, 100, 50 and 10 ml.
   Volumetric pipets, 3, 5, and 20 ml.
   Pipet, calibrated to 2 ml.
   Beaker, 1 l.
4. Magnetic stirrer and Teflon magnetic stirring bars.
5. Thermometers, in 0.1°C increments.
6. pH paper, full range.
7. BOD incubator, capable of maintaining the temperature at 20 ± 0.5°C and excluding all light.
8. Residual chlorine kit.
9. Analytical balance capable of weighing to the nearest 0.0001 g.
10. Toploader balance, capable of weighing to the nearest 0.01 g.

D.3 REAGENTS

The following reagents may be used.

1. *Magnesium sulfate solution*—add 22.5 g of magnesium sulfate heptahydrate ($MgSO_4 \cdot 7 H_2O$) into a 1 l volumetric flask, dissolve, and dilute to the mark with DI water, and mix well.
2. *Calcium chloride solution*—add 27.5 g calcium chloride ($CaCl_2$) into a 1 l volumetric flask, dissolve, and fill up to the mark with DI water, and mix well.

3. *Ferric chloride solution*—add 0.25 g ferric chloride hexahydrate ($FeCl_3 \cdot 6\ H_2O$) into a 1 l volumetric flask, dissolve, and fill up to the mark with DI water.
4. *Phosphate buffer solution*—add 8.5 g potassium phosphate monobasic (KH_2PO_4), 21.75 g of potassium phosphate dibasic heptahydrate ($K_2HPO_4 \cdot 7\ H_2O$), and 1.7 g of ammonium chloride (NH_4Cl) into a 1 l volumetric flask containing 500 ml DI water, dissolve, and fill up to the mark with DI water. Mix well. The pH should be 7.2 without adjustment. The reagent should be discarded if there is any sign of biological growth.
5. *Sulfuric acid, H_2SO_4 concentrate.*
6. *Sulfuric acid, 1 N H_2SO_4*—in a 1 l volumetric flask, slowly add 28 ml of concentrated H_2SO_4 to 700 ml DI water, while stirring. After the solution cools to room temperature, dilute the volume with DI water, and mix well.
7. *Sulfuric acid, H_2SO_4 2 percent*—in a 100 ml volumetric flask, add approximately 50 ml DI water and add into it slowly to 2 ml concentrated H_2SO_4. Mix well. If the mixture is at room temperature, fill up to 100 ml with DI water, and mix well.
8. *Sodium hydroxide, 1 N NaOH*—using a top-loader balance, measure 40 g of NaOH into a 1 l volumetric flask. Add enough DI water to dissolve. Caution: this reaction is highly exothermic and the solution becomes very hot. Let the solution cool to room temperature, fill up to the mark, and mix well.
9. *Sodium sulfite, Na_2SO_3 solution*—weigh 1.575 g of sodium sulfite Na_2SO_3 into a 1 l volumetric flask and add enough DI water to dissolve. Fill up to the volume with DI water. This solution must be made fresh daily.
10. *Potassium iodide, KI, 10 percent solution*—weigh 10 g of potassium iodide (KI) into a 100 ml volumetric flask. Add DI water and mix until dissolved. Dilute to the mark with DI water and mix well.
11. *Glucose–glutamic acid solution*—dry glucose and glutamic acid at 103 to 105°C for 1 h. Cool and store in a desiccator. Weigh 150 mg of dried glucose and 150 mg of dried glutamic acid and add in a 1 l volumetric flask. Add approximately 500 ml DI water and place on the automatic shaker until the chemical is dissolved. Dilute to volume. Prepare fresh before using. The BOD_5 value for the solution should be 200 ± 37 mg per l.
12. *Polyseed inoculum mixture*—add 500 ml of DI water into a 1 l beaker. Add 1 polyseed capsule to the water. Place on a magnetic stirrer until the capsule is dissolved. The contents of the capsule are made from insoluble bran-type material that does not dissolve.
13. *Dilution water*—to each liter of dilution water add 1 ml of each of the following reagents: magnesium sulfate, calcium chloride, ferric chloride, and phosphate buffer. Aerate the solution until the DO content is 8 to 9 mg per l.
14. *Nitrification inhibitor*—2-chloro-6-(trichloro methyl) pyridine (TCMP).

D.4 PROCEDURE

Analysis must begin within 48 h of sampling, preferably within 24 h.

D.4.1 DO Meter Calibration

DO meter calibration includes:

1. Calibrate the DO meter according to the manufacturer's specifications.
2. Check the DO meter performance by measuring the DO content of the dilution water and also determine the DO content by the iodometric titration method (called the Winkler method, discussed in Section D.1.5). The results of the two methods should be close.

D.4.2 Preanalysis Checking

The following procedures are used for preanalysis checking.

1. Check the pH using the full range pH paper:
   - If the pH of the sample is above 2 or below 6, add 1 *N* NaOH solution, see step 8 in Section D.3, continue mixing until the pH is 7.0.
   - If the pH of the sample is below 12 and above 8, add 1 *N* H_2SO_4, see step 8 in Section D.3, continue mixing until the pH is 7.0.
   - If the pH of the sample is below 2 or above 12, take an aliquot of the sample and adjust the pH using NaOH if the pH is below 2 or H_2SO_4 if the pH is above 12.
2. Check the residual chlorine:
   - If residual chlorine is present, add sodium thiosulfate or sodium sulfite, see step 9 in Section D.3 and wait about 15 min to remove the chlorine.
   - Retest the residual chlorine.
3. Temperature adjustment: bring samples to 20 ± 1°C before analysis.

D.4.3 Analysis

The following procedure is used for analysis.

1. Set up and mark the bottle:
   - Set bottles in rows of three.
   - The first row of three bottles is used for the method blank.
   - The second row of three bottles is used for the seed control.
   - The third row of three bottles is used for QC .
     To meet the QC frequency requirements, a QC row must be repeated after every ten samples.
   - Samples.
2. Fill up the bottles in the first row with dilution water to the middle of the ground glass in the neck. Mark the bottles as **Method Blank.**
3. Add 2 ml seed inoculum mixture into each bottle in the second row and fill these bottles with dilution water to the middle of the ground glass in the neck. Mark these bottles as **Seed Control.**
4. Add 5 ml glucose–glutamic acid standard solution (see step 11 in Section D.3) into each bottle in the third row. Add 2 ml of seed to each bottle, and fill up the bottles with dilution water to the middle of the ground glass in the neck. Mark these bottles as the **QC Reference Standard.**
5. Dilute samples according to the origin of the sample:
   - Dilution for industrial waste samples: 0 to 1 percent
     First bottle measure: 0 ml sample
     Second bottle measure: 1.5 ml sample
     Third bottle measure: 3.0 ml sample
   - Dilution for raw wastewater samples: 1 to 5 percent
     First bottle measure: 3.0 ml sample
     Second bottle measure: 7.5 ml sample
     Third bottle measure: 15 ml sample
   - Dilution for effluent samples: 5 to 25 percent
     First bottle measure: 15 ml sample
     Second bottle measure: 30 ml sample
     Third bottle measure: 75 ml sample

- Dilution for surface waters: 25 to 100 percent
  First bottle measure: 75 ml sample
  Second bottle measure: 150 ml sample
  Third bottle measure: 300 ml sample.

6. Add 2 ml of polyseed inoculum into each bottle.
7. Fill the bottles with dilution water to the middle of the ground glass in the neck.
8. If nitrification ihibition is desired, add 3 mg 2-chloro-6-(trichloromethyl) pyridine (TCMP) to each 300 ml bottle before capping. TCMP may dissolve slowly and can float on top of the sample. Some commercial formulations dissolve more readily but are not 100 percent TCMP. Adjust the dosage accordingly.
9. Initial DO reading:
   - After the DO meter is calibrated, place the probe into the neck of each prepared bottle (as described previously) and measured the DO level. Record the readings in the work paper or laboratory notebook (according to the laboratory policy).
   - Stopper the bottles and check each for air bubbles and a water seal. If air bubbles are present, the sample must be prepared again. If a water seal is not present, add a small amount of DI water to the area around the stopper.
   - Place the plastic cap on each bottle and place the bottles into the incubator at 20°C for five days.
10. Final DO reading:
    - After 5 days, remove the bottles from the incubator.
    - Calibrate the DO meter.
    - Read the DO level in each bottle.
    - Record the reading on the workpaper.
    - Pour the samples out of the bottles into the sink. Turn the water on to dilution and rinse.

D.5 CALCULATION

D.5.1 BOD Concentration

$$\text{mg/l BOD} = [(\text{Initial} - \text{Final}) - CF_{SC}] \times DF, \tag{D.1}$$

Where Initial equals the initial DO of sample, in mg per l, Final equals the final DO of sample, in mg per l, CF_{SC} equals the correction factor seed control, SC equals the seed control, and DF equals the dilution factor (Vol_{final}/Vol_{sample}).

D.5.2 Correction Factor for Seed Control

$$CF_{SC} = \frac{(\text{Initial} - \text{Final})(\text{Seed}_{sample})}{\text{Seed}_{SC}}, \tag{D.2}$$

Where CF equals the correction factor, SC equals the seed control, Initial equals the initial DO of seed control, in mg per l, Final equals the final DO of seed control, in mg per l, and Seed equals the amount of seed mixture added, in ml.

D.5.3 Dilution Factor

$$\mathrm{DF} = \frac{\mathrm{Vol}_{\mathrm{final}}}{\mathrm{Vol}_{\mathrm{sample}}}, \tag{D.3}$$

Where DF equals the dilution factor, $\mathrm{Vol}_{\mathrm{final}}$ equals the final volume, and $\mathrm{Vol}_{\mathrm{sample}}$ equals the sample volume.

D.5.4 Alternative Calculation

BOD results may be calculated according to the following formula:

$$\mathrm{BOD}_5,\ \mathrm{mg/l} = \frac{(D_1 - D_2) - (B_1 - B_2)f}{P}, \tag{D.4}$$

Where D_1 equals the initial DO of diluted sample, D_2 equals the final DO of diluted sample, B_1 equals the initial DO of seed control, B_2 equals the final DO of seed control, and f equals the (percent of seed in the diluted sample)/(percent of seed in the seed control).

Report results as CBOD_5 if nitrification is inhibited. Report results as BOD_5 if nitrification is not inhibited.

D.6 QUALITY CONTROL

D.6.1 Method Blank

A method blank (MB) is to be performed once with every sample batch. The MB consists of three bottles of unseeded dilution water that are analyzed as samples. The MB verifies the quality of the dilution water that was used. Ways to improve the quality of the dilution water include storing it in the incubator for at least a week before use and improving the purification of the source water. The cleanliness of the storage bottle is extremely important. Storage bottles are to be thoroughly rinsed three times with DI water, then filled with DI water and placed into the incubator before adding the nutrients, see Section D.3, step 13.

The oxygen depletion of the blank must not be exceed 0.2 mg per l.

D.6.2 Seed Control

The DO uptake of the seeded dilution water should be between 0.6 and 1.0 mg per l.

D.6.3 Reference Standard (Glucose–Glutamic Acid Standard)

A glucose–glutamic acid standard is analyzed once in every ten samples. The BOD value of this standard should be 200 ± 37 mg per l.

The standard recovery must be within ± 20 percent of the true value.

The weekly heterotrophic plate count check is necessary for both the DI water and the dilution water.

In the case of positive results, the system should truly be cleaned up and disinfected.

D.6.4 DO Meter Performance Check

The reading should be checked quarterly against the iodometric titration.

D.6.5 Accuracy

Calculate the recovery on the glucose–glutamic acid standard:

$$\text{percent recovery} = \frac{\text{analyzed BOD value}}{\text{true BOD value}} \qquad \text{(D.5)}$$

Where the analyzed BOD value equals mg per l and the true BOD value equals mg per l.

D.6.6 Precision

Calculate the precision on the results of duplicate samples:

$$\text{RPD} = \frac{(A - B)}{(A + B)/2} \times 100 \qquad \text{(D.6)}$$

and

$$\text{RPD} = \frac{(A - B)}{(A + B)} \times 200, \qquad \text{(D.7)}$$

Where RPD equals the relative percent difference, $A - B$ equals the difference between the duplicate values, and $(A + B)/2$ equals the average of the duplicate values.

D.7 SAFETY

Samples, reagents, and standards processed and/or used by this procedure are to be treated as potential health hazards, and protective measures must be followed. Material Safety Data Sheets (MSDS) for materials used in this test should be available for each person who participates in this laboratory work.

E Determination of Solids

E.1 GENERAL DISCUSSION

Solids refer to matter suspended or dissolved in water and wastewater. Solids may adversely affect water or effluent quality in a number of ways. Water with a high level of dissolved solids is generally not desirable for drinking water purposes and is unsuitable for many industrial applications. Water with a high level of suspended solids is aesthetically unsatisfactory, for example, for bathing. Analyses of solids are important in the control of biological and physical wastewater treatment processes and for assessing compliance with regulatory agency wastewater effluent limitations.

Total solids (total residue) is the residue left in a container after evaporation and drying at 103 to 105°C. Total residue is defined as the sum of the homogenous suspended and dissolved materials in a sample.

Suspended solids (nonfilterable residue) is a portion of the nitrate salts that may be lost.

E.1.1 Sample Collection and Handling

Collect samples in break-resistant glass or plastic bottles, provided that the material in suspension does not adhere to the container walls. Refrigerate samples at 4°C up to the time of analysis to minimize the microbiological decomposition of the solids. Do not hold samples more than 24 h. Nonrepresentative particulates, such as leaves, twigs, insects, and so on must be excluded from the sample.

E.1.2 Sample Pretreatment

Bring samples to room temperature before analysis.

E.2 DETERMINATION OF TOTAL SOLIDS (TS) DRIED AT 103 TO 105°C.

E.2.1 Principle of the Method

A well-mixed aliquot of the sample is quantitatively transferred to a preweighed dish, evaporated to dryness, and dried in an oven at 103 to 105°C to constant weight. The increase in weight over that of the empty dish represents the total solids.

E.2.2 Interferences

Highly mineralized water with a significant concentration of calcium, magnesium, chloride, and sulfate may be hygroscopic and require prolonged drying, proper desiccation, and rapid weighing.

Results for residues high in oil and grease may be questionable because of the difficulty of drying to constant weight in a reasonable time. Disperse visible floating oil and grease with a blender before withdrawing the sample portion for analysis.

Because excessive residue in the dish may form a water-trapping crust, limit the sample to no more than a 200 mg residue.

E.2.3 Apparatus and Material

Evaporating dishes—porcelain, 90 mm, 100 ml capacity; high silica glass (Vycor, a product of Corning Glassworks, Corning, NY) or platinum dishes may be substituted and smaller size dishes may be used if required.)
Steam bath.
Drying oven—for operation at 103 to 105°C.
Analytical balance—capable of weighing to 0.0001 mg.
Desiccator—provided with indicating desiccant.
Muffle furnace—for operation at 500 to 550°C.

E.2.4 Procedure

The following procedure should be followed.

1. Place evaporating dish in a muffle furnace at 550°C for 1 h. Cool and store in desiccator until use. Weigh immediately before use.
2. Transfer a measured volume of a well-mixed sample into a preweighed dish.
3. Evaporate to dryness in a steam bath or in a drying oven. When evaporating in a drying oven, lower the temperature to approximately 2°C below the boiling point to prevent splattering.
4. Dry the evaporated sample for at least 1 h at 103 to 105°C.
5. Cool in a desiccator and weigh.

E.2.5 Calculation

$$\text{mg/l TS} = [(A - B) \times 1000]/\text{sample ml} \tag{E.1}$$

Where A equals the weight of the dried residue plus the dish, in mg, B equals the weight of the dish, in mg, and TS equals the total solids.

E.2.6 Quality Control

E.2.6.1 Method Blank

A method blank is analyzed once at the beginning of every analytical batch and once every 20 samples.

E.3 DETERMINATION OF TOTAL DISSOLVED SOLIDS (TDS) DRIED AT 180°C

E.3.1 Principle of the Method

The method defines filterable residue as those solids which are not retained by a glass fiber filter and are dried at a constant weight at 180°C. A well-mixed sample is filtered through a prepared glass fiber filter into a clean filter flask. The filtrate is transferred to a weighed evaporating dish and is evaporated on a steam bath to dryness. Then it is moved to a 180°C oven and dried until the weight is constant. The increase in dish weight represents the total dissolved solids.

E.3.2 Interferences

Highly mineralized waters containing a considerable amount of calcium, magnesium, chloride, and sulfate content may be hygroscopic and require prolonged drying, proper desiccation, and rapid weighing.

Samples high in bicarbonate require careful and prolonged drying to ensure complete conversion of bicarbonate to carbonate.

Because excessive residue in the dish may form a water trapping crust, select the sample size so that no more than 200 mg of residue remain after drying.

Filter apparatus, filter material, filter preparation, and drying temperature are specified because variations in these parameters have been shown to affect the results.

E.3.3 Apparatus and Material

Glass fiber filter discs—47 mm, without organic binder, 1 mm retention (Whatman grade 934 AH, Gelman type A/E, Millipore type AP40, or the equivalent).
Filtration apparatus—Filter holder and membrane filter funnel (Gelman CMS Cat. No. 192-146 or equivalent) or Gooch crucible 25 to 40 ml capacity with Gooch crucible adapter.
Filter flask—500 ml capacity and vacuum with rubber vacuum tubing.
Drying oven—capable of reaching 180°C.
Porcelain evaporating dishes 100 ml capacity—CMS Cat.No. 077-818 or equivalent; Vycor, a product of Corning Glassworks, Corning NY, or an equivalent product may be substituted.
Analytical balance—capable of weighing to the nearest 0.0001 g.
Desiccator—with color-coded rechargeable desiccant.
Calibrated thermometers—ambient to 200°C.
Steam bath.

E.3.4 Reagents

E.3.4.1 Reference Standard, 500 mg per l

Dissolve 0.5000 g sodium chloride (NaCl) in DI water in a 1 l volumetric flask and fill to the mark. The true TDS value of this standard is 500 mg per l.

E.3.5 Procedure

E.3.5.1 Preparation of Evaporation Dishes

The following procedure is used in the preparation of evaporation dishes.

1. Rinse a clean evaporating dish with DI water and place it in an oven at 180°C for at least 1 h. If volatile solids are to be measured, ignite clean evaporating dishes at 550°C for 1 h in a muffle furnace.
2. Transfer to desiccator and cool for 90 min.
3. Number the dishes with a "sharpie" pen or other heat resistant marker.
4. Weigh the dish on an analytical balance. Record the weight to four decimal places.
5. Once the dishes have been weighed, they can be placed on a tray. If there is going to be a delay before they are used, they must be covered with a lint-free towel to keep airborne

contaminants from falling on them. Alternatively, dishes may be stored in a desiccator until they are needed. Weigh the dishes immediately before use.

E.3.5.2 Sample Analysis

The following procedures are used in sample analysis.

1. Place the filter disc on the filter holder, with the wrinkled side facing upward.
2. Turn on the vacuum and flush the filter with 3 aliquots of 20 ml DI water. Keep the vacuum on for a few seconds to dry the excess water out of the filter.
3. Remove the filter funnel with the rinsed filter and place it on a 250 ml filter flask.
4. Assemble the filter apparatus.
5. Shake the sample well. Be sure to suspend any sediment or particulate.
6. Pour 100 ml of sample into a graduated cylinder. Pour through the filter.
7. Rinse the cylinder with 3 aliquots of DI water using at least 5 ml in each. Pour the rinses through the filter.
8. With a wash bottle, rinse down the sides of the filter holder with DI water.
9. Carefully rinse the filter and any visible residue with water from the wash bottle. Let the suction continue for a minute to dry the excess liquid.
10. Remove the filter with forceps and return it to its marked weighing pan.
11. Transfer the filter from the filter flask to the tared evaporating dish. Carefully rinse the filter flask with three aliquots of DI water and combine the rinses with the filtrate in the dish.
12. If the sample is high in solids, the filter can clog and take a longer time to process the sample. Therefore, if the sample takes longer than 10 min to filter, discard the filter and sample. Start over with a fresh filter and a clean flask. Shake the sample again and pour out a smaller aliquot. The size of the filtered sample should be based on how quickly the filter becomes clogged. If clogging occurs almost immediately, process only 10 or 20 ml. If the sample begins to filter at a normal speed and then, subsequently, begins to slow down, 50 ml would be an appropriate sample size. Any sample that exceeds 2000 mg per l should be reduced to a lower sample volume.
13. The dishes are put on the steam bath to evaporate to dryness.
14. When dry, place the samples in the oven at 180°C. Check the temperature of the oven with a calibrated thermometer. The samples must remain in the oven at least for 1 h.
15. Place the dishes in a desiccator and cool for 90 min.
16. Weigh the dishes with residues on the analytical balance. Record the weight.
17. After the dishes have been weighed, return them to the 180°C oven for 1 h.
18. At the end of the drying time, transfer the dishes to the desiccator, cool for 90 min, and weigh. The result for this second weighing must be within ± 5 mg of the first weighing for each sample. If the weights do not agree within this range, return the samples to the oven and continue the drying/weighing cycle until the weight is constant.
19. If the weights meet the criteria, calculate the concentration. If volatile residue analysis is requested, save the dishes and residues for further analysis.

E.3.6 CALCULATION

$$\text{mg/l TDS} = [(A - B) \times 1000]/\text{sample ml used} \tag{E.1}$$

where A = equals the weight of the dried residue plus the dish, in mg, and B equals the weight of dish, in mg.

E.3.7 Quality Control

E.3.7.1 Method Blank

A method blank is analyzed once at the beginning of every analytical batch and once every 20 samples.

E.3.7.2 Reference Standard

The reference standard is analyzed once with each sample batch and once every 20 samples. The acceptable range of the calculated accuracy is in-house generated.

E.3.7.3 Duplicates

Duplicates are analyzed once with each sample batch and once every 20 samples. The acceptable limit for the calculated precision is established by the individual laboratories.

E.4 DETERMINATION OF TOTAL SUSPENDED SOLIDS (TSS) DRIED AT 103 TO 105°C

E.4.1 Principle of the Method

The method defines nonfilterable residue as those solids which are retained by a glass fiber filter and are dried at a constant weight at 103 to 105°C. A well-mixed sample is filtered through a prepared glass fiber filter, and the residue retained on the filter is dried to a constant weight at 103 at 105°C. The filtrate from this test can be used to determine filterable residue (TDS), EPA method 160.3.

E.4.2 Interferences

Interferences are the same as those listed in Sections E.2 and E.3. Samples high in filterable residue are subject to positive interference. Careful washing of the filter and the residue will minimize this problem.

E.4.3 Apparatus and Material

All the apparatus and material listed in Section E.3 are required except for the evaporating dishes, steam bath, and 180°C drying oven. In addition, weighing dishes are required.

E.4.4 Procedure

E.4.4.1 Filter Preparation

1. Place the filter on the filter holder with the wrinkled side up. Turn on the vacuum and flush the filter with 3 aliquots of 20 ml of DI water. Keep the vacuum on for a few seconds to dry the excess water out of the filter. Remove the filter with forceps and place in an aluminum weighing dish. Dry the filter in the oven for at least 1 h at 103 to 105°C. Place the filters in aluminum dishes and number the dishes with a "sharpie" pen or other heat resistant marker.
2. Weigh the aluminum dish and filter paper on the analytical balance. Record the weight.

3. Once the dishes with the filter paper have been weighed, they can be placed on a tray. If there is going to be a delay before they are used, they must be covered with a lint-free towel to keep airborne contaminants from falling on them.

E.4.4.2 Sample Analysis

1. Assemble the filter apparatus.
2. Using forceps, place the filter on the filter holder. The wrinkled side must be up. Attach the top of the magnetic filter funnel.
3. Shake the sample well and be sure to suspend any sediment or particulate.
4. Pour 100 ml of the sample into a graduated cylinder. Pour the sample through the filter.
5. Rinse the cylinder with 3 aliquots of DI water having at least 5 ml in each. Pour the rinses through the filter.
6. With a wash bottle, rinse down the sides of the filter holder with DI water.
7. Carefully rinse the filter and any visible residue with water from the wash bottle. Let the suction continue for a minute to dry the excess liquid.
8. Remove the filter with forceps and return to its marked weighing pan.
9. If the sample is high in solids, the filter can clog and take a longer time to process. Therefore, if the sample takes longer than 10 min to filter, discard the filter and sample. Use a new filter that was previously prepared. Shake the sample again and pour out a smaller aliquot. The size of the sample filtered should be based on how quickly the filter clogs. If clogging occurs almost immediately, process only 10 or 20 ml of the sample. If the sample begins to filter at a normal speed and then, subsequently, begins to slow down, 50 ml would be appropriate sample size.
10. When all samples have been filtered, place in aluminum weighing dishes in an oven at 103 to 105°C for at least 1 h.
11. The dishes must be cooled in a desiccator for 30 min.
12. Weigh the dishes and filters on the analytical balance and record the weights.
13. Then return them to a 103 to 105°C oven and dry for at least 1 h.
14. After the drying time has elapsed, transfer the dishes and filters to the desiccator and cool for 30 min.
15. Weigh the samples. The results of the second weighing must be within ± 5 mg of the first weighing for each sample. If the weights meet the criteria, calculate the concentration. If the weights do not agree within this range, return the samples to the oven and continue the drying/weighing cycle until the weight is constant.

E.4.5 Calculation

$$\text{mg/l TSS} = [(A - B) \times 1000]/\text{sample ml}, \qquad \text{(E.2)}$$

where A equals the weight of the filter plus the residue, in mg, and B equals the weight of the filter, in mg.

E.5 FIXED AND VOLATILE SOLIDS IGNITED AT 500°C

E.5.1 Principle of the Method

The residue, according to EPA methods 160.1,160.2, and 160.3, is ignited to a constant weight at 500 ± 50°C. The remaining solids represent the fixed total, dissolved, and suspended solids and the weight lost on ignition is from the volatile solids.

E.5.2 Interferences

Negative errors may be produced by the loss of volatile matter during drying. Determination of low concentrations of volatile solids in the presence of high fixed solid concentrations may be subject to considerable error.

E.5.3 Apparatus and Material

Apparatus and material are the same as those listed in Sections E.2.3 and E.3.3.

E.5.4 Procedures

1. Have muffle furnace up to a temperature of 500 ± 50°C.
2. Place residues produced by EPA methods 160.1, 160.2, and 160.3 in the furnace and ignite for 15 to 20 min. However, heavier residues may overtax the furnace and necessitate a longer ignition time.
3. Let the dish or filter paper partially air cool until most of the heat has dissipated.
4. Transfer the sample to the desiccator for the final cooling.
5. Weigh the dishes and the filters.
6. Repeat the cycle of igniting, cooling, desiccating, and weighing until the weight change is less than 4 percent or 0.5 mg.

E.5.5 Calculation

$$\text{mg/l volatile solids} = [(A - B) \times 1000]/\text{ml sample} \quad (E.3)$$

$$\text{mg/l fixed solids} = [(B - C) \times 1000]/\text{ml sample} \quad (E.4)$$

where A equals the weight of the residue plus the dish before ignition, in mg, B equals the weight of the residue plus the dish or filter after ignition, in mg, and C equals the weight of the dish or filter, in mg.

E.6 TOTAL, FIXED, AND VOLATILE SOLIDS IN SOLID AND SEMISOLID SAMPLES

E.6.1 Principle of the Method

This method is applicable to the determination of total solids and the fixed and volatile fractions in such solid and semisolid samples coming from river and lake sediments, sludges separated from water and treatment processes, and sludge cakes from vacuum filtration, centrifugation, or other dewatering processes.

E.6.2 Interferences

The determination of both total and volatile solids in these materials is subject to negative error owing to the loss of ammonium carbonate and volatile organic matter during drying. The mass of organic matter recovered from sludge and sediment requires a longer ignition time than that specified for water. Weigh all samples quickly because wet samples tend to lose weight by evaporating. After drying, residues are very hygroscopic and rapidly absorb moisture from the air.

E.6.3 Apparatus and Material

The apparatus and material are the same as that listed in Section E.2.3.

E.6.4 Procedure

E.6.4.1 Preparation of Evaporation Dishes

1. If volatile solids are to be measured, ignite a clean evaporating dish at 500 ± 50°C for 1 h in a muffle furnace.
2. If only total solids are to be measured, heat the dish at 103 to 105°C for 1 h in an oven.
3. Cool in desiccator and weigh as described in the determination of Section E.2.

E.6.4.2 Sample Analysis

The sample analysis is the same as discussed in Sections E.2 and E.5 with the exception of using the mass of the sample instead of the volume. Preferable sample size is 25 to 50 g.

E.6.5 Calculation

$$\text{percent total solids} = [(A - B) \times 100]/(C - B) \tag{E.5}$$

$$\text{percent volatile solids} = [(A - D) \times 100]/(A - B) \tag{E.6}$$

$$\text{percent fixed solids} = [(D - B) \times 100]/(A - B) \tag{E.7}$$

where A equals the weight of the dried residue plus the dish, in mg, B equals the weight of the dish, in mg, C equals the weight of the wet sample plus the dish, in mg, and D equals the weight of the residue plus the dish after ignition, in mg.

E.6.6 Quality Control

Quality control procedures are the same as those discussed in Section E.3.7.

E.7 SETTLEABLE SOLIDS

E.7.1 Principle of the Method

Measure the settleable solid volumetrically in an Imhoff cone.

E.7.2 Apparatus

Use an Imhoff cone.

E.7.3 PROCEDURE

The following procedures should be followed.

1. Fill the Imhoff cone to the 1 l mark with a well-mixed sample.
2. Let settle for 45 min, gently stir the sides of the cone with a glassrod, or by spinning, and settle 15 min longer.
3. Read the volume of settleable solids in the cone as ml per l.
4. When a separation of settleable and floating materials occurs, do not estimate the floating material as settleable matter.

F Determination of pH

F.1 GENERAL DISCUSSION

Measurement of pH is one of the most important and frequently used tests in water chemistry. Practically every phase of the water supply and wastewater treatment, for example, acid-base neutralization, coagulation, disinfection, and corrosion control, is pH dependent. At a given temperature, the intensity of the acidic and basic character of the solution is indicated by pH.

pH is the negative logarithm of the hydrogen ion concentration,

$$pH = -\log[H^+] \qquad (20.6)$$

The term concentration of the hydrogen ion is written $[H^+]$. The brackets mean mole concentration and H^+ is a hydrogen ion. The concentration of the H^+ is expressed in mol per l.

Water dissociates by a very slight excess into H^+ and OH^- ions. It has been experimentally determined that in pure water

$$[H^+] = 1 \times 10^{-7}$$

and

$$[OH^-] = 1 \times 10^{-7}$$

Therefore,

$$[H^+] = [OH^-]$$

The negative logarithm of 1×10^{-7} is 7.00. Therefore, the pH of pure water is 7.00. Such a solution is neutral and has no excess of H^+ or OH^-. An excess of $[H^+]$ is indicated by a pH below 7.00 and the solution is said to be acidic. pH values above 7.00 indicate an excess of $[OH^-]$ and the solution is alkaline.

F.2 SAMPLE COLLECTION AND HOLDING TIME

Collect samples into glass or plastic bottles. No preservatives are required. Samples must be analyzed for pH within 24 h of sampling.

F.3 POTENTIOMETRIC DETERMINATION OF PH

EPA method 150.1 and SW 9040 are used to determine the pH of water samples and wastes that are at least 20 percent aqueous. SW 9045 is used to determine the pH of solid samples. Nonaqueous liquids are analyzed by SW 9041.

F.3.1 PRINCIPLE OF THE METHOD

EPA method 150.1, SW 9040, and SW 9045 all use an electrometric method to determine the pH of a sample. Samples for EPA method 150.1 and SW 9040 are measured directly after being brought to ambient temperature. SW 9045 requires that the sample first be mixed with DI water or a calcium chloride solution, depending on whether the soil is considered calcareous or noncalcareous. Samples are considered noncalcareous unless specified otherwise.

A pH meter functions by measuring the electric potential between two electrodes that are immersed into the solution of interest. The basic principle is to determine the activity of the hydrogen ions by potentiometric measurement, using a glass electrode and a reference electrode. The most popular, called combination electrode, incorporates the glass and the reference electrode into a single probe. The glass electrode is sensitive to hydrogen ions, and changes its electrical potential with the change of hydrogen ions. The reference electrode has a constant electric potential, see Section 11.1.4. The difference in potential of these electrodes, measured in millivolts (mV), is a linear function of the pH of the solution. The scale of pH meters is designed so that the voltage can be read directly in terms of pH.

A glass electrode usually consists of a silver and a silver chloride electrode in contact with dilute aqueous HCl, surrounded by a glass bulb that acts as a conducting membrane. The hydrogen ion concentration of the HCl solution inside the electrode is constant. Therefore, the potential of the glass electrode depends on the hydrogen ion concentration outside of glass membrane.

The reference, also called a calomel electrode contains elemental mercury (Hg), calomel (Hg_2Cl_2) paste, and Hg metal. This paste is contacted with an aqueous solution of potassium chloride (KCl) solution. KCl serves as a salt bridge between the electrode and the measured solution.

The electrode must be visually inspected every month. The level of the solution should be checked every day and refilled as needed.

F.3.2 Interferences

1. The glass electrode is relatively free from interferences caused by color, turbidity, colloidal matter, oxidants, reductants, or high salt content of the sample.
2. High sodium content samples give higher pH reading interference. It can be eliminated by using a low sodium error electrode.
3. An electrode can be coated with an oily material. It can be cleaned with a detergent solution and rinsed with DI water, or it can be rinsed with dilute HCl (1 : 9), followed by DI water. Soaking the electrode with a 5 percent pepsin solution may also be effective in removing organic deposits.
4. Temperature affects the response of the electrode in two ways. One is that the electrode response varies at different temperature, and the second is that the pH changes at different temperature. The best results are obtained when the temperature of the sample is within 5°C of the temperature of the buffers used to calibrate the instrument. Therefore, samples and buffers are allowed to equilibrate to room temperature prior to analysis. Temperature effects may also be eliminated by using pH meters equipped with a temperature compensator.

F.3.3 Apparatus and Materials

pH meter
pH and reference electrode or combination electrode
Disposable cups, 50 or 100 ml
Analytical balance, capable of weighing to the nearest 0.01 g
Magnetic stirrer
Magnetic stirring bars

F.3.4 Reagents

Electrode filling solution
pH buffer solutions, pH 2.00, 4.00, 7.00, 10.00, and 13.00
Potassium chloride (KCl) ACS grade
pH electrode storage solution

F.3.5 Meter Calibration

1. Check the filling solution level in the electrode. If needed, add more filling solution until the appropriate volume is achieved.
2. Turn on the pH meter. The pH meter requires about 20 min to warm up. The electrodes should be in the storage solution.
3. Set the temperature compensate knob to room temperature.
4. Add pH 7.00 buffer to a disposable cap. The depth of the buffer in the cap should be deep enough to cover the junction spot on the electrode.
5. Remove the electrode from the storage solution and rinse with DI water and dry by gently blotting with a kim-wipe.
6. Place the electrode into the buffer solution and stir gently either by using a magnetic stirrer and stirring bar or by hand-swirling. The rate of stirring should minimize the air transfer rate at the air–water interface of the sample. Be sure that the volume of the buffer is sufficient to cover the sensing elements of the electrode. Read the pH value. If the reading deviates from the actual value (7.00) adjust with calibration knob.
7. Remove the electrode from the pH 7.00 buffer, rinse with DI water, wipe, and place into the pH 4.00 buffer. Read. If the reading deviates from 4.00, adjust to the exact reading by turning the slope knob.
8. Remove the electrode from the pH 4.00 buffer solution, rinse with DI water, wipe, and immerse into pH 10.00 buffer. Read. pH value should be 10.00 or very close. Do not adjust meter at this point. The instrument is now calibrated.
9. Rinse the electrode with DI water and blot with a kim-wipe to remove the excess water. Check the calibrations with the initial calibration verification, by measuring the pH of the pH 4.00, 7.00, and 10.00 buffers. If the readings are correct, rinse the electrode with DI water and wipe.
10. After calibration, but before sample measurement, immerse the electrode in a reference/independent standard. If the reading is satisfactory, the measurement of the samples may start.

F.3.6 Measurement of Aqueous Samples and Wastes Containing Greater Than 20 percent Water

1. Allow the temperature of the sample to equilibrate to room temperature.
2. Pour an aliquot of the sample into the disposable cap or beaker. The depth of the sample should be deep enough to cover the junction spot on the electrode when the electrode is emerged in the sample.
3. Place the electrode into the sample and stir gently by using a magnetic stirrer and stirring bar or by hand-swirling.
4. Read the pH and record.
5. Rinse the electrode with DI water, blot dry with a kim-wipe to remove excess water.
6. After the samples are measured, remove the electrode, rinse, and blot to dry as in step 5.
7. Store the electrode in an electrode storage solution until its next use. Electrode storage solutions are commercially available pH electrode storage solution or may be a pH 7.00 buffer solution. Keeping the electrode in DI water is not recommended.

F.3.7 Measurement of Solid Samples

1. Allow the temperature of the sample to equilibrate to room temperature.
2. Using a two decimal place balance, weigh about a 20 g sample into the cap or beaker used for pH measurement.
3. Add 20 ml DI water to the cap or beaker.

4. Let stand for 30 min, mixing at 5 min intervals.
5. Place the electrode into the aqueous portion of the sample and stir gently either by using a magnetic stirrer and stirring bar or by hand-swirling.
6. Read the pH and record.
7. Repeat steps 5, 6, and 7 in the measurement of aqueous samples procedure.

F.3.8 General Rules

Samples are analyzed in a set referred to as a sample batch. A sample batch for pH consists of ten or fewer actual samples and a duplicate sample. Other QC samples are incorporated into a batch at the frequencies specified in the quality control (Section E.4) section in this methodology.

If a sample reading shows a high alkalinity character, the meter must be recalibrated by using a pH 7.00 and a pH 10.00 buffer instead the pH 7.00 and pH 4.00 buffers as calibration standards. After recalibration, read the sample again.

If field measurements are being made, the electrode(s) may be immersed in the sample stream to an adequate depth and moved in a manner to ensure sufficient sample movement across the electrode sensing element as indicated by drift-free (0.1 pH) sample reading.

Calibration, measurement, and QC data are recorded on the pH working paper.

Because of the wide variety of pH meters, detailed operation procedures are given for each instrument. The analyst must be familiar with the operation of the system and with the instrument functions.

F.4 QUALITY CONTROL (QC)

F.4.1 Duplicate Analysis

One duplicate sample is to be analyzed for every ten samples for each matrix in the sample batch. Duplicate samples are used to determine precision. The calculated relative percent difference (RPD) value should be documented and its value should be within the acceptable range variations determined by individual laboratories.

$$\text{RPD} = \frac{A - B}{(A + B)/2} \times 100 \tag{F.1}$$

or in simpler form

$$\text{RPD} = \frac{A - B}{A + B} \times 200 \tag{F.2}$$

where RPD equals the relative percent difference and A and B equal the values of the duplicate analysis.

F.4.2 Initial Calibration Verification (ICV)

1. ICV is performed immediately after the meter is calibrated. Calibration was previously discussed in Section F.3.5. Read the pH 2.00, pH 10.00, and pH 13.00 buffers. The acceptable ranges are:
   pH 2.00 is 1.90 to 2.10,
   pH 10.00 is 9.95 to 10.05,
   pH 13.00 is 12.90 to 13.10.
2. If the buffer readings are acceptable, the meter is ready for sample measurement.

F.4.3 Continuing Calibration Verification (CCV)

1. CCV is performed once after every ten samples in a batch (if applicable) and once at the end of the sample batch.
2. Read the pH 2.00, pH 4.00, pH 7.00, pH 10.00, and pH 13.00 buffers. The acceptable range for each buffers is as follows:
   pH 2.00 is 1.90 to 2.10,
   pH 4.00 is 3.95 to 4.05,
   pH 7.00 is 6.95 to 7.05,
   pH 10.00 is 9.95 to 10.05,
   pH 13.00 is 12.90 to 13.10.

F.4.4 Calculate Accuracy

Measure the pH of a reference standard with a known value of pH to check the accuracy of the measurement after each ten samples of the same matrix in a sample batch. Express accuracy as percent recovery.

$$\text{percent recovery} = \frac{\text{measured value}}{\text{true value}} \times 100 \tag{F.3}$$

The calculated accuracy value is documented and the value should be within the acceptable range variation indicated by the individual laboratories.

G Bacteriophages

All cellular life seems to be susceptible to viral infection, including bacteria. Viruses that attack bacteria are called bacteriophages or phages, for short.

Like other viruses, phages consist of a protein coat that surrounds a nucleic acid core. We can say each phage consists of a DNA molecule sorrounded by a complex protein coat. The protein coat consists of a head region, that encloses the DNA, and a tail region, that is used to land and attach to a bacterial cell wall.

When a bacterium becomes infected, the phage first attaches to its host's cell wall. Part of the phage then penetrates the bacterial cell's outer covering and takes over the bacterium's metabolic machinery. The phage causes this machinery to replicate viral genetic material, which then forms new phages like itself.

In other words, the phage forces the bacterial host to follow the genetic directions that the phage injects into it. Eventually, what remains of the host cell busts open, releasing a swarm of phage particles. Some of these may come in contact with other bacterial cells and then repeat the process.

Bacteriophages may also serve as useful indicators of fecal contamination. The presence of coliphage (bacteriophages that infect and replicate in coliform bacteria) in water shows good correlation with water quality and coliform bacteria because they are more resistant to water treatments and disinfection. The numbers of the total and fecal coliforms are determined as follows.

G.1 COLIFORM COUNTS

G.1.1 Total Coliforms

$$\log y = 0.627(\log x) + 1.864 \qquad \text{(G.1)}$$

where y equals the total coliforms per 100 ml and x equals the coliphages per 100 ml.

G.1.2 Fecal Coliforms

$$\log y = 0.805(\log x) + 0.895 \qquad \text{(G.2)}$$

where y equals the fecal coliforms per 100 ml and x equals the coliphages per 100 ml.

Bibliography

Alcamo, *Fundamentals of Microbiology,* 4th ed., Benjamin Publishing, 1994.
Baron, Peterson, and Finegold, *Diagnostic Microbiology,* 9th ed., Mosby Publisher, 1994.
Black, *Microbiology, Principles of Applications,* 2nd ed., Prentice-Hall, Englewood Cliffs, NJ, 1993.
Brock, Smith, and Madigan, *Biology of Microorganisms,* 4th ed., Prentice-Hall, Englewood Cliffs, NJ, 1984.
Corps of Engineers Method 3-373, CE-81-1, 1981.
Csuros, M., *Environmental Sampling and Analysis for Technicians,* CRC/Lewis Publishers, 1994.
Csuros, M., *Environmental Sampling and Analysis Lab Manual,* CRC/Lewis Publishers, 1996.
EPA method 405.1, *Methods for Chemical Analysis of Water and Wastes,* EPA 600/4-79-020, 1983.
Ewing, Differentiation of Enterobacteriaceae by biochemical reactions,
Greenberg, Clesceri, and Eaton, *Standard Methods for the Examination of Water and Wastewater,* 18th ed., American Public Health Association, 1992.
Ingraham and Ingraham, *Introduction to Microbiology,* Wadsworth Publishing, 1995.
Johnson and Case, *Laboratory Experiments in Microbiology,* 4th ed., Benjamin Publishing, 1995.
Kelly and Post, *Basic Microbiology Techniques,* 3rd ed., Star Publishing, 1989.
McKane and Kandel, *Microbiology,* 2nd ed., McGraw-Hill, New York, 1996.
Microbiological Methods for Monitoring the Environment, EPA, 8-78-017, 1978.
Miller, T., *Environmental Science: An Introduction,* Wadsworth Publishing, 1986.
Mitchell, R., *Environmental Microbiology,* Wiley-Liss Publishers, 1992.
Semmelweis, I., The cause, concept, and prophylaxis of childbed fever, 1861.
Smith, *Principles of Microbiology,* 10th ed., Times Mirror, Mosby, 1985.
Solomon and Berg, *The World of Biology,* 5th ed., Saunders Publishing, 1995.
Standard Method 5210, AWWA, 18th ed.
Standard Methods for the Examination of Water and Wastewater, 18th ed., APHA. AWWA. WPCF, 1992.
Sullivan, T. F. P., *Environmental Law Handbook,* Government Institution, 1995.
Talaro and Talaro, *Microbiology,* Brown Publishing, 1996.
Tortora, Funke, and Case, *Microbiology,* Benjamin Publishing, 1998.

Index

A

Acanthamoeba, 204
Accuracy, calculating, 305
Acid-base neutralization, 301
Acidophiles, 47
Actinomycetes, detection of, 271–275
  actinomycetes, 271
  determination of actinomycetes density, 272–275
    calculation and reporting, 274–275
    counting, 273–274
    medium, 273
    plating method, 272
    preparation and dilution of samples, 272
    preparation of soil samples, 272–273
    procedure, 273
  streptomycetes, 271–272
Adenosine diphosphate (ADP), 30, 37
Adenosine monophosphate (AMP), 37
Adenosine triphosphate (ATP), 30, 36, 37
Adonitol, 259
ADP, see Adenosine diphosphate
Aerobic respiration, 37, 42
Aeromonas, 255
Aerotolerent microbe, 49
Agar
  brain heart infusion, 247
  media, 171, 239
    M-Endo, 222
    procedure using, 223
  plate
    incubation of, 182
    inoculation of, 182
    preparation of, 181
  preparation, 210, 215
  slant(s)
    culture, characteristics of, 179
    inoculate, 179
    preparation of, 178
  starch-casein, 273
  temperature, testing, 211
Agrobacterium, 219
Air
  density plates, 106, 118
  filtration, 65
  microorganisms found in, 63
  quality, 106
Alcohols, 55
Algae, 12
Alkali-azide reagent, 284
Aminoglucosides, 271
Ammonium sulfate, 131
AMP, see Adenosine monophosphate
Anabolic reaction, 29
Anaerobic respiration, 37
Analytical balance, 155
Animalcules, 2
Anthrax, 6
Antibiotics, 8, 9
Antiseptics, 51
Apoenzyme, 31
Arabinose, 259
Asepsis, 51
Aseptic techniques, basis of, 5
Aspergillus amstelodami, 64
ATP, see Adenosine triphosphate
Atrazine, 79
Autoclave, 125, 153, 235
Azide dextrose broth, 249
Azotobacter, 26, 60

B

Bacillus
  anthracis, 18
  stearothermophilus, 175
  subtilis, 64
Bacteria, 11
  alternate pathway of glycolysis by, 38
  basic shapes of, 16
  denitrifying, 62, 77
  detection of pathogenic, 205
  disease, causative agent of, 7
  fecal coliform, 87
  Gram-negative, 20
  growth, 45
  methylotrophic, 60
  nitrifying, 61
  nitrogen-fixing, 60, 70
  pathogenic, 89
  rod-shaped, 219
  in soil, 67
  sulfate reducing, 62
  sulfur oxidizing, 63
  thermophilic, 69
Bacteria, iron and sulfur, 265–270
  enumeration, enrichment, and isolation of iron and sulfur bacteria, 270
  iron bacteria, 265–267
  sulfur bacteria, 267–270
    common forms of sulfur bacteria, 267–268
    identification of sulfur bacteria, 269–270
Bacterial numbers, estimation of by indirect methods, 193–200
  biochemical reactions and enzymatic tests, 194–200

catalase test, 194
citrate utilization test, 199–200
clumping factor test, 194–195
decarboxylase test, 200
IMViC tests, 198
indole test, 197–198
methyl red and Voges-Proskauer test, 198–199
motility test, 200
nitrate reduction test, 195–196
oxidase test, 196
rapid urease test, 198
rapid identification systems, 200
turbidity, 193–194
Bacteriophages, 307
fecal coliforms, 307
total coliforms, 307
Balances, types of, 155
Balantidiasis, 89
Beaches
natural bathing, 205
sample collection from bathing, 143
Beggiatoa alba, 269
Benthic zone, 70
Benzoates, 81
Benzoic acid, 54
BGLB, see Brilliant green lactose bile
BHI broth, see Brain heart infusion broth
Biochemical cycles, 57
Biochemical oxygen demand (BOD), 283–290
apparatus and material, 285
calculation, 288–289
alternative calculation, 289
BOD concentration, 288
correction factor for seed control, 288
dilution factor, 289
procedure, 286–288
analysis, 287–288
DO meter calibration, 286
preanalysis checking, 287
quality control, 289–290
accuracy, 290
DO meter performance check, 290
method blank, 289
precision, 290
reference standard, 289–290
seed control, 289
reagents, 285–286
safety, 290
scope and application, 283–285
carbonaceous vs. nitrogenous BOD, 283
dilution requirements, 283
interferences, 285
sample collection for Winkler method, 284
sampling and storage, 284
summary of method, 284
Biogenesis theory, spontaneous generation vs., 2
Biohazard control, 115
Biological Stain Commission, 129
Bioreactors, 82
Bioremediation, of organic contaminants, 79–84
advantage of bioremediation, 79
application of bioremediation, 80–84
biological control of groundwater pollution, 83–84
degradation of synthetic chemicals in soil, 82–83
modern microbiological concepts and pollution control, 84
solid and hazardous waste bioremediation, 81–82
wood preservative industries, 81
biotransformation by subsurface microorganisms, 80
development of bioremediation, 80
disadvantage of bioremediation, 80
objective of bioremediation, 79
Blow-out pipets, 165
BOD, see Biochemical oxygen demand
Brain heart infusion (BHI) broth, 247, 248
Brilliant green bile broth, 222
Brilliant green lactose bile (BGLB), 228
Bromthymol-blue indicator, 169
Broth
culture tubes, 180
media, 239
bacterial growth in, 176
growth patterns on, 176
procedure using, 223
Brucella, 219
Buffer, 47
Bunsen burner, 220

C

Calcium
chloride solution, 285
propionate, 54
Calibration stock, 125
Campylobacter jejuni, 85
Cancer, 8
Candida albicans, 96, 204
Carbenicillin, 234
Carbohydrate catabolism, 37
Carbolic ac id, 55
Carbon cycle, 58
Casitone, 241
Catabolic reaction, 20
Catalase test, 194, 195
CCV, see Continuing calibration verification
Cell wall, 19
atypical, 20
damage to, 21
structures internal to, 21
Centrifuge, 123
CERCLA, see Comprehensive Environmental Response Compensation and Liability Act
CFUs, see Colony forming units
Chain of custody form, 148
Chemical agents, 65
Chemicals, storage of, 111, 115
Chemoautotrophs, 35
Chemoheterotrophs, 35, 58
Chemotherapy, 8, 9
Chemotrophs, 35

Chlorine, 55
Chlortetracycline, 272
Chromatophores, 21
Chromatum okenii, 270
Citrate, 199, 259, 260
Citric acid cycle, 39, 40
Citrobacter, 88, 255
Cladosporium, 64
Clostridium
  botulinum, 91
  perfringens, 92
  tetani, 91
Clumping factor test, 194
Coccobacilli, 17
Coliform
  colonies, 224
  counts, 307
  determination, flow chart for total, 232
  estimates, 220
  group, 87
  monitoring requirements, 98
  organisms, 255
  test, 206
  verification outline for total, 226
Coliform, determination of fecal, 237–244
  application for soil, sediment, and sludge samples, 244
  definition of fecal coliform group, 237
  delayed incubation MF method, 240–242
    application, 240
    culture media, 241–242
    counting and recording colonies, 242
    equipment and glassware, 240
  membrane filter method, 237–240
    application, 237
    counting and recording colonies, 239–240
    culture media, 237–238
    equipment and glassware, 237
    procedure, 238–239
    verification, 240
  most probable number method, 242–244
    application, 242
    calculation and reporting, 244
    equipment and glassware, 242
    media, 242
    procedure, 242–243
  MUG test, 244
Coliform, determination of total, 219–236
  analytical quality control, 236
  application for soil, sediment, and sludge samples, 235
  delayed incubation by MF method, 226–227
    application, 227
    equipment and glassware, 227
    culture media, 227
    procedure, counting, and recording, 227
  determination of *Klebsiella*, 233–235
    apparatus, 234
    counting colonies, 235
    culture medium, 234–235
    procedure, 235
    verification, 235
  membrane filter technique, 219–222
    application, 219–220
    culture media, 221–222
    equipment and glassware, 220–221
    set up of filtration apparatus, 222
    suggested sample volumes for membrane filter total coliform, 222
  membrane filtration procedure, 223–226
    counting and recording colonies, 223–225
    procedure using agar media, 223
    procedure using broth media, 223
    verification, 225–226
  most probable number method, 227–233
    application, 228
    calculation, 231–233
    culture media, 228–229
    dilution water, 229
    equipment and glassware, 228
    procedure, 230–231
Colony(ies)
  counting, 213, 223, 275
  forming units (CFUs), 106, 185
  spreading, 214
Completed test, 188
Composting, 82
Compound light microscope, 158, 159, 160
Comprehensive Environmental Response Compensation and Liability Act (CERCLA), 82
Confirmed test, 188, 230, 251
Contact hazard, 113
Continuing calibration verification (CCV), 305
Conversion factors, see Units and conversion factors
Copper, 56
Corrosion control, 301
Corynebacterium diphtheriae, 25
Counterstain, 191
Cowpox blisters, 7
Coxsackieviruses, 93
Crenothrix polispora, 266, 68
Cryptosporidium, 85, 103
Culture media, 171–182
  bacterial growth in media, 176–182
    bacterial growth in agar slant cultures, 178
    bacterial growth in liquid media, 176–178
    bacterial selection by sugar fermentation, 178–180
    obtaining pure culture, 180–182
  culture media and culture, 171–176
    chemically defined media, 171
    complex media, 171
    pH check of media, 175
    preparation of media, 172–174
    selective and differential media, 172
    sterile media from commercial sources, 176
    sterilization of media, 174–175
    storage of culture media, 175
    storage of dehydrated culture media, 172
Cyanobacteria, 58, 60, 69
Cycloheximide, 274
Cytochrome
  oxidase, 196
  system, 41

Cytophaga, 69
Cytoplasm, 23

D

Dark field microscopy, 161
DDT, see Dichloro-diphenyl-trichloroethane
Decarboxylase test, 200
Dechlorinating agent, 140
Decolorizer, 190
Deep freezing, 52
Degerming, 51
Depressuization, 125
Desiccation, 53
Desulfovibrio, 268
Detection method, rapid, 203
Dialysis, 22
Dichloro-diphenyl-trichloroethane (DDT), 82
1,2-Dichloropropane, 79
Diffusion, 22
Dilution(s)
  bottles, 167
  preparation, 210, 216
  requirements, 283
  serial, 183, 184
  single, 183
  water, 184, 222, 248, 286
Dimethylamino-cinnamaldehyde, 197
Dioxin, 81
Diplobacilli, 17
Discharges, sample collection from, 146
Disease
  Germ Theory of, 6
  outbreaks
    swimming associated, 86
    water-borne, 85
Disinfection, 51, 301
  by-product regulations, 100
  effluent, 104
  intermediate, 105
  requirements, for public water supplies, 99
Disposable pipet, 165
Dissecting microscope, 158
Dissolved oxygen (DO), 283, 286, 290
DNA, 12
DO, see Dissolved oxygen
Drinking water
  disinfecting municipal, 55
  laboratories, 137
  quality, 96
  sampling frequency for, 141, 142
Dulcitol, 259
Duplicate analysis, 304
Durham tube, 179, 195
Dye, primary, 189

E

Echoviruses, 93
EDB, see Ethylene dibromide
EDP, see Entner Doudoroff Pathway
EDTA, see Ethylene diamine tetra acetic acid
EEC, see European Economic Community
Effluent(s)
  chlorinated, 203
  disinfection, 104
Ehrlich, Paul, 9
Electric vacuum pump, 157
Electron transport chain, 36
EMB, see Eosine methylene blue
Embden–Meyerhof Pathway, 38
Endocytosis, 23
Endoenzymes, 194
Endospore formation, 27
Energy production, 34
Entamoeba histolytica, 85, 94, 95
Enterobacter aerogenes, 130, 132, 133
Enterobacteriaceae, 255–263
  API 20E system, 260–263
  BBL OXI/FERM TUBE II, 260
  coliform organisms, 255
  differentiation of, 255–256
  ENTEROTUBE II, 256–259
    indicating positive reactions, 257
    operation of, 256–257
    ordering information, 257–259
    reading, 257
Entner Doudoroff Pathway (EDP), 38
Environment, microorganisms in, 57–78
  aquatic microbiology, 70–74
    factors affecting microorganisms in aquatic environments, 71
    freshwater environment, 70
    marine environments, 71–72
    pathogens in water, 73–74
    water pollution, 72–73
  biochemical cycles, 57–63
    carbon cycle, 58–59
    nitrogen cycle, 59–62
    phosphorus cycle, 63
    sulfur cycle, 62–63
    water cycle or hydrologic cycle, 57
  microorganisms found in air, 63–65
    controlling microorganisms in air, 65
    factors affecting indoor microbial levels, 65
    sources of indoor pollution, 64–65
  microorganisms in soil, 66–70
    abiotic factors influencing microorganisms in soil, 69
    actinomycetes, 68
    bacteria, 67–68
    components of soil, 66–67
    cyanobacteria, algae, protista and viruses, 69
    fungi, 68
    importance of decomposers in soil, 69–70
    soil pathogens, 70
  sewage treatment, 75–78
    primary treatment, 75
    secondary treatment, 75–76

septic tanks, 77
tertiary treatment, 76–77
water purification, 74
chlorination, 74
filtration, 74
flocculation, 74
Environmental Protection Agency (EPA), 97, 100
Environmental samples, collecting and handling of for microbiological examination, 139–151
sample collection from different sources, 141–148
collecting potable water samples, 141–142
collecting samples from river, stream, late, spring or shallow well, 143
collecting samples from sediments and sludges, 146–147
marine and estuarine sampling, 146
sample collection from bathing beaches, 143–146
sample collection from domestic and industrial discharges, 146
soil sampling, 147–148
sample identification, 148–149
sample transportation, preservation and holding time, 149–151
discard samples, 151
holding time, 150–151
laboratory custody procedure, 150
sample transportation and preservation, 149–150
sampling, 139–140
chelating agent, 140
dechlorinating agent, 140
sample containers, 139
sampling procedures, 140
sampling program, 139
type of samples, 139
Environmental samples, microbiological quality of, 85–106
actinomycetes, 96
air, 106
fungi, 96
groundwater, 100–101
classification of groundwater, 100
groundwater standards, 101
iron and sulfur bacteria, 94–96
microbiological parameter of sanitary quality on environmental samples, 88–89
coliform bacteria group, 88–89
heterotrophic plate count, 88
monitoring microbiological quality, 85–88
indicator bacteria, 87–88
recovery of pathogens from environmental samples, 87
standards on microbiological quality, 88
waterborne disease outbreaks, 85–87
pathogenic microorganisms, 89–94
pathogenic bacteria, 89–92
pathogenic protozoa, 93–94
viruses, 92–93
regulations for drinking water quality, 96–100
disinfection by-product regulations, 100
disinfection requirements for public water supplies, 99–100
minimum coliform monitoring requirements, 98–99
monitoring agencies, 97–98
monitoring requirements, 98
standards, 99
soil and sediment, 105
surface waters, 101–103
surface water classification, 101
surface water quality standards, 103
Surface Water Treatment Rule, 102–103
wastewater, 103–105
industrial wastewater, 103
minimum treatment standard of domestic wastewater effluents, 104–105
permit for industrial and domestic wastewater effluents, 104
Environmental sources, 234
Enzyme(s), 31
action, mechanisms of, 32
inhibition of, 32
naming, 32
tests, 194
Eosine methylene blue (EMB), 228
EPA, see Environmental Protection Agency
Escherichia coli, 10, 193, 255
Estuarine sampling, 146
Ethyl alcohol, 55
Ethylene
diamine tetra acetic acid (EDTA), 140
dibromide (EDB), 79
oxide, 54
European Economic Community (EEC), 97
Eutrophication, 73
Evaporating dishes, 292, 293
Exocytosis, 23
Exponential notation, 277–278
adding and subtracting numbers in, 278
multiplying and dividing numbers in, 278
numbers greater than 10 in, 277
numbers less than 1 in, 278
Eye
protective device, 110
wash fountain, 110

F

Facultative aerobe, 49
Facultative anaerobe, 49
FAD, see Flavin–adenine–dinucleotide
Fecal coliform, 136, 201, see also Coliform, determination of fecal
bacteria, 87
colonies, verification of, 241
count, 239
group, 237
MPN test, 243
test, 204
Fecal contamination, indicators of, 307
Fecal streptococcus, see Streptococcus, determination of fecal

Fermentation, 6
bacterial selection by sugar, 178
organic compounds produced by, 43
tubes, 168
Ferric chloride solution, 286
Ferrobacillus ferrooxidans, 265
Filter(s)
flask, 293
nitrocellulose, 186
systems, trickling, 76
Filtration apparatus, 238
First Aid Charts, 112
Flagellates, 93
Flammable hazard, 113
Flatworm, 12
Flavin–adenine–dinucleotide (FAD), 36
Fleming, Alexander, 9
Flocculation, 74
Fluorescent microscopy, 162
Food(s)
supplies, preserving household, 46
use of desiccation to preserve, 53
Foraminiferans, 72
Formaldehyde, 54
Fossil fuels, 58
Frame sampler, 143
Frankia, 60, 271
Freezers, 126
Fresh water ecosystems, 101
Fungi, 11, 68, 96

G

Gallionella, 265, 266
Galvanometers, 193
Gas(es)
production, 259, 260
storage of, 111, 115
vacuoles, 26
Gastroenteritis, 74, 87, 89
Gelidium, 171
Geosmin, 271
Germicide, 51
Germination, 26
Germ Theory of Disease, 6, 7
Giardia lamblia, 85, 93, 103
Giardiasis, 74
Glassware
cleaning laboratory, 169
specifications for, 220
sterilization, 169
Global warming, 59
Glucose–glutamic acid standard, 289, 290
Glycocalyx, 18
Glycolysis
end product of, 43
intermediates of, 39
Graduated cylinders, 166
Gram, Hans Christian Joachim, 20
Gram stain, 189
preparations, 231
procedure, 191, 192
Greenhouse effect, 59
Groundwater
classification of, 100
standards, 101

H

Halophile bacteria, 48
HAV, see Hepatitis A virus
Hazardous Agent Charts, 112
Hazardous waste bioremediation, 81
Hazard warning signs, 110
Health hazard, 113
Heavy metals, 55
Helminths, developing, 13
Hemophilus influenza, 64
Hepatitis A virus (HAV), 93
Herbicides, 79
Heterotrophic plate count (HPC), 88, 202, 209–217
plate count from soils and sediments, 216–217
calculation and reporting, 217
moisture determination, 216
pour plates, 217
preparation of dilutions, 216
spread count, 217
pour plate method, 209–214
counting, 213–214
dilution preparation, 210–211
equipment and material, 209
media preparation, 210
plate preparation, 211–213
preparation of agar, 210
spread plate method, 215–216
counting and reporting, 216
equipment and material, 215
media and agar preparation, sample dilution, 215
preparation of agar plates, 215
procedure, 215–216
Holding time, 150
Hook, Robert, 2
Hot air oven, 123, 125, 153
HPC, see Heterotrophic plate count
Humus, 67
Hydrogen peroxide, 55
Hydrolase, 32
Hydrologic cycle, 57
Hypertonic solution, 22
Hypotonic solution, 22

I

ICV, see Initial calibration verification
Imhoff cone, 299
IMViC test, 198
Incubator temperature control log, 120, 121
Indicators, 129

Indole
formation, 259
test, 197, 257
Indoor pools, disinfected, 204
Industrial wastes, 73
Infections, pioneers in control of, 50
Initial calibration verification (ICV), 304
Inoculating needles, 157
International System of Units (SI), 279–280
derived SI units, 279
derived SI units with special names, 279
prefixes, 280
useful conversion factors, 280
Iodine, 55
Iodophor, 55
Iron bacteria, see Bacteria, iron and sulfur
Irrigation, 147
Isomerase, 32
Isopropyl alcohol, 55

K

Kemmerer sampler, 143, 144
Kingdom system, of biological classification, 11
Klebsiella, 88, 219, 233, 255
Kovacs method, 196
Krebs cycle, 37, 39, 41

L

Lab guard safety label system, 113, 114
Laboratory
bench areas, 117
cleanliness, 118
custody procedure, 150
facilities, 107
glassware, cleaning, 169
incubator, 154
pure water, 169
records, 151
standard safety practices in, 109
Laboratory, safety in environmental microbiology, 107–115
laboratory facilities, 107–108
laboratory safety considerations, 108–111
electrical precautions, 111
general handling and storage of chemicals and gases, 111
general laboratory safety rules, 108–109
standard safety practices in microbiological laboratory, 109–110
summarized safety check list for environmental microbiology laboratories, 111–115
administrative considerations, 111–113
biohazard control, 115
disinfection/sterilization, 115
emergency precautions, 115
handling and storage of chemicals and gases, 115
laboratory equipment, 114
personal conduct, 114
Laboratory equipment and supplies, in environmental microbiology laboratory, 153–169
chemical and reagents, 169
laboratory equipment, 153–162
autoclaves, 153–154
balances, 155
hot-air sterilizing oven, 153
incubators, 153
inoculating needles or loops, 157
line vacuum or electric vacuum pump, 157
membrane filtration equipment, 156
microscope, 157–162
optical counting equipment, 155–156
pH meter, 154–155
refrigerator, 156
laboratory glassware, 162–169
cleaning laboratory glassware, 169
dilution bottles, 167
fermentation tubes and vials, 168
glassware sterilization, 169
graduated cylinders, 166
petri dishes, 162
pipets, 162–166
safety trap flask, 167
thermometers, 168
vacuum filter flask, 166
Laboratory quality assurance and quality control, 117–137
analytical quality control procedures, 133–136
measurement of method precision, 134
membrane filter method verification, 135–136
performance sample, 134
quality control in routine analysis, 133–134
reference sample, 134
interlaboratory quality control, 136–137
quality assessment, 117
quality assurance, 117
quality control for laboratory equipment and instrumentation, 119–127
autoclave, 125
balances, 119
centrifuge, 123–124
hot-air oven, 125
incubators, 126
membrane filter apparatus, 123
microscope, 127
pH meter, 120
refrigerators and freezers, 126
safety hood, 126
thermometer and temperature-recording instruments, 119
UV sterilizer, 121–122
water bath, 126
water deionization unit, 120–121
spectrophotometer, 127
quality control of laboratory pure water, 130–133
checks and monitoring criteria, 130
test for bacterial quality, 130–133
quality control of laboratory supplies, 127–130
chemicals and reagents, 128
culture dishes, 128

culture media, 129–130
dyes and stains, 129
glass, plastic, and metal utensils for media preparation, 128
glassware, 127
membrane filters and pads, 129
sterility check on glassware, 128
records and data reporting, 136
requirements for facilities and personnel, 117–119
laboratory bench areas, 117
laboratory cleanliness, 118–119
personnel, 119
ventilation, 117
walls and floors, 118
Lactic acid, 43
Lactose, 259
Lancefield's group D streptococcus, 245
Lauryl sulfate broth, 221
Lauryl tryptose broth, 188, 221, 242
Legionella pneumophila, 64, 91
Legionnaire's disease, 91
Leptospira, 89
Leptospirosis, 87
Ligase, 32
Light microscope, 158
Limnetic zone, 70
Lindane, 102
Line vacuum, 157
Lipid inclusion, 26
Lister, Joseph, 6, 7, 50
Littoral zone, 70
Living systems, characteristics of, 2
Lyase, 32
Lyophilization, 52
Lysine decarboxylase, 259

M

Magnesium chloride, 222
Manganese sulfate reagent, 284, 285
Mannitol, 262
Marine environments, 71
Marine sampling, 146
Material safety data sheets (MSDS), 111, 112, 290
MB, see Method blank
Media
bacterial growth in broth, 176
from commercial sources, 176
holding times for prepared, 175
pH check of, 175
preparation of, 172
reaction, 261
selective, 172
sterilization of, 173, 174
Membrane filter (MF), 129
fecal coliform, 237
method
delayed incubation by, 226
verification, 135
procedure, delayed, 249
system, 174
technique, 186, 201, 219
advantages of, 187
limitations of, 187
outline of, 187
Mercurochrome, 56
Mesophiles, 46, 69
Mesosomes, 21
Metabolic pathways, 30, 31
Meter calibration, 303
Method blank (MB), 289
Methyl red (MR), 198, 199
Methylotrophic bacteria, 60
Metric system, see International system of units
MF, see Membrane filter
Microaerophile, 49
Microbes
cold loving, 46
heat loving, 47
moderate temperature, 46
Microbial growth, 45–56
chemical methods of microbial control, 53–56
alcohols, 55
aldehydes, 54
ethylene oxide, 54
halogens, 55
heavy metals, 55–56
organic acids, 54
oxidizing agents, 54–55
phenol, 55
surfactants or surface acting agents, 53–54
chemical requirements, 48–50
carbon, 48
nitrogen, sulfur, and phosphorus, 48
organic growth factors, 49–50
oxygen, 49
trace elements, 48
control of microbial growth, 50–51
antiseptics, 51
asepsis, 51
degerming, 51
disinfection, 51
germicide, 51
pasteurization, 51
sanitization, 51
sterilization, 51
death phase, 45–46
Microbial growth, direct measurement of, 183–192
dilutions, 183–185
dilution water, 184–185
prompt use of dilutions, 184
serial dilution, 183–184
single dilution, 183
direct microscopic counts, 191–192
membrane filter technique, 186–187
advantages of, 187
limitations of, 187
outline of, 187

most probable number method, 188–189
calculation and reporting of MPN values, 189
completed test, 188
confirmed test, 188
presumptive test, 188
plate counts, 185–186
pour plate method, 185–186
spread plate method, 186
streak plate method, 186
staining procedures, 189–191
Gram stain, 189–191
preparation of bacterial smears, 189
Microbial growth, physical methods of microbial control, 52–53
desiccation, 53
filtration, 52
high heat, 52
low temperature, 53
radiation, 53
Microbial growth, physical requirements, 46–48
osmotic pressure, 48
pH and buffers, 47–48
temperature, 46–47
Microbial metabolism, 29–43
carbohydrate catabolism, 37–43
aerobic respiration, 42
alternate pathway of glycolysis by bacteria, 38
fermentation, 43
glycolysis, 38
Krebs cycle or citric acid cycle, 39–40
respiratory chain or electron transport system or cytochrome system, 41
transition reaction, 38–39
energy production of organisms, 35–37
formation of ATP during respiration, 36–37
formation of ATP during fermentation, 37
oxidation–reduction, 36
phosphorylation as ATP formation in biological system, 36
Microbial metabolism, enzymes, 31–34
inhibition of enzymes, 33–34
meaning of metabolism, 29–30
energy transport through ADP and ATP, 30
first and second law of thermodynamics, 29–30
mechanisms of enzyme action, 32
metabolic pathways, 30–31
naming enzymes, 32
nutritional classification of organisms, 34–35
temperature and pH effect on enzymes, 32–33
Microbiological quality of environment, methods for analyzing, 201–207
detection of pathogenic bacteria, 205–206
detection of soil microorganisms, 206–207
moisture determination, 206–207
most probable number method, 207
plate count, 207
sample collection, 206
sample handling and storage, 206
method for microbiological examination of recreational waters, 204–205
natural bathing beaches, 205
swimming pools, 204–205
whirlpools, 205
methods and techniques, 201–203
detection of stressed organisms, 203
heterotrophic plate count, 202
method for fecal coliforms, 201–202
method for fecal streptococcus, 203
method for total coliform group, 201
rapid detection methods, 203–204
Microbiology
aquatic, 70
Golden Age of, 5
history of, 2
modern developments in, 10
Microbiology, scope and history of, 1–13
brief history of microbiology, 2–5
aseptic techniques, 5
microscope and microbes, 2
spontaneous generation vs. biogenesis theory, 2–5
chemotherapy, 8–10
antibiotics, 9–10
first synthetic drugs, 9
diversity of microorganisms, 11–13
algae, 12
bacteria, 11
fungi, 11–12
protozoa, 12
multicellular parasites, 12–13
viruses, 12
Golden Age of Microbiology, 5–8
fermentation and pasteurization, 6
germ theory of disease, 6–7
vaccination, 7–8
microbes and microbiology, 1–2
basic characteristics of living systems, 2
general concepts of microbiology, 1
investigation of microorganisms, 2
size of microorganisms, 1
unicellular and noncellular organisms, 1
modern developments in microbiology, 10
naming and classifying microorganisms, 10–11
Microorganisms
control of in air, 65
diversity of, 11, 13
investigation of, 2
major groups of, 1
Microscope(s)
bright field, 163
compound light, 158, 159, 160
dissecting, 158
Leeuwenhoek's, 3
parts of, 157
phase contrast, 163
quality control functions for care of, 127
rules for using, 161
scanning electron, 162, 163
stereo, 158
transmission electron, 162, 163
types of, 163

Microscopic counts, direct, 191
Microscopy
  dark field, 161, 163
  electron, 162
  fluorescent, 162, 163
  oil-immersion, 160
  phase-contrast, 161
Mohr pipet, 164, 165
Mordant, 190
Most probable number (MPN)
  index, 201, 232
  method, 188, 207, 222, 242
  table, 231
Motility test, 200
MPN, see Most probable number
MR, see Methyl red
MSDS, see Material safety data sheets
Muffle furnace, 292
Multiple-tube fermentation method, 202
Multitest systems, 200
Mycobacterium tuberculosis, 6
Mycoplasma pneumonia, 20

N

NA, see Numerical aperture
NAD, see Nicotinamide–adenine–dinculeotide
National Bureau of Standards (NBS), 119
National Institute of Standards and Technology (NIST), 119, 153
National Pollution Discharge Elimination System (NPDES), 104, 146
NBS, see National Bureau of Standards
Needham, John, 4
Neisseria meningitidis, 64
Nephelometry, 193
Nicotinamide–adenine–dinculeotide (NED), 36
NIST, see National Institute of Standards and Technology
Nitrate reduction test, 195
Nitrification inhibitor, 286
Nitrogen
  cycle, 59
  sources, 35
Nitrosomonas, 61
Nowalk agent, 93
NPDES, see National Pollution Discharge Elimination System
Nucleic acids, 1, 12
Numerical aperture (NA), 159

O

Obligate aerobe, 49
Obligate anaerobe, 49
Ocean water, discharging to open, 104
Oil-immersion microscopy, 160
Organisms
  destruction of, 51
  identification of, 13
  nutritional classification of, 34
Ornithine decarboxylase, 259, 262, 263
Osmosis, 22
Osmotic pressure, 22, 48
Outdoor pools, disinfected, 204
Oxgall solution, 248
Oxidase test, 196
Oxidation–reduction, 36
Oxidizing agents, 54
Oxidoreductase, 32
Oxygen, microbes using molecular, 49
Oxytetracycline, 272

P

PAHs, see Polycyclic aromatic hydrocarbons
Parasites, multicellular, 12
Pasteur, Louis, 4, 5, 20
Pasteurization, 6, 51
Pathogens
  expelled by humans, 64
  recovery of from environmental samples, 87
  in water, 73
PCBs, see Polychlorinated biphenyls
Pellicle, 177
Penicillin, bacterial growth of, 9
Penicillium notatum, 9
Peptidoglycan, 20
Performance sample, 134
Pesticides, 81
Petri dish
  handling of, 182
  with loose fitting lids, 162
  pipetting sample into, 212
  solid agar in, 171
  with tight fitting lids, 163
Petri plate, 171
Petroff-Hausser counting chamber, 191
Pfizer selective enterococcus (PSE) agar, 249, 250, 252
pH, determination of, 301–305
  general discussion, 301
  potentiometric determination of pH, 301
    apparatus and materials, 302
    general rules, 304
    interferences, 302
    measurement of aqueous samples and wastes, 303
    measurement of solid samples, 303–304
    meter calibration, 303
    principle of method, 301–302
    reagents, 302
  quality control, 304–305
    calculation of accuracy, 305
    continuing calibration verification, 305
    duplicate analysis, 304
    initial calibration verification, 304
  sample collection and holding time, 301
Phase-contrast microscopy, 161
Phenols, 81
Phenylalanine deaminase, 259
Phosphate buffer solution, 286
  stock, 184
  working, 185

Phosphorus cycle, 63
Phosphorylation, oxidative, 36, 38
Photoautotrophs, 35
Photoheterotrophs, 35
Pili, 19
Pipet(s), 162
  blow-out, 165
  disposable, 165
  measuring, 164
  Mohr, 164, 165
  serological, 164
Pipettors, 166
Plate(s)
  counts, 185, 207
  crowded, 253
  preparation, 211
  uncountable, 214
Poliovirus, 74, 89, 93
Pollutants, biodegradability of volatile organic, 83
Pollution control, 84
Polychlorinated biphenyls (PCBs), 81
Polycyclic aromatic hydrocarbons (PAHs), 80, 81
Polysaccharide granules, 26
Polyseed inoculum mixture, 286
Pontiac fever, 87
Pools, disinfected indoor, 204
Potassium
  iodide, 286
  sorbate, 54
Pour plate method, 185, 209, 251
Preanalysis checking, 287
Presumptive test, 188, 230, 251
Primary dye, 189
Procaryotic cells, 15–27
  cell wall, 19–21
    atypical cell wall, 20
    damage to cell wall, 21
    Gram negative bacteria, 20
    Gram positive bacteria, 20
  morphology of bacterial cell, 15–17
    bacilli, 17
    cocci, 17–18
    spiral bacteria, 16–17
  nucleoid in, 24
  procaryotic and eucaryotic cells, 15
  structures external to cell walls, 18–19
    flagella, 19
    glycocalyx, 18
    pili, 19
  structures internal to cell wall, 21–27
    cytoplasm, 23–24
    endospores, 26–27
    inclusion, 25–26
    movement of material across membranes, 22–23
    nuclear area, 24
    plasma or inner membrane, 21
    ribosomes, 24–25
Profundal zone, 70
Protoplasm, 1
Protozoa, 12, 93
PSE agar, see Pfizer selective enterococcus agar
Pseudomonas, 219
  aeroginosa, 205
  dermatitis, 86
Psychrophiles, 46
Pyruvic acid, 43
  derivatives, 39
  formation of acetyl CoA from, 40

Q

QA, see Quality assurance
QC, see Quality control
Quality
  assessment, 117
  assurance (QA), 117, see also Laboratory quality assurance and quality control
Quality control (QC), 117, 304, see also Laboratory quality assurance and quality control
  analytical, 236
  criteria, for prepared media, 130
  interlaboratory, 136
  in routine analysis, 133
  tests, analytical, 137
Quaternary ammonium compounds, 54
Quebec colony counter, 155, 213

R

Radiation
  ionizing, 53
  nonionizing, 53
  ultraviolet, 65
Radioactive substances, 73
Radiolarians, 72
Radionuclides, 102
Rapid indole test, 197
Rapid urease test, 198
RCRA, see Resource Conservation and Recovery Act
Reactivity hazard, 113
Reagents, 131
Reference sample, 134
Refrigerators, 122, 126, 156
Reproducible Organisms Detection And Counting (RODAC), 118
Resolving power (RP), 159
Resource Conservation and Recovery Act (RCRA), 82
Respiratory chain, 41, 42
Rhizobium, 60
Ribosomes, 24
RNA, 12, 25
RODAC, see Reproducible Organisms Detection And Counting
Root systems, of higher plants, 67
Rosolic acid solution, 238
Roundworm, 12
RP, see Resolving power

S

Safe Drinking Water Act (SDWA), 97
Safety
  hood, 126
  trap flask, 167

Salmon, Daniel, 90
Salmonella, 89, 219, 255
Sample(s), see also Environmental samples, collecting and handling of for microbiological examination
  analysis, 294, 296
  aqueous, 303
  collection, 206, 291
  containers, 139
  dilution, 215
  discard, 151
  identification, 148
  label, 150
  log sheet, 148, 149
  performance, 134
  potable water, 141
  reference, 134
  sludge, 235
  transportation, 149
  types of, 139
Sanitization, 51
S. avium, 246
S. bovis, 245
Scanning electron microscope, 162
SDWA, see Safe Drinking Water Act
Sedimentation rate, 25
Sediments
  bottom, 146
  procedures for, 254
Seed control, 288, 289
Selective agents, 129
Semmelweis, Ignaz Philipp, 50
S. equinus, 245
Septic tanks, 77
Serological pipet, 164
Sewage
  raw, 222
  soluble components of, 77
  treatment, 75, 78
  water, 85
S. faecalis, 246
S. gallinarum, 246
Shigella, 219, 255
Shigellosis, 74, 89
SI, see International System of Units
Silver, 56
Silvex, 83
Simmons citrate agar, 198, 199
Slide coagulase test, 194
Sludge(s)
  disposal on into receiving waters, 147
  procedures for, 254
  samples, 235, 244
  separation of from water, 297
  system, activated, 76
Smallpox epidemics, 7
Soap, 53
Sodium
  benzoate, 54, 241
  citrate, 131
  sulfite, 286
  thiosulfate, 140
Soil(s)
  bacteria, decomposition of organic materials by, 77
  components of, 66
  degradation of synthetic chemicals in, 82
  heaping, 82
  microorganisms in, 67
  moisture determination, 206
  pH, 69
  plate count from, 216
  procedures for, 254
  suspensions, 217
Solids, determination of, 291–299
  determination of total dissolved solids dried at 180°C, 292–295
    apparatus and material, 292, 293
    calculation, 292, 294
    interferences, 291, 293
    principle of method, 291, 292
    procedure, 292, 293–294
    quality control, 292, 295–296
    reagents, 293
  determination of total solids, 291–292
  determination of total suspended solids, 295–296
  determination of total suspended solids at 103 to 105°C, principle of method, 295
    apparatus and material, 295
    calculation, 296
    interferences, 295
    procedure, 295–296
  fixed and volatile solids ignited at 500°C, 296–297
    apparatus and material, 297
    calculation, 297
    interferences, 297
    principle of method, 296
  general discussion, 291
    sample collection and handling, 291
    sample pretreatment, 291
  settleable solids, 298–299
    apparatus, 298
    principle of method, 298
    procedure, 299
  total, fixed, and volatile solids in solid and semisolid samples, 297–298
    apparatus and material, 298
    calculation, 298
    interferences, 297–298
    principle of method, 297
    procedure, 298
    quality control, 298
Sorbic acid, 54
Sorbitol, 259
SPC, see Standard plate count
Spectrophotometer, 127, 193
Spherotylus natans, 266, 267
Spirillum, 26
Spontaneous generation
  biogenesis theory vs., 2
  dispute of theory of, 4

Spore formation, 26
Sporulation, 26
Spread plate method, 186, 215
Stains, microbiological, 129
Standard plate count (SPC), 209
Staphylococcus aureus, 10
Starch-casein agar, 273
Stereo microscope, 158
Sterilization, 51, 52
Sterilizer
  acid–anionic surface acting, 53
  UV, 121, 220
Stormwater runoff, 147
Streak plate method, 180, 186
Stream, sampling large, 145
Streptobacilli, 17
Streptococci, fecal, 136
Streptococcus, determination of fecal, 245–254
  delayed procedure, 249
  enterococci of fecal streptococcus group, 246
  fecal coliform and fecal streptococcus, 245–246
  membrane filter technique, 246–249
    application, 246
    counting and recording colonies, 248
    culture media, 247–248
    equipment and glassware, 246
    procedures, 248
    verification, 248–249
  most probable number method, 249–251
    calculation, 251
    culture media, 249–251
    procedure, 251
  pour plate method, 251–254
    counting and recording, 252–254
    media, 252
    procedure, 252
  procedures for soils, sediments, and sludges, 254
Streptococcus pneumoniae, 18
Streptomycetes, 96
Substrate–enzyme complexes, 33
Sugar fermentation, 178
Sulfanilamide, 241
Sulfur bacteria, see Bacteria, iron and sulfur
Surface
  acting sterilizers, 53
  water classification, 101
  Water Treatment Rule (SWTR), 102
Survey monitoring, 240
Suspended solids, 291
Svedberg unit, 25
Swimming pools, 204, 222
SWTR, see Surface Water Treatment Rule
Synthetic chemicals, 73, 82
Synthetic drugs, 8, 9

T

2,4,5-T, see 2,4,5-Trichloro-phenoxyacetic acid
Temperature
  control log, 120, 121, 122, 123
  -recording instruments, 119
Test
  catalase, 194, 195
  citrate utilization, 199
  clumping factor, 194
  coliform, 206
  completion, 230
  confirmed, 188, 230, 251
  decarboxylase, 200
  enzymatic, 194
  fecal coliform, 204, 243
  IMViC, 198
  indole, 197, 257
  methyl red, 199
  motility, 200
  nitrate reduction, 195
  oxidase, 196
  presumptive, 188, 230, 251
  rapid indole, 197
  rapid urease, 198
  regular indole, 197
  slide coagulase, 194
  Voges-Proskauer, 198, 257, 258
Tetracycline, 272
Tetramethyl-*p*-phenylenediamine dihydrochloride, 196
Thermodynamics, law of, 29, 30
Thermometer(s), 119
  accuracy of, 168
  calibration chart for, 124
Thermophiles, 47
Thiobacillus ferrooxidans, 35, 265
Thiodendron mucosum, 269
Thiospirillum jenense, 270
THMs, see Trihalomethanes
Top loading balance, 155
Total solids (TS), 291
Total suspended solids (TSS), 104, 295
Toxic wastes, 203
Trace elements, 48
Transferase, 32
Transmission electron microscope, 162
Treponema pallidum, 9
2,4,5-Trichloro-phenoxyacetic acid (2,4,5-T), 83
Trihalomethanes (THMs), 74
Tryptophane
  broth, 198
  deaminase, 263
TS, see Total solids
TSS, see Total suspended solids
Typhoid fever, 74, 89

U

Units and conversion factors, 281
  conversion from metric measures, 281
  conversion to metric measures, 281
Urea, 259, 262
UV sterilizer, 121, 220

V

Vacuum filtration, 297
van Leeuwenhoek, Anton, 2, 3
Verification procedures, 225
Vibrio cholera, 85, 89
Vibrios, 16, 17
Virchow, Rudolf, 4
Viruses, 1, 12, 92, 307
VOCs, see Volatile organic compounds
Voges-Proskauer test, 198, 257, 258
Volatile organic compounds (VOCs), 82
Volatile solids, 296
Volcanic activity, 58
Volumetric pipets, 162
Volutin, 25

W

Wallemia sebi, 64
Wastewater, industrial, 103
Water
  bath, 126
  classification, surface, 101
  contamination, drinking, 94
  cycle, 57
  deionization unit, 120
  dilution, 184, 242, 286
  discharging to ocean, 104
  diseases transmitted by drinking, 86
  drinking, 141, 142
  ecosystems, fresh, 101
  laboratory pure, 169
  pathogens in, 73
  pollution, 72, 75
  purification, 74
  quality
    check of laboratory pure, 131
    regulations for drinking, 96
    test, 133
  rules in counting and reporting for potable, 225
  samples, collecting potable, 141
  sludges separated from, 297
  supply(ies)
    disinfection requirements for public, 99
    distribution networks, 234
    reservoir, sampling, 145
Wells, sampling from, 142
Whirlpools, 205
WHO, see World Health Organization
Winkler bottles, 284
Wood preservative industries, 81
World Health Organization (WHO), 97, 99

Y

Yersinia
  enterocolitica, 89, 90
  pseudotuberculosis, 90

Z

Zinc chloride, 56